Race and Curriculum

Race and Curriculum
Music in Childhood Education

Ruth Iana Gustafson

RACE AND CURRICULUM
Copyright © Ruth Gustafson, 2009.
Softcover reprint of the hardcover 1st edition 2009 978-0-230-60840-5

All rights reserved.

First published in 2009 by PALGRAVE MACMILLAN® in the United States—a division of St. Martin's Press LLC, 175 Fifth Avenue, New York, NY 10010

Where this book is distributed in the UK, Europe and the rest of the world, this is by Palgrave Macmillan, a division of Macmillan Publishers Limited, registered in England, company number 785998, of Houndmills, Basingstoke, Hampshire RG21 6XS.

Palgrave Macmillan is the global academic imprint of the above companies and has companies and representatives throughout the world.

Palgrave® and Macmillan® are registered trademarks in the United States, the United Kingdom, Europe and other countries.

ISBN 978-1-349-37532-5   ISBN 978-0-230-62244-9 (eBook)
DOI 10.1057/9780230622449

Library of Congress Cataloging-in-Publication Data is available from the Library of Congress.

A catalogue record of the book is available from the British Library.

Design by Scribe Inc.

First edition: July 2009

To my former students and to Duwayne Hoffman,
their teacher and mentor

The most dangerous and most prevalent illness of our time ... strikes at our core ... by which we orient ourselves to take in what we need and close against what is dangerous to us, and most prevalent because everyone is subject to it, more or less. ... Every exchange with the world is either a match, of mutual benefit, or a mismatch, of advantage to one, and disadvantage to another.

—*James P. Gustafson, The Great Instrument of Orientation*

Contents

Acknowledgments		ix
Prelude		xi
1	Fabricating the Future Citizen: The Ballads of a Nation	1
2	The Child as Charmed Victim: Early Vocal Instruction and the Social Distinctions Conferred by Disease	17
3	Making Daily Life Sublime: Verse and Rhythm "Never to Abase or Degrade"	35
4	Bacchanalian Chaos, Degenerate Hymns, Public Music Instruction, and the Discursive Fabrication of Whiteness	61
5	Ranking the Listener, Disciplining the Audience	81
6	Goodbye Darwin: Music Appreciation and Musical Publics	103
7	Reason, Ventriloquism, and National Music Memory Contests	123
8	The Listening Body and the Power of the Good Ear	145
9	Aural Icons and Social Outcasts: Beethoven, Lincoln, and "His Master's Voice"	165
10	Rethinking Participatory Limits: From Music Standards to Hip-Hop	183
Notes		205
Bibliography		239
Index		259

Acknowledgments

The energy and resources to pursue this subject came to me when I was a graduate student under the guidance of my mentors and colleagues at the University of Wisconsin–Madison. Much of the research material presented here is based on my PhD dissertation and more recent readings in intellectual history and music. I have tried to avoid the formalities of a thesis as much as possible by expanding the field of investigation to subjects I see as vitally related to schools without losing the excitement I experienced when I was with students playing music. For comments on chapters and sections of the book, I am grateful for the keen eyes of Jim Gustafson, whose intellect and endurance are a labor of love and patience; Thomas Popkewitz has been a great friend and mentor on theoretical matters of history writing, the Enlightenment, and its limitations and paradoxes. Julia Eklund Koza, my former thesis adviser, is still, fortunately, an adviser on substantive matters in this book and a close friend. I thank her for her labors on my projects over the last decade. Jinting Wu, whom I met through Tom Popkewitz's Wednesday Group discussions, has been a close friend and source of inspirational courage in her pursuit of the problems of representation of subjects, difference, and the future. Gloria Ladson-Billings, Dory Lightfoot, and Ruth Latham contributed to the formation of my thought with regard to racialism, equity, and equality both through their writing and their personal journeys. Although it is impossible to mention all the writers I have admired in my field, I am especially indebted to the work of Bernadette M. Baker in curriculum history, and the writings of Sylvia Wynter, Joyce E. King, Noah Sobe, Ronald Radano, and Eric Lott on the subjects of nation building, education, race, and music. Friendships I made through the Wednesday Group have meant a lot to me over the years as we not only argued but also shared so many intellectual and personal points of view. For those hours and hours of companionship, I want to thank so many, especially Dory Lightfoot, Marie Brennan, Thomas Popkewitz, Lynn Fendler, and Noah Sobe. I am grateful for support of the Spencer Fellowship in Education Program as well. Finally, I want to acknowledge my husband, Jim Gustafson, for his guidance on living through the trials and joys of family while engaged with that other force in my life, academia. Jim has taken this path before me, and from him I have learned how to clear the hurdles and perversities of institutional life while preserving the heart of my convictions.

Prelude

This book is an attempt to come to terms with the near 100 percent attrition rate of African American students from public school music programs across the country. In most school districts, black students rarely commit to the traditional class offerings such as music history, theory, band, orchestra, and chorus, and if they do, they often drop out. The issue is not new in itself, but it has not been approached from the perspective I offer here—a historical exploration of racialism in school music. In many school music programs, the absence of African American music boils down to a rejection of the embodied musical culture that African Americans (and others) identify with. Yet, what makes this problem especially challenging, as I point out throughout the book, is that it is complicated by the very opposite and paradoxical popular reception of black music (over more than two centuries) by a public that consumes, appropriates, and sometimes vilifies entertainment associated with black culture. How do I explain so obvious a contradiction as black nonparticipation in school music without resorting to the conclusion, which is simplistic and distorting, that the schools and music teachers are racist?

Research into this problem is limited. Although there is writing about inequity in music education, there has been, so far, no systematic attempt to look into the racialist underpinnings that have unconsciously (I use Joyce E. King's term "dysconsciously")[1] supported the pedagogy and selections of material in the curriculum. The primary aim of this book is not to assign blame, but to analyze how and why the curriculum's central tenets have historically translated a broad range of erudite and popular ideas about race into music pedagogy. Following a strategy akin to that of Eric Lott's *Love and Theft*, in which the author links blackface minstrelsy to white working-class social anxieties, my book traces a genealogy or a branching "family" tree in music education's past to the governance of urban populations, entertainment venues, mass immigration, and economic changes in the northern industrial cities. Similar themes continue to find a perch in public school music instruction today.

My interest in this subject grew out of my observations, my teaching, and my doctoral dissertation, but primarily it was my experiences in music classrooms that made me want to write this book. "Children were shown icons of ascending and descending melodies. Teshawn, an African American first grader, sings the melodies accurately, by imitation, when the teacher plays them on the piano, but refuses to engage in the question and answer session regarding their visual representation. Teshawn began instead to quietly improvise a vocal line. Mrs. Prentiss (name changed), the teacher, asks Teshawn, 'Can you listen first, Teshawn?' Teshawn fidgets and then starts to make arm motions to the music, looking in my direction for approval. I nod and she re-engages with the activity in progress" (R. Gustafson 1991, 10). [scr1]With the exception of feminist studies in music, education, and linguistics studies such as Shirley Heath's *Ways with Words*, there was, in the early 1990s, very little written on the marginal status of very young black children in classrooms—especially on the subject of the absence of African American music and African American students from music programs.

Race and Curriculum recounts how it is that children with a disposition like Teshawn's have come face-to-face with a racialist template in music pedagogy. Teshawn's encounter provides a clearly defined starting point for looking into northern European aristocratic traditions, pedagogy, and curriculum documents that flesh out a template for the ideal listener and singer. The template matches a relatively motionless body, a reverent demeanor, and a minimizing of gesture. Historical ironies abound, since what poses as superior emerged from a complex web of social and musical intermixing.[2]

My focus on school music as a site of compliance and conflict in forming musical difference may appear, at first, to load the school and teachers with undeserved blame for the current situation. As I try to make clear throughout the book, curriculum and pedagogy derive from a broad array of racialist ideas and aesthetic tenets in music teaching. Making up the formal "foundations" of the field, these inscriptions of race have profound, dysconscious effects on the everyday task of the teacher. Her training, with all of the history of the field that implies, added to the conditions of schooling, inform the possibilities of her teaching. Another cautionary note: the curriculum and pedagogy are not the only factors in black underachievement and attrition, but they play a crucial part. One of my aims is to provide a sense of how music instruction governs body, diction, and affect to produce participatory limits.

Entrainment

The question of how music entrains each one of us in different ways is at the heart of my narrative. When I use the word *entrain*, I mean the way we react to music, with reference to the interaction of sound, memory, body motions, and gestures. These are sometimes beyond words, but they are the feelings and signs that link us to various tastes in music and social groups (Berthelot, 1991). Contemplating our own, or another's, entrainment lets us into a world structured by cultural history and one that has its own intimate meanings. As individuals watch others, they interpret motions as familiar or strange, either like their own values or different from them.

It is my observation that, insofar as musical responses are understood as part of the modern self, each person harbors a sovereign sense of his own entrainment, as much as over musical experience as over his language, carriage, and feelings. With the exception of mimicry, a subject that complicates the notion of subjective integrity in important ways, it is the child's *sense* of sovereignty that counts in her assent to, or withdrawal from, a musical activity. This can grow into a crisis for that child in school.

> I was sitting in the back of a second grade general music class where the teacher, a young man about thirty-five years old, is teaching some of the "elements" of music by playing notes on the piano and asking his students to tell the difference between a series of rising notes and a pattern of descending notes. One of the students, an African American, who I will call David, has been humming along with the notes played on the piano. After each pattern was played, the teacher asked, "Who has good listening ears?" David did not raise his hand and when the teacher called on him with a variant of this question, he turned his face toward the window at the back of the class, counting himself out of the group with "good listening ears" who had raised their hands. This was surprising to me at first, since I had heard him softly humming the pitches just played. Most striking, none of the other African American children in the class responded either. As class ended I asked myself, what did David think Mr. Taylor meant by listening ears? (R. Gustafson 1991, 12)

In other classrooms that school year, I noted similar incidents. This was not what teachers were hoping for as they devoted great effort to reengaging students in listening lessons when they noticed withdrawal, usually to no effect. I wanted to find out if my hunch about the "good ear" and similar pedagogical prompts was relevant to the general pattern, which, I concluded, was not David himself, but concerned whatever it was that made "who has good listening ears?" a plausible device for focusing attention on music in a particular way.

Mr. Taylor (name changed) was representative of the group of teachers I had occasion to observe. He was from a white middle-class background and had worked in the school for many years before it was integrated. Integration came about through busing from de facto segregated neighborhoods to cross-city schools. Howard Taylor, like many music teachers in the district, was eager to present a curriculum that would involve children of many backgrounds. He had planned a multicultural repertoire for listening lessons that he felt suited the classes he would teach. Most of my observations of the early grade levels occurred during the first year through the fifth year of this plan. I deliberately chose schools where music educators at the university in the district were asking particular music teachers to pilot a new approach to teaching musical "elements" as discrete, testable entities.

The pattern of participation of African Americans seemed to revolve around more than any one aspect of the curriculum. What's more, it was obvious that music was just one of many school subjects in which

I might have observed similar demographic divisions. In spite of these considerations, I felt strongly that music was a special case in which the involvement of the body discloses more about cultural dispositions than one could observe in other subjects. I found, for one thing, that lessons on marking rhythm became increasingly conservative as the school year progressed. This point was brought home to me in a vivid way whenever students were asked to mark rhythm by clapping or playing percussion instruments. What I came to recognize was the highly disciplined aspects of two traditions in those classrooms. There is the traditional Western split between body and mind. Within the musical culture called black or African American, the body is the mediator and an inextricable part of what music is and what musical thinking is.

Racial Labels and Racial "Essence"

There is a very real danger in the discussion of race and music of making race a determining characteristic and a biologically "real" category. In my narrative, I try to make it clear that racial labels have been produced through the effects of racism as well as by popular and academic theories of race that designate difference in a society bifurcated by notions of racial origin.[3] When I use the term *race* in the context of school music, indicating African American, Caucasian, or any other racial category, I am reading its historical influence as a category that divides individuals and groups.[4]

Race, then, is a dubious marker, but it cannot be avoided as it allows us to identify patterns of projection, exploitation, persecution, and discrimination that have had enormous effects on various populations. *Black music* is a term that speaks to the racialist way in which we refer to musical types, but it also champions work of crucial symbolic value to both African American culture and the multiracial reality of American culture. Between acknowledging that symbolic value and steering a course away from absolute racial difference, this book attempts to piece together the historical role that notions of race play in judging musical entrainment in school music.

Going against the grain of a prevailing view of neutral, universal foundations, my reading of this history of music education proposes that musical values associated with whiteness took their social heft from comparisons to abject blackness.[5] Interpreting the curriculum as both a conveyor and reinventor of racial difference requires what Ronald Radano describes as hearing "black music's power not in race per se but in the wild fluctuations from sameness to difference that racial ideologies have constructed. From this mode of hearing, we identify, finally ... a kind of musical-textual double speak that claims for music the unities and incommensurabilities of blackness and whiteness, at once" (Radano 2003, 13).

Thinking along similar lines, I set about investigating the ethnic, racial, and religious anxieties as they rippled through the society at various points in music education. Still, there is a thin line between situating ethnic, national, and racial identities as causal and recognizing their function as *imaginaries*, in sanctioning or barring particular forms of music and dance. For example, notions of blackness and Celtic parentage in the early nineteenth century alerted educators to dangerous differences in musical practices. Throughout this book, I point out that a fabricated whiteness has been erroneously taken as straightforward and real. In music pedagogy, whiteness takes the shape of the cultivated persona, as it continually projects from itself what is cast as "black," forming a semantic shorthand for "high" and "low" musical and so-called citizenly dispositions.

Good Listening Ears

I suspect that asking about good ears is a very widely used ploy to get children to listen and to identify what they are hearing. However, even while we know that styles of rhetoric used in schools depend on cultural affinity and that differences between the child's background and the institution can be great, the idea of clashing rhetorical understandings begs the question of what the good ear symbolizes in relation to goodness (or a misrecognition of it). There was something about the partiality for the ear that bore closer examination, I thought, and because I had gone through ear training instruction, I recognized its history in the training of music teachers. Insofar as this book attempts to get closer to the significance of anatomical detail, it looks into the mind–body split in music teaching and listening. Pursuing this anatomical question, it seemed plausible that what is at stake is not miscommunication between teacher and student but a whole system of thinking about music and the body that coincides with the racial ranking of musical dispositions. Looking at David's situation from this angle, the good ear took the class conversation away from his humming response toward a focus on listening in silence. David was not prepared to give up singing for a good ear at this tenuous stage of involvement with music instruction. This does not mean that ear training is a bad thing, but that it makes music an affair of the ear alone—with the consequence that much less merit is attached to the performance of voice and the body's gestures.

A Rhythm Lesson

One of the most common methods of teaching music today is to ask students to mark the rhythm of a musical phrase by clapping on the strong beats that alternate with weak or counter beats. When I observed David and Teshawn's classes, I noticed the exercise often divided classes into two groups, those who marked both strong *and* counter beats with many kinds of body movement, including finger snapping and foot tapping, and a group that marked strong beats only with hand claps. This occurred whether the music was classical, folk, or a popular song. What was striking about the musical differences was that they followed demographic lines, with African Americans marking counter beats. In one class, students were to use handclaps in the usual way to mark strong beats. They were marching around the room to Tchaikovsky's *The Nutcracker*. One student was marking counter beats, eliciting a shake of the head from the teacher, after which he sat down and put his head in his hands for the rest of the class. I am not asserting that students of color have *inherent* musical differences, yet it was apparent that the teacher's preference left him dejected and confused, since the entrainment with which he was familiar would have recognized the aptness of his rhythmic gestures. This does not mean that every classroom is like this. There are teachers who forego the regular curriculum and follow their students' own musical interests. One example I observed was a teacher who took gestural cues from his students, so that a variety of rhythmic accentuation with any rhythmic pattern took place. However, his ability to retain students of all backgrounds in his program was not valued as highly by his colleagues as the production of star ensembles in the classical mode. The classroom incidents I cite are specific moments of inclusion and exclusion. Some might object to the fact that I base my hypothesis on relatively few events. Yet, I believe they represent, in a nutshell, the aesthetic differences that compel high attrition rates. It is not so much the particulars of a curriculum or method that are at issue, but deeply ingrained attitudes toward the body in motion that need to be accounted for in historical terms. Historical bias, I argue, insinuates itself into the music classroom as a seemingly neutral set of values. While this book does not provide a formula that will reverse the situation, it offers a reading of public school music teaching that questions the present's "rightness" and inevitability.

Since the 1990s, in spite of the Brown desegregation mandates, many schools have been reverting to internally (re)segregated enclaves matching cultural and demographic differences. Academic tracking is one dimension of resegregation patterns. The most knotty problem is belonging in school when differences of language, tastes, and comportment are significantly far from the school's culture and assimilation is not an option. Cultural loyalty is an investment and form of love and, in this sense, participation hinges on a mutual recognition of qualities that are part of the child's *social* being and heart, inextricable from the child's sense of "sovereignty'" in musical entrainment. Central to well-being is the recognition that the body needs in order to open to other experiences.

Race Theory and Curriculum History

Several years after my observations of early elementary school classrooms, I began to investigate the "dysconscious" effects of the historical construction of race-based images of beauty, intelligence, and culture. What I came to understand as dysconscious, according to Joyce E. King, is a lack of awareness of racial preferences packed into the seemingly neutral logic of aesthetic judgment. Taking up this idea of dysconscious racial judgments,

I have attempted, in this book, to show how these judgments *produce different kinds of children*. I give special weight to the word *produce*, since the child that undergoes music instruction acquires an array of labels and a bona fides that give her an identity of relative advantage or disadvantage in the eyes of the school. This is, in a sense, a biography that maps her future. Norms represented as neutral aesthetics outweigh overt acts of willful racism in that they pass without notice. For certain, racist ideas are abundant, but it is the more scientific and seemingly equitable precepts that produce the template of the music student as a racial type.

Until recently, music was about as perfect an arena as any in which to encounter a set of purely aesthetic and enchanted practices devoid of material or political aims. Here, I have undertaken a different orientation for music education history in turning toward sociology, cultural studies of the body, and the cultural history of the curriculum. It is a cross-disciplinary approach that, with few exceptions, had been dropped from music studies for decades. Fortunately, the idea that music is reflective of society has been renewed by the recent, exciting work of numerous scholars. As I apply some of their ideas to music instruction, I have drawn inspiration from a new collective sense that music is far from being *just* music. In a serendipitous coincidence of place and time, I am much indebted to Thomas Popkewitz's work on the cosmopolitan child and to Julia Koza, who has worked toward building a critical perspective on music education. Their work has been foremost in my thinking about how school music became an affair of the head that produced the "reasonable" versus the abject citizen.

Michel Foucault's seminal idea is that knowledge is a form of power.[6] Turning that insight in the direction of music education has meant considering many concerns: health regimens, theories of vocalization, fears of immigration, public exhibitions, racialist forms of entertainment, and rhetoric—as components of power. This "gnarled history," a phrase Eric Lott uses in relation to the early nineteenth century, leads us to consider black absence from music classrooms as an effect of these historical practices. I write about these gnarled events as a pedagogical cleavage of body from mind that makes Teshawn and David's musical disposition appear deficient.

1
Fabricating the Future Citizen
The Ballads of a Nation

I begin with the body and its expression of cultural nobility. In a contrast between listeners that has been illustrated in a lighthearted way in a *Peanuts* cartoon strip, we have two contrasting views of embodied responses to music. The first window of the cartoon has Schroeder telling Lucy he has a recording of Brahms' Fourth Symphony. When Lucy asks what he is going to do with it, he replies that he will listen to it at home. Lucy asks whether he is going to dance. "No, just listen."[1] Flabbergasted, Lucy asks if Schroeder will march, whistle, or sing. Schroeder says again that he is only going to listen.

Schroeder's approach toward music, in contrast to Lucy's, comes from the traditions of "cultivated" listening. Passed down from the traditions of body comportment prescribed for audiences from the nineteenth century to the present, school music listening lessons have historically favored relative immobility as opposed to movement. This type of posture is also represented as that of the "thinking" individual. In Auguste Rodin's sculpture, *The Thinker*, the figure is seated and posed with his head bowed in contemplation. Critics have linked Rodin's statue to a profundity associated with the reflection of reason.[2]

The Thinker and *Peanuts* illustrate the model of cultural nobility that has taken precedence in elite society and in the historical process of the formation of school music. "One has only to think of the extraordinary value nowadays conferred on the lexis of 'listening' by the secularized . . . versions of religious language. As the countless variations on the soul of music and the music of the soul bear witness . . . all concerts [classical or 'art' music] are sacred . . . 'Insensitivity' to music doubtless represents a particularly unavowable form of materialist coarseness" (Bourdieu 1984, 19). As I pointed out in the prelude to this book, dispositional divides in classrooms present a unique opportunity to consider how claims of music's "universal" appeal are not universal at all. Differing responses represent conflicting cultural affiliations and the valuing of particular kinds of musical activity with limited appeal to any demographically mixed collective, such as a public school classroom.[3]

This chapter takes the *Peanuts* cartoon and its representation of contrasting musical dispositions as a jumping off point for a historical inquiry into patterns of cultural recognition or misrecognition of music and personhood in public music instruction in the United States. Its chief focus is on analyzing documents related to curriculum through the social realities of popular and intellectual life, beginning with the distinctions of personhood that were being etched in the postrevolutionary period up to the Civil War. The period is characterized by historians as exhibiting the strains and tensions of the judicial, racial, and economic contentions that led to that war. Conflict often took the form of high-minded rhetoric of battles between the civilized and barbaric peoples. This rhetoric also worried civic actors and school boards, so much so that in early vocal curriculum documents (1825 through the 1850s), the main rationale for initiating public vocal instruction was to extend the principles of reason and civilization to populations who appeared to lack those virtues. The distinction between barbaric groups and worthy citizens was consistently made through racialist doctrines and popular prejudices. In the antebellum years, racial anxiety was aimed at Irish Catholics and freed slaves, living primarily in northern cities, who would be the "uncivilized" school population.

Whiteness pervades models for singing and listening, although it rarely makes an appearance as an explicit rubric in school music methods. Rather, whiteness is a "cobbled together" ideal, absorbed from both nineteenth-century and twentieth-century notions of racial destiny in which characteristics of "races" were important to early nation building and public education.

One way to conceptualize the fabrication of racial personae is to see them as entwined with the qualification for citizen status on many scales and across many cultural practices. For example, there are the large-scale ideas of musical merit that Thomas Jefferson describes in his *Notes on the State of Virginia*. On the smaller scale of the classroom, one finds that school songbooks reiterate notions of musical or racial stereotypes. Similarly, whiteness and blackness inflect the registers of school songs and vocal production itself through everything from treatises on singing to the methods teachers learn in preparatory courses. Interrelating these scales provides a window onto the dysconscious[4] comparisons rife in particular teaching protocols and curriculum outlines. Elucidating the way racial ideals were articulated in the early decades of public music instruction is the central concern of this chapter.

The Civilizing Process and Early Vocal Instruction

Standards for the reasoning citizen and his cultural nobility have deep roots in the early nineteenth-century mission to civilize the child. Asa Fitz's *A Child's Songbook*[5] contains verses to teach how reason would supplant unreason (corporal punishment) in disciplining animals:

> Then I will never beat my dog
> Nor ever give him pain,
> But good and kind
> I'll be to him and he'll love me again. (Fitz 1819, 6)

This song exemplifies the kind of object lesson through which the child would not only learn to restrain the use of force but would also learn to put much greater emphasis on reason, language, and bodily constraint.[6]

Singing's potential to teach the self-governance invoked by the Declaration of Independence was an important factor in convincing school boards to support public music instruction. Lowell Mason, the first public school music teacher in the United States, invokes the spirit of self-governance in the following injunction to music teachers as surrogate lawmakers: "LET ME HAVE THE MAKING OF THE BALLADS OF A NATION AND YOU MAY MAKE THEIR LAWS."[7] Similarly, testifying on behalf of public vocal instruction in the Massachusetts public schools, Horace Mann, state secretary of education, commented: "If the subject of school-books is important... [it] can hardly be less so [with songbooks]... They will constitute a part of moral nature... regard and sympathy for domestic animals; consideration and benevolence towards every sentient thing, whether it flies, or creeps or swims; all filial, all brotherly and sisterly affections; respect for age; compassion for the Sick" (Mann 1844/1982, 153). Order and reason would be achieved through singing. Horace Mann wrote, "[With singing]... the schools of Prussia are kept in such admirable order... with so rare a resort to corporal punishment."[8] He illustrated his point with a vignette about Napoleon's pacification of the Egyptians: "[Music would] mollify and subdue the hearts of the people, to make them yielding and receptive... to gain that conquest over their feelings by his arts, which he had achieved over their power by his arms" (Mann 1844/1982, 151).

Boston educators believed that music's charm would mold the child *from inside* rather than through corporal punishment. According to this vision, public school vocal instruction promised to distinguish bona fide citizens from those who, a priori, lacked the requisites for liberty. According to the major social institutions of the times, those viewed as fit for citizenship displayed the qualities of reason and civilization on the body through dress, comportment, and social etiquette.[9] In the first half of the nineteenth century, the scale of comparison for citizenship was explicitly racial. Frederick Douglass's *Narrative of the Life of Frederick Douglass, An American Slave* describes the two poles of existence: between the civilized and the abject figure of the slave and between the individual at liberty who uses reason to conduct his own affairs and the quasi-animal status of the descendants of Africans.[10] "We were all ranked together at the valuation. Men and women, old and young, married and single were ranked with horses, sheep and swine" (Douglass 1845/2001, 18). Between the poles of slave and free citizen there was a continuum of human types with various degrees of social recognition: "The savage now permanently reminds the civilized, not in a noble way, but as threat of cultural failing, of how an absence of 'elite' culture itself would look... For the child of civil society and those cast into the category of savage life, the consequence... was enormous" (B. M. Baker 2001, 382). Public music instruction in this period can be understood not simply in terms of a set of texts and pedagogues but as an institution that caught fire when schools were seen as places that would counteract the uncivilized, savagely inspired, social chaos in the crowded and newly industrialized cities.[11] Boston was described as Satan's seat by one minister, who asserted that in the crowded section of town "there are three hundred [people] wholly devoid of shame and modesty" (Schultz 1973, 28). In the preface of Lowell Mason and Elam Ives's early songbook, *The Juvenile Lyre*, published in 1831, Lowell Mason points out that many of the pieces in the collection were translated from Prussian songs. The latter were thought to convey the highest form of social discipline.[12] Music's molding of body and mind permeated the Prussian system of education of the early 1830s in which teachers used songs as pedagogical and disciplinary routines.[13] According to Horace Mann, "The universal practice of music in most of the schools of the German states, for a long series of years, is an experiment sufficient of itself to settle the question of its utility ... Prussian teachers rarely have occasion for resorting to coercive measures; and thus the Prussian schoolroom becomes the abode of peace and love ... The whole country, indeed, is vocal with music ... Pervading all classes, it softens and refines the national character" (Mann 1844/1982, 147).

However, American ideals and circumstances were different, and while the order of the Prussian classrooms Mann visited made a deep impression, Lowell Mason's methods and song collections were to distinguish the future citizen's comportment from the unrestrained irrationality of the twin scourges of tavern and revival meeting, which were distinctly American social ills.[14] "[Music] ... tends to improve the heart. This is its proper and legitimate and ought to be its principal object ... Its effects in softening and elevating the feelings, are too evident to need illustration. There is something in the nature of musical tones, viewed in their pure and simple ... state which is truly heavenly and delightful ... music of such a character [should] become universal throughout the nation" (Mason 1834/1982, 130). In agreement with Mason's intentions, a committee appointed by the Boston School Committee describes how the sentiments of the songs would teach virtue: "In music the very image of vice and virtue is perceived" (Boston School Committee 1837/1982, 134). Mason writes in the *Manual for the Boston Academy of Music* that "in this way amusement may be blended with instruction; and cheerfulness, happiness and order introduced into the family and into the school ... We will give a few examples of [songs] to illustrate ... 'O come to the garden, dear mates of the school, And rove through the bowers so fragrant and cool ... We'll cull all the sweetes [sic] to make a bouquet, to give to our teacher this warm summers day ...' Now who does not know that such an exercise tends to unite the hearts of the children, and to make them love and obey their teachers" (Mason 1834/1982, 130). In the 1830s, the first waves of mass

Irish Catholic immigration were arriving on the eastern seaboard. At this time, the Irish were considered a non-white (Celtic) race unto themselves and their religion represented regression to autocratic rule and religious superstition. What deserves emphasis here is that public vocal instruction served as a response to social conditions and was presented to the general public as primarily an issue of governance in the pastoral sense. "The bells worn by the sheep [in Southern Germany] are tuned to the common chord, so as never to make a dissonant sound . . . there arises from the earth, an exhalation of music . . . we have evidence nearer home of the beneficial effects of music in schools" (Mann 1844/1982, 148). Singing was the basis of order and the organization of it as a school subject made it relevant to every child's personhood and being. This development was different from the notion that children were born into their destiny—peasant or aristocrat. By the early nineteenth century, the child took his potential from the Enlightenment view, which was that an aptitude for music became a universal quality. The change accompanied the humanist idea that mankind reveals itself through a knowledge of the individual's physical being and aptitudes.[15] No longer would a child simply possess a "faculty" for mental activity—each child was the *object* of instruction, making singing in school an exhibition of the capacity for speech, reading, arithmetic, and reason in general: "Every child can vary the tones of his voice; and if he receives early instruction, it will be as easy for him to learn to sing, as to learn to talk or read" (Mason 1834/1982, 127). "Vocal music furnishes the means of intellectual exercises. All musical tones have mathematical relations . . . Music furnishes problems sufficient to task the profoundest mathematical genius" (Mann 1844/1982, 149). Music promoted the convivial, democratic society in which "happiness" for all was the aim: "Substantially, then the voice and the ear are universal endowments of nature and thus the means of enjoying the delights and of profiting by the utilities of music are conferred upon all" (Mann 1844/1982, 145). Nineteenth-century Boston School Committee reports represented vocal instruction as an activity that would lead *all* children to happiness and self-governance in school and state. Universal methods drew out "a natural repugnance between music and fear, envy, malevolence" (Boston School Committee 1837/1982, 140). "Let all parents understand that every pure and refined pleasure for which a child acquires a relish, is, to that extent a safeguard and preservative against a low and debasing one" (Boston School Committee 1837/1982, 140). "Natural . . . musical sound can only produce good, virtuous and kindly feelings . . . happiness, contentment cheerfulness, tranquility—these are the natural effects of music . . . connected intimately with the moral government of the individual . . . melody is concerted action . . . 'Where Music is not, the Devil enters'" (Boston School Committee 1837/1982, 138). If abilities are universal, how does one explain lack of ability? Simply put, the premise of universal ability characterized those who were unresponsive to singing's value and development as deficient. Exceptions to the rule of universal ability signaled deviance that showed itself through different tastes and dispositions. This made it possible to articulate a human typology using both pious and secular registers. In both, and, ultimately, in the more dominant secular rhetoric, the shepherd or flock imagery would save those who had yet to grasp "universal" truths. For music instruction, this meant seeing difference as abject and unnatural: "The exceptions are not inherent in the nature of things, but only punishments for our infraction of the Physical Laws and . . . [this] calamity of privation shall be wholly removed . . . [by vocal instruction]" (Mann 1844/1982, 145). Classrooms were divided along the lines of those who found "pure and refined" pleasure in singing and those who did not. School songbook writers took on the mission of removing deprivations in others. Hence, it is common to find rhetoric that compares human types in a general way in Boston school board testimonies and singing manuals of the early nineteenth century.[16] While aspiring to reach all children and asserting the universality of ability, the pastoral attitude and rhetoric simultaneously etched lines of exclusion granting recognition to those children who *already resemble* the ideal. This dysconscious ordering of schoolchildren was produced through what were read as marks of unreason in demeanor, speech, and musical interests. In this respect, race, as I discuss further on, was one of many factors in the classification of student's abilities and dispositions embodied in the verse and aesthetic principles of singing lessons. Early public music instruction inscribed a prototype of pedagogical comparisons, for example, in the Boston School Committee testimonials that dealt in various different guises in notions of the "good ear" and the worthy listener.

The process of reworking subject content with matters that appear to have psychological and social urgency has been compared to a kind of alchemy in which a school subject converts disciplinary knowledge into the gold of civic virtue.[17] While it is not the deliberate intention of most teachers to brand students as unqualified, the comparative principles within pedagogy produce categories of social worthiness. That the comparisons also answer to racial and other demographic groupings often escapes notice. In this connection, it is worth examining the disposition that made David's "singing along" a violation of the pedagogical protocol. Drawing on a similar system of reason about singing, we see in the curriculum documents of early nineteenth-century vocal instruction, in David's case, that the habitus of the child stands in sharp contrast to the curricular principles. These principles insert an overriding concern for using musical distinctions as characteristics of a universal musical reasoning in the worthy future citizen and her supposed opposite. The linkage of distant times and spaces to the present moment does not mean they are the same, but it does imply that the process of organizing the school curriculum is heavily influenced by social anxieties with important consequences for schoolchildren. If, two centuries ago, it was the Negro and the uneducated masses of immigrants in both rural and urban locales who were seen as social inferiors without civic virtue, early vocal instruction contributed a number of ordinances and qualifications with respect to the *particular* musical expression of the model citizen. Parallels to the present circulate within schooling in similar ways, in that the political climate produces various internal and external threats.

The curriculum's part in this drama is to articulate the scale of national concerns at the local level. Although we may think terms such as *at risk* and *cultural disadvantage* are fairly recent, their use in schooling, and particularly in music, emerged, in the United States, from the pact between social elite members such as Lowell Mason and Horace Mann in the earliest years of public schools. The social strata identified with singing and drinking in taverns, street ballads, and, as Mason put it, those who participated in the degenerate vocalizations common in revival meetings, would be outstripped by the trained voice of the ideal citizen who armed himself with reason, morality, civilization, and musical taste.[18] Based on what was considered "high art," rather than on popular song, the child associated with the cultivated style would resist the seduction of (unreasonable) religious movements, and of entertainment such as minstrelsy and magic shows. The line between religious ecstasy and "refined" religious observance was particularly important for the genteel sector associated with established churches and public education. This stratum of society treated ecstatic spiritualism as a serious threat to the rule of reason necessary to civic government and the republic as a whole.

Social unrest swelled the ranks of disgruntled common men and women as disappointment with the social and economic hierarchy in the 1830s and 1840s kept poor farmers poor, new immigrants marginal, and the propertied and landed elite disproportionately wealthy. The alliance of black and white revivalist sects, the popularity of "uncultured" forms of entertainment, and the mass immigration of Irish Catholics stirred the fears of the upper class in the northern cities. Popular enterprises and spectacles, such as blackface minstrelsy, although attended and enjoyed by people from all social classes, were considered degenerate. True religion was a private, dignified, and a basically individual affair. Given the fear that more aggressive religious recruitment would engulf the establishment churches, religious conversions and revival meetings appeared to threaten the principles of rule by secular reason. Reason was not godlessness but required the subordination of beliefs to secular law. Aimed at making the three branches of government the arbiters of justice and progress, musical elites and civic leaders (the two affiliations often overlapped) gave public music instruction its salvific mission to compare individuals, assign them biographies, and project their future in terms of how well the citizen aligned herself with reason.[19]

The appearance of early vocal instruction occurred at a historical moment when postrevolutionary economic insecurity, debates over slavery, and fears of Catholic dominance were in the air. In the decades that followed the first public music classes in Boston, the lyrics approved for school songbooks intensified their emphasis on patriotic and genteel themes so that the combination of music and social instruction shaped a political anatomy for the recognition of the bona fide citizen type. Images of the ideal singer-citizen circulated in musical periodicals, meetings, and performances of the Handel and Haydn Society and through pedagogical texts such as Thomas Hastings's *A Dissertation on Musical Taste*.[20] Built on admiration for German music and the principles of the canon of common practice in the northern European tradition, the genteel music societies and their publications were, not surprisingly, active in supporting public music instruction.[21] While music education's traditional histories stress the improvement of hymn singing as the aim of early public school music instruction, it was equally the case that social concerns, chiefly immigration and school integration in the large cities, provided the support for singing instruction to improve and select the voice, taste, and comportment of the worthy individual and distinguish her from the rest. Members of the genteel music societies envisioned public music instruction as a conduit for propagating good taste and comportment in the mass population. In *Spiritual Songs for Social Worship*, published in 1831, Lowell Mason collaborated with Thomas Hastings on a hymnal whose aim was to resist the many publications of "insipid, frivolous, vulgar and profane melodies" (Mason and Hastings 1831, i), referring particularly to revival hymns and tavern ballads. Driving this outlook was the fear of popery among the governing elite of Boston, which increased in proportion to Irish Catholic immigration. The Irish were also closely identified with a racial difference related to blackness and were even in particular musical circumstances, such as minstrel shows, referred to as the "smoked Irish" (Lott 1993, 94–95) in the hybrid adaptation of Irish musical instruments, singing, and dancing styles. Tensions between particular secular (for the most part Protestant) ideals, revivalism, and Catholic values ran high. Bias against freedmen and Catholics in Boston, New York, and other cities in the North fueled the clamor for economic progress and tension in the political atmosphere about how best to contain the revolutionary energy emanating from the lowest classes: freedmen, the Irish, and potentially those who would be freed by the abolitionist movement that was so prominent in Boston in particular. Consciousness of race, social class, and religious creed deepened in a milieu where upheaval was in the air: "Everyone who was not free [not working for himself] was presumed to be a servant ... Dependency was now equated with slavery ... 'By Freemen,' wrote John Toland, 'I understand that men of property or persons are able to live of themselves.' 'There are,' stated simply by John Adams, 'but two sorts of men ... freemen and slave.'

Such a stark dichotomy collapsed all the delicate distinctions and dependencies and created radical and momentous implications for Americans" (Wood 1992, 179). Always politically unstable and increasingly untenable, the categories of free individual and slave presented daunting challenges to the moral and legal authority of city and nation and of the new western territories.[22] Political challenges gave new meaning to the deeply embedded Calvinist ethos once focused on sin in the church community. By 1830, sinning was located in the obstacles that would hinder the building of a unified nation with a manifest destiny to civilize the world.[23] For John Adams, at the time of the War of Independence, liberty was rhetorically framed as the gift of the religious communities of New England.[24] In contrast, his son John Quincy Adams wrote that secure liberty came only through economic progress, through the erasure of "barbaric" creeds, and through mutual civility.[25]

The shift in emphasis from religious covenant to the secular structure of social life transmuted old assumptions of lines of authority and heritage into new political values. Thomas Paine had written that hereditary government meant the hereditary ownership of a people "as if they were flocks and herds" (Paine 1792/1945, 364). He also wrote about the necessity of abolishing slavery. In general, this strong belief in the liberty of the individual was the underpinning of the schism and contradiction in the egalitarian spirit as it continued to prize a particular type of persona. Negative racial, linguistic, and religious tenets were projected onto non-Protestants and non-whites as if they were the very obstacles to equality.[26] Grace, once given by God, was now fully invested in economic achievement through the attributes of a charmed human type who showed the virtue of reason and taste catalogued in the secular, British manner by Shaftesbury.[27] Music instruction, as part of a general education, was to work a physical transformation on the student by making "inner" reason socially recognizable: "But you must have a good Master ... that can teach what is graceful and becoming and what gives a Freedom and Easiness to all the Motions of the Body" (Locke 1693/1982, 88). The play and effect of this ethos operated selectively to certify only those conversant with gentility, with values that were highly sectarian and racially prescribed. Consequently, the parameters for what were considered noble qualities for the American citizen and the music learner in the early part of the nineteenth century were set in relief against a backdrop of codifications of human types. As I will discuss in later chapters, the public school music instruction's traditional foundations approach has not included an analysis of these social parameters, factors that account for the large differences between the aesthetics of school songbooks and the dynamism of the social situation outside the school.

Popular Song, Public Music Instruction, and the Jeremiad

Songs used in the public schools were very often selected from collections of popular ballads and hymnals. While harmonic and rhythmic difficulties were simplified in the school songbooks, they often shared similar subject matter with some types of popular songs. A major difference was their omission of the immensely popular satirical ballads and the provocative, racially destabilizing songs sung in blackface.[28] These were highly irreverent with respect to high culture, sexual mores, and manners and were enjoyed in many settings outside the school.

One satirical song circulating around the middle of the century made fun of the polymath, Goethe; it was entitled, "Give My Chewing Gum to Gerty."[29] Other popular tunes, representative of those omitted from school songbooks, were similar to the following:

> I cannot eat, I loath my meat,
> I feel my stomach failing me,
> Steward, hasten, bring a basin,
> What the Deuce is ailing me. (Anon., cited in Levy 1971, 88)

During the decades when public vocal instruction was gaining support, there was a sharp upturn in music publishing of all sorts for the home, theater, and social gatherings. By the 1830s, with the increasing sales of pianos to the middle class, many publications of popular songs were sold as sheet music. One of these songbooks included "A Favorite Old Song on Mortality, Think and Smoke Tobacco, Made Agreeable and Pleasing to all Classes From the King to the Beggar" by Joseph Gear, printed by John Ashton and Company Pendleton of Boston in 1836. Songs, such as this one, celebrating smoking flaunted irreverence by paraphrasing the Bible and flying in the face of medical advice that smoking was bad for one's health:

> This Indian weed now wither'd quite,
> Though green at noon,
> Cut down at night,
> Shows thy decay:
> All flesh is hay,
> Thus think and smoke Tobacco
> Thus think and smoke Tobacco. (Gear, cited in Levy 1971, 30)

A broad cross section of the middle class purchased and enjoyed these sacrilegious and irreverent ballads as well as European art music. Yet, if tastes were mixed in this respect, strict distinctions were made between entertainment and what was appropriate for church and public music education. This was consistent with the elevated style associated with northern European art music and was the main authorization for school songs: "In connection with singing in the schools ... whether joyful ... or grave and solemn [it] should be of elevated character" (Mason 1847, ii). In this sense, the vocal curriculum of the 1830s was to be, inseparably, an achievement in music literacy[30] and an achievement in spiritual deliverance from barbarism. School music, with the exception of tunes such as "Yankee Doodle," represented, in part, the deep gulf between the cultivated and vernacular tradition: "On the one hand there continues a vernacular tradition . . . essentially unconcerned with artistic or philosophical ideals, a music based on established or newly diffused raw materials; a popular music in the largest sense . . . On the other hand there grows a cultivated tradition of fine-art music significantly concerned with moral, artistic or cultural idealism . . . based on continental European raw material and models, looked to rather self consciously, an essentially transatlantic music . . . by no means widespread throughout the populace" (Hitchcock 1969, 44).

The tradition of oratory and writing that promoted public schools and music instruction was closely related to a type of Protestant sermon, the jeremiad, urging the development of aptitudes necessary for spiritual and material progress. The jeremiad, writes Sacvan Bercovitch, was endemic to local and national politics: "[It] gave the nation a past and future a sacred history, rendered its political and legal outlook a fulfillment of prophecy, elevated its 'true inhabitants,' the enterprising European Protestants who had immigrated within the past century or so, to the status of God's chosen, and declared the vast territories around them to be their chosen country" (Bercovitch 1978, 140–41). Lowell Mason's dictum, "Let me have the making of ballads," shared aspects of Abraham Lincoln's style, with its repeated cadences of the first person plural or *anaphora*:[31] "Let reverence for the laws . . . Let it be taught in schools . . . in seminaries, in colleges . . . Let it be preached . . . Let it become the political religion of America and let [all people] sacrifice unceasingly upon its altars" (Lincoln 1838/1967, 11–21). The urgency behind the jeremiad with respect to vocal instruction and the nation was that it coincided, in Boston, with the emergence of Irish Catholic immigration and what was regarded as the resulting threat to the social order. Lyman Beecher (1835/1977), one of the best-known ministers in early nineteenth-century New England, saw it as everyone's business to further education, especially when, in his view, Irish Catholic immigrants would precipitate authoritarian rule. Beecher wrote and preached in *A Plea for the West* that the only response to the need to save civilization from Catholic autocracy was to demand, with great urgency, that the public be educated. "[Our] nation is blessed with such experimental knowledge of free institutions, with such facilities and resources of communication . . . for the free unembarrassed application of psychical effort and pecuniary and moral power to evangelize the world" (Beecher 1835, 10). Horace Mann crafted a jeremiad with promises of music's unique qualifications as a civilizing tool, one in which singing would restore agreement where there was conflict. "Harmony of sound produces harmony of feeling. Can it have escaped the observation of any reflecting man, when present at a public concert . . . Competitors in business; rivals, almost sanguinary, in politics; champions of hostile creeds; leaders of conflicting schools in art or philosophy—in fine a collection and assortment of contrarities [*sic*] and antagonism; and yet the whole company is fused into one by the breath of song!" (Mann 1844/1982, 153).

Leaving his position as lawyer for the state of Massachusetts in 1837 to head up the state's public education system, Horace Mann is reported to have said, "Let the next generation be my client."[32] Like many civil servants and educators of this period, Lowell Mason and Horace Mann saw themselves as "the constant guardians of virtue" (Mann, cited in Brooks 1952, 181). Initiating a new way to certify future citizens in publicly funded schools, they spoke for a secular theology in which schools and music societies would remake the child into a more reasoning character than he or she would be otherwise. The transformation was nothing short of salvation (with the tacit proviso that many were not eligible for this redemption). Several Boston School Committee members echoed Mann's sense of the sacred mission in music instruction: "There . . . [are many] thousand common schools in this country . . . [these] will mould the character of this democracy . . . If vocal music were generally adopted as a branch of instruction in these schools, it might be reasonably expected that in two generations we should be changed into a musical people . . . Thereby you set in motion a mighty power which silently, but surely in the end, will . . . refine and elevate a whole community" (Boston School Committee 1837/1982, 141). Public school jeremiads, like the speech of Abraham Lincoln, turned the biblical lexicon and rhetoric to civic and institutional aims. It was an all-inclusive secularism in which the consecration of civic and state laws and actions followed particular contours and legal definitions of personhood prevalent at that time.

The notable result of the historical shift from religious institutions to schooling, with music instruction as an organ of socialization, was that it made new distinctions about the individual's musical life and participation possible. Interchangeable civic, religious, and racial registers qualified or disqualified the child as a bona fide political actor—that is, a citizen.[33] Music instruction was to develop musical skill, not for its own sake but for the good of the community, defined as the social group with genteel taste. Yet, the mechanisms of comparisons of musical dispositions went unnoticed as they were hidden in the proposition of universality offered by the characteristic melodies and lyrics of school songbooks. How melody and lyric posed as universal ideals is the next subject to be addressed.

School Songbooks and the Fabrication of Biographies

Considering singing as a recognition or hailing of the merit of a person's entrainment is a substantially different way of organizing the relation between music, teachers, and students than we find in traditional histories of music education. In standard professional texts, the objects of study and pedagogy function as markers of social standing. In contrast, the concept of entrainment, as I use it here, is an attempt to disrupt the normalization of music and human hierarchies embedded in curriculum and in traditional histories of the field.[34] That is to say, the persona portrayed in a song provides an occasion, on the one hand, for hailing the singer, or, on the other hand, for demonstrating a failure to make connections between the song, the student, and her embodied musical disposition.[35]

The alignment of singer, song, and social niche is complex. Differences between singers cover a range of experiences and associations that defy strict correlations between an individual's social class, musical disposition, and aspirations.[36] Because the idea of enacting social recognition in these school songs has not been systematically addressed in the field of music education, I will look more closely into these dynamics through a random selection of school songbooks published in the middle to late nineteenth century.

Song texts and melodies, more than the manuals, rationales, and testimonies to the school boards, provide a view of the fabrication of the biographies of child singers. Images of these songsters serve as models for future citizens—that is, for those who are hailed by or who exhibit a familiarity with the sentiments and musical sensibility of this period.[37] Exemplars of songs from the 1830s to the 1880s reveal the frequent pairing of images of the nation with school, teacher, friends, domestic scenes, and landscape.[38] For some, the language of the verses intensified autobiographic memory, while for others, a different set of images would be a better match.[39] Performing identities triggers a range of memory and also reconstructs memory. In this sense, school songs perpetuate, distinguish, and enshrine ideal images, while there is a counter process of omitting or forgetting, similar to the process of drawing a map of the nation and of citizenship in musical terms.[40]

The most common topics of songs sung in school in the nineteenth century were nature and social life.[41] Song verses emanated from popular stereotypes and pseudoscientific tracts on race that divided the population into phylogenetic types with respect to musical disposition. This is not to suggest that human types were systematically described in curriculum documents. More accurately, vocal instruction drew students to compare themselves to particular images of home, genteel mores, and patriotic sentiments. This placed them at points along a continuum from the barbaric to the civilized, forming a template for the genteel life—depicting the home, work, and social life of the bona fide citizen.[42] Many songs echoed the slogans of various moral militias, such as temperance leagues, composed to reclaim the errant, reform the alcoholic, and save the family.[43]

This side of public music instruction locates the selective hailing of the singer-citizen,[44] which fabricated future citizens with the virtues characteristic of a particular segment of the population.[45]

Reiterated performances of imagery have a cumulative affect,[46] evoking the world of the song and no other, and making the singer's world, however momentarily, the world that *is*.[47] Songs published in four decades of school songbooks demonstrate this idea, in that the allure and charm of the citizen ideal is constantly being shaped and rehabilitated in response to the older canon and in response to the social anxieties of the urban, industrial cities of the North. The preface by Lowell Mason and Elam Ives's in *Juvenile Lyre or Hymns and Songs*, notes that singing was the perfect device for establishing a pastoral calm in the classroom. Songs such as "Come Children, to the Garden We'll Go," "The Morning Call," "Spring Flowers," "Little Wanderer's Song," "Humble Is My Little Cottage," and "The Thunder Storm" paint a bucolic scene, reflecting the Romantic Movement's tropes on childhood and nature. In "The Morning Call," for instance, children are the flower gathering "we" of the song, bringing lilies, jasmine, and roses to the schoolroom.[48] These images fabricated a particular experience of the middle-class city child and, by implication, characterized the cities and towns as bucolic. However, only privileged parts of Boston had flowering gardens. The area outside Beacon Hill was crowded, noisy, decaying, every spare inch of land covered with shanties and refuse. This was an area settled by immigrants where "brick warehouses and small manufacturers edged streets once lined with trees . . . Ann Street [the main thoroughfare] had become a synonym of vice and intemperance."[49] By evoking the vision of Boston as a garden, the floral images in the songs elide the slums with a heavenly scene, hailing the experiences of the few as belonging to Paradise and the garden city. In a similar way, typical songbooks such as *The School Chimes*, *The Silver Bell*, and *The Golden Wreath* contained titles such as "Cheerily, Cheerily," "Come Boys Be Merry," "Glad Hearts," "Oh Merry Bells," "Love of School," "Oh what a Happy Group," and "Love of Friendship," creating a space where the school pupil was "free from care" and "full of glee." *A Choice Collection of Favorite Melodies, Schools, Seminaries, Select Classes and Etc* reiterated middle-class social values as the prerequisite for learning. In another songbook, the performative function of the pronoun "we" links domestic tranquillity with the peace of nature as the normal situation:

> Our haunts shall be
> Nature's own beautiful bowers
> Our gems shall be
> Nature's own beautiful flowers. (Bradbury 1847, 82–83)

"Our haunts" are pastoral imaginaries that represent the child-singer in a bucolic, unencumbered world. By all accounts, and particularly in Horace Mann's records of city life, living conditions were unsanitary ad extremely poor for most of Boston's inhabitants.[50]

Moreover, as reflections of the Romantic style, a particular register of social thought, connected to Immanuel Kant's concept of a cosmopolitan society built on reason, linked the universal sense of beauty to the moral dimension of experience: "The literary character and moral sentiment of the poetry [of a school songbook] which children learn, will have an abiding effect upon them through life—or rather it would be more correct to say, they will constitute a part of their moral nature—In sentiment it should inculcate all kindly and social feelings; the love of external nature—the love of country and that philanthropy which looks beyond country and hold all contemporaries and all posterity in its wide embrace" (Mann 1844/1982, 153). The pastoral verses inscribe images that will have an "abiding effect" according to Mann, on those who base their moral sentiments on cosmopolitan principles. Such sentiments were to unite singers in social values, yet they also identified types of schoolchildren in relation to experience and disposition. With no mention of religion, the secular songbook relied on bucolic themes to encompass different creeds, identifying "nature" as the sacred. The pastoral prototype for the school songs far outweighed in number other kinds of music available from publishers and in live performances in the environment. These songbooks tied the home to a pastoral landscape, ranging from the peaceful domesticity of the gentle hearth to a noble nation with its rocky coast and rugged imagery.[51] The overall pattern that emerges from the lyrics of the songbooks of the period establishes an ideal portrait—snug, secure, restful—of a home that upholds secularism in all matters of the state and a mild Christianity as the faith of the patriot. These were unquestioned universals and they had the effect of normalizing *genteel* social settings, domestic circumstances, and expression as the trappings that would identify the worthy citizen.

School songbooks' romantic, literary character sharply differentiated them from the musical practices of the street and ordinary folk.[52] Omitting the popular, including items that were transcribed from "Indian" and Negro music, school song collections purveyed a homogeneous image of countryside, people, and nation bound by mutual sentiments of love and admiration. The twin ideals of domestic tranquility and the pastoral fabricated the child's future and, through reiterated performance, lent them solidity as the exemplary sentiments and physical charm of meritorious personhood.[53] Imbued with genteel values, school songs were moments that intensified social differences, deploying, unwittingly, a politics of inclusion and exclusion. Like the classroom incidents recalled in the prelude to this chapter, the intersecting registers of music instruction and comportment reinscribe distinctions between students even as they are creating ideals of overcoming class and racial conflict. At the time, the myriad social institutions and practices of, for example, phrenology and ethnology churned out an industry of literature on human types that sharply differentiated the Irish Catholic and the "Negro" or "African" from the white.[54] Underlying the urgency of these comparisons was the constitutional conflict in which the debates on slavery and the Civil War heightened the fragility of white social privilege. Thus, the atmosphere in which school songbooks were published was highly charged with religious and racial contentions; these fed fears of miscegenation and the disintegration of civil society that erupted in accusations of sexual violation of white women. In the early nineteenth century, these fears were manifest in public school segregation debates and in the discursive web formed by ethnological journals and the popular press.⁵⁵ In these venues, whiteness was a marker of moral strength, rationality, and virility. Inflected with a racial dimension, the effort to prove manliness drove the enlistment of black men who were attempting to garner a living as well as higher status through military service: "During the Civil War, 180,000 Black men enlisted in the Union Army . . . they understood that enlisting was their most potent tool to claim that they were men and should have the same rights and privileges . . . They understood that the only way to obtain civic power was through gender—by proving that they, too, were men" (Bederman 1995, 21).

Patriotic songs in the school canon of songs fortified the qualities of citizenship, duty, patriotism, and manliness as aspects of whiteness. This was especially the case for school songbooks, which intermingled a domestic sense of the pastoral with patriotism as a normative scenario. As the contest between civilization and barbarism took shape between the Union and its enemies, the song collections published in the North were part of a discursive web connecting whiteness, civic duty, and domestic virtue.⁵⁶ *The Nightingale: Songs, Chants and Hymns Designed for the Use of Juvenile Classes, Public Schools and Seminaries* (Perkins and Perkins, 1866) outlines a typical collection of themes that make sacrifice for the nation a familiar identity, equating citizenship with the recognition of manliness granted only to white heroes.⁵⁷

In the decades that followed the Civil War, there was a new discursive intensification of race in tandem with evolutionary theories on child development and the growth of civilization and music. Complexity and refinement were the hallmarks of the evolutionary superiority of Western music.⁵⁸ Self-cultivation in the arts was coterminous with progress. As these changes registered themselves in curriculum literature and music texts, the school subject of music bore a unique responsibility, specific to music, for recognizing merit in the future citizen. He now displayed the comportment and advanced racial "inheritance" as the product of evolution through his participation in singing and ethnic identity.

The continual references to European vocal methods and the proliferation of songbooks with models for civic comportment were texts and practices that provided a contrast to "savage," particularly non-white, elements in the population and to those who would not merit recognition. A fuller treatment of this subject requires a survey of the nineteenth-century cultural themes of degeneration and the racial anxieties enfolded in the weakening of the Puritan heritage. The arc of degeneration left a heavy imprint on school singing and on the curriculum's notions of what "music" was. In some sense, the idea of the fall was inscribed in the idea of progress itself and the recuperation of mankind from a savage state by overcoming those obstacles embodied in uncivilized groups such as Native Americans, African Americans, and Irish Catholic Americans. In the next chapter, I discuss the relation of degeneracy to the medical-biographical aspects of the music curriculum, focusing on how health regimens furthered the reading of human types as carriers of contagion or victims of contamination.

# 2

The Child as Charmed Victim

Early Vocal Instruction and the Social Distinctions Conferred by Disease

> Natural magic ceased to belong to the [West], but it persisted for a long time in the interaction of beliefs . . . That [these beliefs] do not possess the formal criteria of a scientific form of knowledge does not prevent them from belonging, nevertheless, to the positive domain of knowledge.
>
> —Michel Foucault, *The Order of Things*

General health occupied a sizable place in the Boston School Committee testimonials for vocal instruction in the 1830s and 1840s. That period is marked by the transition from older phrenologic and homeopathic practices in medicine to the laboratory science practiced first in Germany. The role that medicine at this time played in shaping the general kindergarten through twelfth grade music curriculum has been underestimated and largely overlooked. Medical concepts were prominent in Horace Mann's support for vocal instruction: "Good blood [enriched by the oxygenating effects of singing] . . . gives more active and vigorous play to all the organs of absorption, assimilation and excretion" (Mann, 1844, 149). Mann's ideas were considered unscientific by the early twentieth century, when historical accounts of music education began to appear, but they have considerable significance for the reception of and support for public music instruction.

Ideas about the circulation of oxygenated blood and lung exercise established a way of reasoning about the young singer as a specific human type, and Mann's medical concerns tell us something important about the role of health in relation to recognizing a particular type of person as the pinnacle of civilized humanity. These also bring up the anxieties about disease that circulated along with racial descriptions of the typical contagious individual and her victim.

The Vocal Curriculum and Circulation of the Blood

Cholera epidemics in New York and Philadelphia in the 1830s and the ongoing toll of tuberculosis affected a sizable portion of the population in the larger cities in the eastern United States. Representing the possible destruction of the American city and cultural life in the early nineteenth century, the possibility of preventive regimens made public vocal instruction alluring to a health-minded population. In the larger picture, concern for the health and disease of one's fellow citizens resonated with the cultivation of a disinterested civic virtue that was necessary to build a strong republic, a type of virtue that was favorable to economic progress and self-government.[1]

The hygienic concerns appearing in several of the Boston School Committee reports were underpinned by various intellectual and scientific authorities of the day.[2] T. Kemper Davis, a member of the Boston board, cited John Locke's belief that music was a healthful release from the wearying effort of study.[3] In Lowell Mason's "Manual" (1834), one of the stated major effects of vocal instruction was improving the health of young girls by strengthening the chest and fortifying resistance to consumption. Citing Benjamin Rush, a prominent medical authority, Mason writes, "[Singing] has a salutary operation in soothing the cares of domestic life."[4] These statements presented the ideal woman as a homemaker, an image that overlooked the labor performed by women in other spheres and social classes.

Preventives in the form of quack medicines, but also exercise regimens, were the stock in trade of doctors and surgeons. One of the era's most popular treatises on consumption was based on a theory of the positive effects of inhalation treatments or lung exercise on tuberculosis patients in a pneumatic institute. Another widely read work on tuberculosis presented the central thesis that fresh air should be continually inhaled in order to replenish the blood supply with oxygen and to counteract the miasmic conditions in the cities that in themselves were thought to be the cause of tuberculosis.[5]

Horace Mann cited physiological science and the advice of a number of prominent physicians to establish that singing promoted good digestion and nutrition.[6] He wrote that poor lung exercise and diet led to degeneration of the blood, describing the risks that epidemic disease posed to civil society and the latest regimens for the stimulation of circulation as a prime preventive: "Vocal Music promotes health. It accomplishes this object directly by the exercise that it gives to the lungs and other vital organs and indirectly by the cheerfulness and genial flow of spirits, which it is the especial prerogative of music to bestow. Vocal music cannot be performed without an increase in action of the lungs ... and [this] necessarily causes an increase in action of the heart and of all the organs of digestion and nutrition" (Mann 1844, 148–49). Similarly, Lowell Mason wrote that singing was responsible for low rates of consumption among Germans: "The music master of our academy [reference to Gardiner, with no citation in Mason's text] informs me that he had known several instances of persons strongly disposed to consumption, restored to health by the exercise of the lungs in singing. About twenty percent of all deaths that occur are caused by consumption" (Mason, 1834b, 130). The Boston School Committee asked, "But why cite medical or other authorities to a point so plain? It appears self-evident that exercise in vocal music, when not carried to an unreasonable excess, must expand the chest and thereby strengthen the lungs and vital organs" (Boston School Committee 1837, 141).

Phrenology and *Wissenschaft*

Several of Horace Mann's annual reports to the Boston School Committee reflect some of the most advanced early nineteenth-century medical protocols practiced in the German universities. *Wissenschaft*, or laboratory science, as it was called then, codified disease processes through symptom observation rather than through the framework of the conjecture, the four humors, skull shape, or other means that had dominated medicine for centuries.[7]

> The singer brings a greater quantity of air in contact with the blood. Hence the blood is better purified and vitalized. Good blood gives more active and vigorous play to all the organs ... The better these functions are performed, the purer and more ethereal will be the influences which ascend to ... a well-formed brain, when it is supplied with healthful and oxygenated blood. The scientific physiologist can trace the effects of singing from the lungs into the blood; from the blood into the processes of nutrition ... and finally from the whole vital tissue into the brain ... there may be various conspiring or disturbing forces tending to aid or defeat the result, but still from beginning to end the connection between cause and effect is ... as a broad line running across a black surface. (Mann 1844/1982, 149)

Phrenology, practiced in those decades alongside scientific physiology, provided a system of knowledge based on a reading of abilities and propensities through analysis of the skull. Practitioners and adherents included Mann and many in the elite circle around him who referred to the well-formed brain as the beneficiary of singing instruction. In addition to diagnosis from features of the skull, phrenology used physique and skin tone to diagnose illness. These areas predisposed the individual to disease in different ways, depending upon race and class, among other factors. The elaboration of diagnostic terms in vocal instruction provided the schoolchild with a calculable medical future and a biography based on appearance. The overly studious type was a sure mark for early death: "In visiting schools I have found it a common occurrence that when the hour of recess arrives ... some half dozen pupils, with pale faces and narrow chests and feeble frames, will continue bending over their desks, too intent upon their lessons to be aroused by the joyous shouts ... from abroad ... Alas ... those children are victims of an overactive brain and that every such disproportionate mental effort is a cast of the shuttle that weaves their shrouds"(Mann 1843, 36).

Phrenological biographies emphasized exteriorly visible symptoms and signs as prophesies of physical constitution, especially with regard to tuberculosis, and as Mann wrote, with regard to those who were both pale and frail.[8] While the thrust of these remarks was aimed at all schoolchildren, the remarks also located the child within a white population that was seen as having an "inherent" susceptibility and being more vulnerable to carriers of the disease, who were pictured in the popular literature as members of the laboring class, dark-skinned, or Catholic. In this manner, the qualities of paleness, thinness, and studiousness refer to a phenotype that stood in sharp relief to the stereotypes of Native Americans, blacks, and the Irish Catholic (Celtic "race") immigrants.

Further on in his *Sixth Annual Report*, Mann interjects a sense of emergency into his presentation to the school board through a quotation of the mortality rates reported in a British study of occupational hazards.[9] Arguments through statistics were part of the new *Wissenschaft* methods practiced in advanced laboratories in Germany. Mann used the appeal and prestige of the new science to demonstrate the need to save the nation through school courses on physiology as well as the pulmonary benefits of music instruction.[10] His interest in physiology was consistent with the arguments he later made in the "Report for 1844" for vocal music's healthful effects.[11] In this report, he used the jeremiad's rhetorical devices to link vocal instruction to the medical sciences but also to tie vocal pedagogy to the urgency of saving the population and the republic.

Perhaps the most important consequence of the populational reasoning enfolded in Mann's phrenologic and laboratory-based prescriptions was that in posing physical causes for disease and in tracing the effects of singing in the respiratory system, Mann was also arguing against the religious notion, still strong among in the population at large, that disease was a punishment for sin.[12]

Phrenology, most infamous as the study of skull shape with regard to character and personality traits, also outlined complex systems of exercise and diet that overlapped with a growing middle-class white concern about the degeneration of the race. As popular medical advice, phrenologic regimens were popular among cosmopolitan elites and the middle class who had interests in classical music, music education, dietary regimens, and exercise.[13] One tenet of phrenology held that it was hygienically wise to avoid exciting or extreme activity such as exposure to atmospheric changes, excess in eating and drinking, mental fatigue, impure air and too sedentary activity. In agreement with these warnings, vocal music's effects were seen as moderating the tendency of the body to overheat during the usual recess in school routine. Touching on this concern, the educators involved in vocal instruction wrote that music added just the right kind of excitement to social events, especially music that was not too boisterous and stimulating. Singing the right music prevented the boisterous overheating suggestive of promiscuity. This connects the physical disposition of the child discursively to warnings about dance and degenerate sectors of the population.[14] In addition, the fear of boisterousness forms a link or discursive chain from the health regimens popular among the genteel class and its aspirants to the distinctions with regard to body comportment made in vocal instruction that would mark some children as meritorious, others as less so. A letter published in the *Boston Musical Gazette* on the subject of music classes at the Hawes School illustrates the connection between genteel mores, singing instruction, and moderation in exercise. As an example of the way the school subject of music undergoes an alchemical mixing of health concerns, scientific precepts about the body, and genteel values, the letter states little about music itself, except that school attendance seemed to improve on days when there was music. Focusing instead on the moral and hygienic effects of singing, there is a particular concern about maintaining control over girls' behavior, related to their sexual and reproductive roles. This emphasis on gender at the end of the letter underlines the overall theme of the preservation of a genteel race and coincides with the notion that the sexual role of genteel white women ensured the continuance of civilization: "The children are agreeably employed; and we are certain that they are innocently employed ... Of the great moral effect of music there can be no question ... It is a relief to the wearisomeness of constant study ... It excites the listless ... It is with moderate physical exertion ... [and it is] preferred to boisterous, overheating and sometimes dangerous play ... [thus preventing] an excess that almost unsexes [the girls]" (Harrington and Harris 1838, 7).

Horace Mann, outlining a corollary to the idea of too much excitement, spells out the consequences of excessive constraint, which endangers the middle class in the city: "With the decline of physical vigor, the natural desire for physical effort ceases; animal spirits are not generated ... [That which remains is] ... inertia ... The offspring of such parents inherit their feebleness. In the price-current, too, of certain classes of society, clothes are more valuable than health ... children must not be allowed to sport ... Because boys and girls were made for shoes and stockings ... There is a tendency in cities to a rapid deterioration of the race ... The body shrinks, the limbs droop and pine; the size of the brain diminishes" (Mann 1843/1950, 148). Mann's criticism of sedentary activity is addressed broadly to the relatively small portion of Boston that belonged to the white middle class. We see the ready associations to a visible whiteness that pictorially surface in his criticism of an imbalance of values: "clothes are more valuable than health ... children must not be allowed to sport." Mann also sounds the note of fear in the "rapid deterioration" (Mann 1843/1950, 148) of his generation's physical well-being, but he also fears for the future, as the enfeeblement of body and brain were, as Mann and many believed, inheritable.[15]

In this manner and in similar arguments put forward in prefaces to school songbooks, vocal instruction represented a nonfatiguing, yet invigorating, exercise, an alchemical mixture of vocal activity and physiological precepts that produced the gold of the genteel body in its masculine and feminine prime. School songs facilitated the division of children into categories of domestic tranquility as against the stereotyped images of social inferiors. As this type of idealization continued throughout the century, numerous songs pictured the sacredness of a peaceful home threatened by forces outside the sacred circle of the home:

> The hearth of home is beaming,
> With rays of holy light;
> And loving eyes are gleaming,
> As fall the shades of night;
> And while their steps are leaving
> The circle pure and bright. (Giffe 1875, 74)

The orchestration of this domestic scene inscribes a kind of civility that speaks to genteel values and to vulnerability: a sharp contrast to the images of the home life of the poor that alluded to moral and physical degeneracy during this same period. These were consistent with the atmosphere of concern over the social obstacles that threatened the progress of the republic. "[The Fort Hill District] ... was a perfect hive of human beings, without comforts and mostly without common necessaries ... [crammed] together like brutes without regard to sex or age or sense of decency ... grown men and women sleeping together in the same apartment, and sometimes wife and husband, brothers and sisters, in the same bed ... under such circumstances, self-respect, fore-thought, all high and noble virtues soon die out ... disorder, intemperance and utter degradation reign supreme (Report of the Committee on Internal Health 1849, Boston, cited in Schultz 1973, 222–23).

Home versus Tavern: Cholera and the Drinking Poor

The words of school songs painted ideal domestic scenes urging the schoolchild to take up the reform of the poor. Ironically, while public school music appeared to offer the possibility of health and happiness to *all* through vocal exercise, the school songs created enclosures around the ideal home and its typical inhabitants as those who would reform *others.*

The issue of school truancy divided children along several lines, since truancy was associated with immigrants, tavern-going families, and crime. One Boston city document numbers some 963 truants in the category of truants, with a significant number of them listed as indentured servants. Children who ranged in age from six to sixteen were described as raised for prostitution, thievery, and alcoholism, in the words of Josiah Quincy, Jr., mayor of Boston in the early nineteenth century.[16]

Intemperance was associated with the spread of cholera.[17] As the song "The Brewer and the Goat" makes clear, the schoolchild whose family could be linked to tavern going was in no position to be a right-reasoning citizen:

> The brewer had tried it,
> His customers too;
> How silly it made them, they partially knew;
> And hence thought the Brewer
> It seems so at least,
> We'll see if we can't be No Worse than a beast. (Giffe 1875, 74)

While lung disease had a starring role in the reports on vocal instruction in Boston, the cholera epidemic of 1832 was an important, if less prominent, school board issue in the minds of the civic fathers. Some blamed taverns for spreading cholera. The epidemic intensified the classification of immigrant versus native born, and white versus black. Such distinctions churned through school songs, heightening the importance of teaching "the right" or righteous child to display the zeal to reform others:

> Be our zeal in heaven recorded
> With success on earth rewarded,
> God speed the right,
> God speed the right. (Mason 1856, 44)

Cholera evoked images of the plague and was perceived as a massive threat to civil order on both sides of the Atlantic. There was little known about the infection.[18] To make matters worse, the similarity of the first stage of cholera to more ordinary sicknesses raised the level of hysteria. Exaggerated rumors prevailed over medical advice. The belief that the disease appeared to travel from port to port convinced people who could afford to do so to flee the coastal cities. Doctors attempted to educate people to the fact that cholera was passed on in contaminated food and water and had to be ingested by prospective victims in order to cause the disease. While the threat of mass death eventually led, in midcentury, to the regulation of populations through public sanitation and quarantine, in the 1830s, physicians and civic leaders were limited to the pastoral role of comforting and advising citizens.[19]

The epidemic had long-lasting effects in prompting cities to keep statistical records establishing a correlation between food, water, and disease. These statistics embodied a scientific authority that was to transcend superstitions and religious belief. However, the algorithms that compared groups in the population, specifically the U.S. census records of each decade, also inscribed notions of inherent difference, since the conditions under which people live overlaps with ethnic, religious, and racial distinctions.[20] Margo

Anderson wrote, "It is the census that triggers increased (or decreased) resources for a geographic region, and thus it is the census that has been used to illustrate the virtues or vices of particular regions, peoples or ways of life in America" (M. Anderson 1988, 22). Statistical calculations translated demographic divisions into scientific nomenclature.[21] The result was a form of racial and social demonization that attributed the cholera epidemic to poor blacks and Irish Catholics, or those who could not travel out of harm's way and who succumbed to cholera in disproportionate numbers. The popular belief was that the disease came from stagnation of the air in "miasmic" urban eras, where sin, unwholesome diets, lasciviousness, and intemperance stereotyped the lives of the poor. Taverns, especially, were seen as breeding grounds of infection.

Public and private institutions disseminated a mixture of popular misconceptions and medical advice. School temperance songs decried the proliferation of taverns from which the French wine industry distributed its poison,[22] the substance that wrecked health and home. Drinking was a sign of playing fast and loose with cholera. The school singer took up the mission of rescuing others, bearing his duty contentedly in temperance ballads such as "Cheerily, Cheerily" from W. T. Giffe's *The New Favorite*. In this song, a band of schoolmates perform as a "temperance army," cheerful souls who equate sobriety with liberty: "Sing again the merry charm, we are free." While temperance songs were performed in other places, their inclusion in the school repertoire underlines the continuity between the moral societies, a budding public health consciousness, and the activation of a template for the self-governing, vigilant citizen.[23]

Temperance songs made the singing lesson the province of the domestically virtuous, where both singer and song deployed reason against alcohol. Another temperance ballad identifies abusers as follows: "They that tarry long at the wine, they that go to seek mixed wine," and "Who would be a drunken sot, the worst of miseries . . . See how many bosoms bleed!" The song "Wine is a Mocker" asks, "Who hath babbling? Who hath red eyes? Who hath wounds without cause" (Bradbury and Sanders 1841, 136–37). In these types of verses, the singer as the adjudicator of merit appointed himself the guardian of others and himself as well. His role was to reform drinkers and thereby save society from the ravages of disease purportedly caused by drinking:

> Raise the cry in every spot,
> Touch not—taste not—handle not!
> Who would be a drunken sot! (Giffe 1875, 137–38)

Songs sung in school were important as instruments of governance that brought moral values to the home where parents were deemed lacking or especially resistant, an administrative function noted in the records of the Boston public schools.[24] Thus the registers of civil administration, vocal pedagogy, and health were mutually reinforcing.[25]

The child's future was fabricated through a reading of his or her physical comportment and bodily characteristics but also through a reading of the family and home. This brought music pedagogy into the center of schooling's alchemical fabrication of a national home life and the mystique of the meritorious persona.

Medical Prescriptions, Consumption, and the Pale Complexion

In the nineteenth century, a charmed comportment and appearance were closely linked to the disease of consumption. The illness was believed to strike only the white, genteel class. The bacterial causes of consumption, or tuberculosis, as it was later called, were unknown until the late nineteenth century. It was not until the twentieth century that other groups in the population were recognized as vulnerable to the disease. Medicine provided a rationale for vocal methods as both an intellectual and a physiological practice to achieve the full potential of an ideal Anglo-Saxon or Teutonic individual.[26]

Horace Mann's image of the well-formed skull and brain resonated with the ranking of mentality according to a racial hierarchy. The racial nuances emanating from the school songbooks—where a particular complexion, comportment, temperance, and affect spelled future health or ill health—made students legible along two lines: the self-governing and the ungovernable.[27]

When Horace Mann was thirteen, his father died of what was thought to be consumption; then his first wife succumbed to the disease after one year of marriage.[28] Mann argued passionately for measures that would make a difference in mortality rates: "In our climate . . . about twenty percent of all the deaths that occur, are caused by consumption; and this estimate includes infancy and childhood as well as [adulthood] nearly one half of all the deaths that occur, are caused by this terrible [consumption] alone."[29] These would be the deaths of the children with "pale skin" and "narrow chests," children who overexerted themselves mentally and remained sedentary when other children went out to recess.[30] As inducements for vocal music instruction, both Mann and Mason's prescriptions borrowed a page from the body culture movement, in which singing would enlist a proactive capacity to take an interest in the body as a *project to perfect* and to ensure its long life.[31] Lowell Mason wrote in his *Manual for the Boston Academy of Music* that singing is one aspect of a system of regimens that "aspires . . . to develop Man's whole nature"(Mason 1834b, 130). Horace Mann argued that vocal exercise produced a superior human type: "[This person] . . . developed into the flower and fruit of cheerfulness [enabling] a prolonged life, just as easily and as certainly as a skilful manufacturer can trace a parcel of raw material which he puts into his machinery . . . until it comes out at last, a finished and perfect product" (Mann 1844/1982, 149). The body, as a functioning machine, used, and was liable to *use up*, its supply of energy; in the set of tenets associated with phrenology, the life quantum was sometimes referred to as a vitalism or life force. Mann's vocal prescriptions are derivative from vitalism and often combined with other physiological diagnoses.[32] The benefit of vocal instruction from this vantage point was in the regeneration from a state of expended energy to bodily and mental renewal.[33] "The simple and calm delights of music restore the energies that have been wasted by toil, revivify the spirits languishing with care" (Mann 1844/1982, 153). Speaking of music's regenerative qualities, Lowell Mason wrote that "[Music] is almost the only branch of education . . . whose direct tendency is to cultivate the feelings . . . Our systems of education generally proceed too much on the principle that we are intellectual beings, not susceptible of emotions, or capable of happiness" (Mason 1834/1982, 129).

According to the profile, the consumptive was usually female, fair, sedentary, and sensitive, a person whose life habits of low exertion led to underinflation of the lungs and poor oxygenation. In the decades that followed the first public singing instruction, dozens of school songbooks appeared with lyrics reflecting popular images of consumptives as white, saintly, and frail. The phenotype in medical terms was also the romantic heroine, recognized through blushing, as, for example, in the song "A Maiden Fair" in *The Cyclone of Song*: "A rosy blush her cheeks did flush" (Leslie 1888, 46). The rosy cheeks had a double function: not only did they signify the subject as the "chosen" of grace, they also indicated that the individual was vulnerable. Likewise, a rosy complexion or flushing was a condition that might indicate fever, one of the signs of consumption. The proper musical activity would have restorative effects, especially where there was physical or mental exhaustion; it would provide the affective basis necessary for the mind to refresh itself and restock the "energy and electric celerity"[34] that was protection against disease.[35] *The School Harp's* preface states: "A lively, exhilarating song is like an electric shock, which will suddenly arouse both the physical and intellectual powers into renewed and invigorated action" (Bascom 1855, iii).

The "bright," "merry," and "fair" qualities of the future citizen in school songs also connoted superior respiratory and circulatory systems. A theory of hematological inferiority among blacks was disseminated from the work of S. A. Cartwright, a nineteenth-century medical doctor who presented papers on his observations of the circulatory process among slaves at medical meetings. Cartwright believed that blacks had a circulatory system that polluted the blood with carbon dioxide, creating mental deficiency.[36] The disorder, called dysthesia, received support from a network of physicians and scientists who subscribed to polygenist views of racial, mental, and physiological inferiority. Another medical theory posited that the low rate of consumption among blacks was due to lesser mental acumen and less sensitivity to the environment.[37] These were but two of a wide range of biological comparisons that relegated blacks to the lowest rung in the racial hierarchy. The image of rosy cheeks functioned as part of an assembly of whiteness that contrasted with images of darkness in school song lyrics.[38] Where darkness appeared, it was invoked as a sign of danger, vice, and bestial impulse. For example, the song "Sleep, baby Sleep" sets the child up as the white lamb that is rescued by a shepherd:

> Thou old black dog, so fierce and wild,
> Adown there falls a sweet dream on thee
> Baby, sleep . . .
> The full, fair moon is the shepherdess. (Giffe 1875, 74)

"Sleep Baby Sleep" is typical of many school songbooks' selections in its use of darkness to contrast the "fair" with disease and moral degradation as the prevalent dangers.[39] Where the dark or light tropes in the song match racial stereotypes, the health of the "fair" is the hope of the health of the nation.

Many songs were instructions for the future citizen to distance himself from particular areas in the city. Some areas of the city were rumored to be sources of mass contagion for cholera and other infectious diseases; these were low-rent districts where immigrants and freed slaves lived.[40] In several school songbooks, there are selections that describe the streets as breeders of vice and sources of contagion where susceptibility to illness was read through skin color, build, and affect. Whole neighborhoods were stigmatized as diseased;[41] night in the street was imagined as the time of day when sickness and evil snatched their victims from the streets. As a song from *The New Favorite* songbook phrases it:

> Let fate be dark or bright;
> At home no dart can wound thee,
> Then don't stay late tonight. (Giffe 1875, 74)

Physical and racial typologies obliged individuals to regulate their activities and life expectations. The upstanding were secular guardians of others and *themselves*, where virtue was increasingly defined as avoiding the immoral behavior that led to ill health.[42] The song "Your Neighbors O'er the Way" reads as follows: "You judge their actions every day, You know when they've gone wrong" (Leslie 1888, 46). But the song's main thrust is self-vigilance:

> You say and say it truthfully,
>   "I'm sure I know myself,
> I never had to struggle with some
>   wicked hidden elf." (Leslie 1888, 46)

The hailing of the future citizen as the prospective consumptive was a double recognition of a particular *medical* type attached to the charmed appearance of personae who exemplified whiteness. The obsession with the tropes of light and dark is, of course, dispersed throughout Western art, in effect comprising a genealogy of the use of color to symbolize cultural taboos. Toni Morrison discusses "blackening up" and "whiting out" (Morrison 1992, 87) as literary strategies of sexual prohibitions and titillation to forward the plots of American novels in the late nineteenth and early twentieth centuries. Eric Lott notes this propensity as central to blackface entertainment, which degrades color while participating in the vicarious thrills of miscegenation, among other taboos, in that period. By calling attention to the diction of song lyrics, I am attempting to describe how racialist themes in medicine, gender, physiology, physique, and skin tone make their way into the vocabulary of comparisons between the civilized and the barbaric, separating the afflicted and threatened members of the Anglo or Teutonic racial group with "the long-flowing graceful ringlets . . . rosy cheeks and coral lips . . . that noble gait" (Charles White, cited in Gould 1996, 73–74) from those categorized as the contaminated degenerates around them. According to the literature and popular myths that took the Nordic or Teutonic type as the ideal, this was the *only* body type that could be worked on through self-discipline, shaped, and protected from contagion.

Happiness, Facial Expression, and Ranked Merit

There were also affective components to the charmed appearance. The typical victim of consumption lacked "gayety," an indication that the disease was thought to visit the melancholy personality. According to Horace Mann, "gayety" eroded the opportunities for disease to take hold. The disease-prone type could be merry and bright, fortified against contagion, for the moment, but his affect, if changed, would invite disease. Consequently, happiness or gloom played a large role in fabricating the biography of the child and inscribing a character that was fit or unfit for his role as citizen.[43] School songs activated the worthy to keep watch over those posing risk, hence the reiteration of the imperative of cheerfulness in many of the school and friendship songs:

> Come away, Come away,
> Like a merry boy
> With a tug and a pull and a smile. (Bradbury and Sanders 1831, 110–11)

Vocal instruction was also linked to the pursuit of happiness as a republican ideal. Lowell Mason's preface to *The Normal Singer* reads as follows: "This book, if it be indeed normal, must be right, or a book in which the principles of song are treated according to their true relation to the great work of education, or to human improvement, goodness and happiness" (Mason 1856, iii). Other school songs reiterated cheerfulness; for example, from C. E. Leslie's *Cycle of Songs*: "You know the work he has to do is always gladly done."[44] As Mann stated, vocal instruction's physical effects were instrumental in achieving cheerfulness through increased circulation. "The scientific physiologist can trace the effects of singing from the lungs into the blood; from the blood into the processes of nutrition ... and finally from the whole vital tissue into the brain ... to be there developed" (Mann 1844/1982, 149). This was consistent with the old religious covenant that bound the individual to do God's will; secular pursuits of happiness, as a key idea in the Declaration of Independence, came as liberty merited by work. The circle of duty and happiness was intimately connected with the government of the individual and civil society, in which "new patterns of governing in the nineteenth century ... focused power on persons in everyday life."[45]

The preface to *The Young Choir* school songbook states that the constant use of music "inspires and cheers ... [as an efficient] contributor to cheerfulness and good nature" (Bradbury and Sanders 1831, iii). "Brightness" and "fair," recalling the description in "School Song" from *The School Chimes* (Baker and Southard 1852, 112), were words that characterized the school cohort. Similarly, a song about giving a chance to a young boy who is diligent in everything, with its explicit connection between happiness and duty, illustrates the idea that cheerfulness was a predictor of health. "Giving the boy a chance" meant viewing him as the meritorious citizen, the one who would do his work cheerfully, marry and begin a household, and be defined through the body's affects, acts, and attitudes. In popular ballads, blackness was childlike, self-deprecating, and sexually promiscuous.[46]

The type of "gayety" portrayed in the songbooks examined in this study included an occasional humorous song; for the most part, "gayety" was circumscribed within an overall tone of reverence for school, teacher, and landscape. A happy face on songsters was markedly different from the painted grin of blackface in minstrel performances. While the future citizen was to keep himself in a state of robust "gayety" to fulfill his obligations,[47] the stereotype of the Negro minstrel represented a carefree, *unreasonable* indifference to his misfortune.

Another prominent theme in school songbooks was the downward trajectory of the idle scholar who would become the dissolute disease carrier. Industriousness marked the biography of those who would follow a trajectory of good fortune and health. An individual to watch was the idle schoolboy who ruined his chances for happiness, fell into gloom and, from there, developed bad health habits:

> This idle boy no comfort had,
> His face was gloomy, dull and sad,
> And happiness flew at his touch. (Fitz 1846, 82)

A related premise was that happiness and good habits would erode in bad company. "School, Sweet School" encouraged the child to stay away from the street child:

> If in the streets with bad children we play,
> We never should learn to be happy or good,

School, sweet school,
There's no place like school. (Peabody 1830, 3)

The rules for comportment inscribed in school singing compelled distinctions about social life, emotions, and the way to exhibit worthy aspirations for the future. The verse above and others like it typify ideas about the dangers of cities, identifying and hailing the child of merit in moral terms that connected goodness with the white subject. Both explicitly and implicitly, the imaginaries of race regulated the behavior of the population as a whole, yet reckoned the life and death of the white subject in radically different ways.

When Rosy Faces Pale and Cheer Fades: A Consumptive's Death

Reposing opposite the specter of the cholera carrier and her victim was the ailing but well-comported figure of the consumptive approaching death. Her form became the center of literary tropes in novels and fictional genres of the eighteenth through early twentieth centuries. She was portrayed as dignified but frail, with pale white skin. The consumptive charmed onlookers or filled them with pathos; she was not a repulsive figure, nor was she frightening.[48]

A chivalric tone also pervaded the writing on consumptive death: virtue appeared as pure, disembodied, feminine nobility in the wan, quiet, melancholy, and spiritually uplifted persona. The degenerative process of consumption appeared to regenerate the soul, inscribing the chosen body, even in the spitting of blood, fevers, and delirium, with a martyred attractiveness.[49] Although consumption in the family was a catastrophe and the disease was tragic in and of itself, the ascetic appearance of the consumptive—the victim's frailty and paleness—became an emblem of giftedness. A common image in the poetry adapted for school singing was the "fair lady," as, for example, in the song "Two Angel Hands," where the singer clasps the "wee, white hands" (Leslie 1888, 32) of a girl or woman on her deathbed. Traditional lament songs, such as tributes to war dead, had made the culturally noble death a chivalric occasion in which the "gallant" manners of a particular social class were the Victorian and the Romantic style stock-in-trade:

> We will strew their graves
> With sweet flowers;
> They shall bloom
> Graves enshrining honored men,
> Gallant boys low in the tomb. (Giffe 1875, 44)

One study of the literary images of tuberculosis describes a predilection for sainthood: "Lady Barbara's small pox comes as a punishment and renders her outward appearance as ugly as her inner character. Lady Lucy, whose soul is innocent and pure develops a disease [consumption] that does not destroy her looks, though it takes her life" (McMurry 1985, 105). By contrasting the two diseases, the author of the study was able to highlight the peculiar nature of consumption in the nineteenth century: physical dissolution accompanied by moral and aesthetic transcendence. Similar descriptions of consumptives occupy diaries, novels, short stories, and theater and opera of the eighteenth and nineteenth centuries. These depictions cohere around a pale, thin body type, with curiously little substance to the attributions of saintliness and wisdom except for their appearance.[50] This was the predominant image of the consumptive in the school songbooks examined for this study published between 1819 and 1912.

Stereotypes of the consumptive's ennobling features appeared in marked contrast to those of suffering victims of diseases such as cholera, yellow fever, and typhoid. These diseases were associated with stereotypes of the lower-class Irish Catholic and African American and, later in the century, with southern and eastern Europeans.[51] Dark-skinned, stocky, and broad-chested persons were not represented in the literature extolling the consumptive.

The image of the typical consumptive conferred a spiritual dimension on the Teutonic or Nordic by making whiteness the chosen site for this disease. The ennobled death of the "fair" was both a convention and a conferring of nobility through an appeal to whiteness. One verse from the song "He Doeth All Things Well" from *The Silver Bell* reads,

> She was a lovely star
> Whose light around my pathway
> Amid this darksome vale of tears . . .
>
> That star went down in beauty
> Yet it shineth sweetly now
> In the bright and dazzling coronet
> That decks the Savior's brow. (C. Butler 1869, 203–4)

The last verse from the song compared the deceased with a ray of moonlight and a pearl; another compared the woman to the whiteness of a winter scene. "She died in beauty! Like the snow on flowers dissolved away" (Mason 1856, 86). In the song "Little Dora," the ethereal quality of the lost child is matched by the piling on of metaphors of lightness:[52]

> Yes, we'll meet in the morning, darling,
> And our tears will be wiped away,
> And we will lose this shadow from our hearts
> In the dawn of an endless day. (Giffe 1775, 42–43)

True to Victorian conventions, women's deaths were memorialized as the passing of a pure and white figure, where "lilies" and "snow" were metaphors for a brief and innocent life. The whiteness not only evokes the archetype-ideal but also conforms to the popular image of the consumptive.

Even more common was the death of a mother taken in childbirth. Memorial songs and laments on this subject were a significant category in the school songbooks written between 1830 and 1865. It was a common tragedy among all classes. "I Have No Mother Now" in Giffe's *The New Favorite* songbook is a lament of this type, combining grief with the specificity of the twin deaths of mother and infant: "Yet not alone she lieth / One an angel is there" (42–43). Yet songs like the ones quoted above, bounded by genteel values and subject matter, did not include the loss of child or mother through slave sale and brutality. Some depictions of slave life appeared in poetry, for example, in John Greenleaf Whittier's 1838 poem, "The Farewell of a Virginia Slave Mother to Her Daughter Sold into Southern Bondage." In his *Narrative of the Life of Frederick Douglass, An American Slave*, Frederick Douglass describes the disappearance of his grandmother, whose time of death and burial was obscured by her slave status. Toward the end of her life she was put out in the woods in a makeshift hut where her "grave was at the door." This ending provides a harsh contrast to the elaborate funerals of white people. "She saw her children, her grandchildren and her great grandchildren, divided like so many sheep without being gratified with the small privilege of a single word, as to their or her own destiny" (Douglass 1845/2001, 40). Rites of death and burial are accompanied by religious and social customs that feature, in bold relief, the social rank of the individual. Following long-standing practices, the obscurity of the death of slaves and their frequent disappearance through sale are memorialized in songs such as "Nellie Gray," a song that has been attributed to an abolitionist from Ohio. Its central verse describes Nellie leaving Kentucky for the Georgia cotton fields.

Aristocratic Comportment, Physical Grace, and Images of Whiteness

School songs depicted ideal whiteness as the comportment of the aristocrat dancing a minuet.[53] Modeling a gradual change between stillness and motion, the minuet exemplified the height of grace of the moving body until about 1820, when the waltz gradually overtook it.[54] "The bodily *Bildung* [cultivation] and the sense of propriety which was produced . . . furnished the balls of the high society with a special excitement, because of the minuet, the then dominant dance" (Lempa 1999, 271). Litheness and a serene facial expression set dancers of the minuet apart from others, as similar qualities did the consumptive.[55] Even as the minuet began to disappear from the scene in the early nineteenth century, its poised and deliberate gestures—similar qualities of bodily grace associated with the Nordic type—were displayed in the forms of greeting and manners mentioned in school songbooks.[56] For example, in "Excursion Song" there are greetings to nature: "With joy once more we hail thee, O lovely rural scene!" In "My Native Hills," we read the phrase "Give home its tribute tear"; the song

"Our Daily Task" includes the line "Nature's praises sing, her welcome's warm and willing." The rhetoric was not simply composed of Victorian conventions of address; it was accompanied by a whole set of codes for genteel behaviors among acquaintances and intimates. Sentiments about nature and friends, the modeling of manners, inscribed modes of distinction between the genteel and the common types of greetings. As previously discussed, "The Song of Friendship" in Baker and Southard's *School Chimes* instructs the singer in the ritual of wishing good health. Similar to the preventive effect of cheerfulness in relation to disease, genteel manners inoculate the right-living against the corrosive effects of idleness and rowdiness. School songs carried the performances and enunciations through which each case could be compared and interpreted. Facial expressions— "gayety" or gloom, plus comportment—were interleafed with images of the "fair" as those at risk from others. These performances demanded that the singer be as much person-to-be-rescued as rescuer. In this sense, school songs embodied multiple missions as object lessons in etiquette and hygiene; medical and moral governance materialized as inseparable from music instruction, as a program of lifesaving exercise that accentuated the millennial zeal for a constant watchfulness over contagion and corruption. As vocal instruction offered disease awareness and lessons in comportment that overlapped racial divides, it also compelled divisions between those who would participate in the benefits of democracy and those who could not.

The mission to conquer alcoholism and disease ran parallel to the nation's special destiny of material and moral progress.[57] The systems of reason deployed from medicine, literary traditions, social etiquette, and object lessons in music satisfied a "common sense" written into the Declaration of Independence—all, provided they met the standards for the worthy, had the right to the pursuit of happiness.[58] Occurring in the decades when the ethos of a natural equality put pressure on elites to acknowledge that they were fundamentally on the same political footing as their compatriots,[59] aristocratic forms of manners were pressed into service as part of the school's civilizing mission. In this sense, the fabrication of a medical biography heightened social distinctions that were being erased on other fronts.

I began this chapter with a quotation from Michel Foucault, in which he noted the conflation of magical thinking with modern, scientific theories. In what followed, I interwove the older religious and literary themes from school board testimonials with the lyrics of school songs and literature on health regimens for the prevention of contagious disease. By mapping this unevenly rational and irrational terrain, as we look back on it, I have tried to translate several very disparate enterprises, appealing to old and new ways of thinking about human types, into a pattern of distinctions that anticipated the racial divides I observed in public school music instruction. From the archives and the repeated imagery of darkness against whiteness, a former, contested world reveals itself in its projections of contagion onto freedmen and Irish Catholics in the northern cities. In Chapter 3, I will pursue this "magic" of cultural ideas and their dark or light modality in school songs, to trace how they fabricate degenerate groups in the population. This is a language with semantic ties similar to those of the fair complexioned as he or she follows the charmed life.

3
Making Daily Life Sublime
Verse and Rhythm
"Never to Abase or Degrade"

In 1837, when a select panel of citizens filed their report to the Boston School Committee recommending public vocal instruction, their main concern was how singing lessons would introduce fine music to schoolchildren and contribute to character building by staving off the degenerative influences of other forms of music: "[Reprehensible music] ... would call forth the sentiments of a corrupt, degraded and degenerate character" (Boston School Committee 1837, 35). As the preface to Lowell Mason's *Songbook of the School Room* states, ten years after the committee report, "Songs ... should ever ... not abase or degrade ... [nor songbooks include] any song corrupted by mean human experience" (Mason 1847, ii). The purpose of moral uplift encouraged inclusion of many songs like "Our Pleasant School," in which school resembles the Garden of Eden.

> Where do children love to be
> When the summer birds we see,
> Warbling praise on every tree?
> In our pleasant school. (J. Green 1852, 23)

This romantic outlook was inextricably bound to school ballads through imagery and diction. In this chapter, I analyze the diction and rhythm of school songs, relating their frequent portraits of childhood innocence to the Romantic Age's ideals of freedom and to the underlying grid of racial comparisons that circulated in intellectual and popular discourse. Much of this material included language inappropriate for the school canon even though salacious, subversive, or satirical material played to audiences from all walks of life in theaters and taverns.[1] The refined society delineated by the school songs emanated from an intellectual milieu in which pseudoscientific tracts on race divided the population into gradations of animal or man and enlightened or primitive. This is not to suggest that human types were systematically described in curriculum documents or songbooks; nor were the songs always about degeneracy. Rather, as I attempt to establish, vocal instruction's mission of producing future citizens of the new nation, through singing, depicted the life of school, home, work, and nation as infused with genteel tastes that were litmus tests for the true citizen.[2]

The Trope of Degeneration

The word *degenerate* is a combination of *de* (from), and *genus* (kind, seed, or origin). *Degenere*, in Latin, the root of the English word degenerate, means to decline from a particular kind of being or thing to a lower form—physical, mental, or moral. The concept has had an active life in comparative notions of the progress of societies and peoples: "Degeneration . . . permeates nineteenth century thought with a model (or a series of models) for decline and images of decay . . . [It is] embedded in biological models and images, but its import . . . [is that it] overwhelmed the purely biological character of the paradigm. It borrows or subverts other terms, such as decadence, but it remains for the nineteenth century the most frightening of prospects" (Chamberlin and Gilman 1988, i–xi). Insofar as there was an acknowledged proclivity in the nineteenth century for the motif of degeneration to penetrate intellectual enterprises, degeneration provided "a palette with which to color religion, the arts, anatomical science, fiction political institutions and . . . music—in effect all of what were considered essential to human 'civilization' (Chamberlin and Gilman 1988, xiv). One of the most degenerate types of song was the minstrel ballad. It was performed in a multitude of settings, including theaters, but is conspicuously absent from song collections used in schools up to 1900. "Opossum Up a Gum Tree" was sung during one performance at The Negroes' Theatre in New York in the 1820s:

> Opossum up a gum tree
> On da branch him lie;
> Opossum up a gum tree,
> Him tink no one is by.
> Opossum up a gum tree,
> Nigger him much bewail . . .
> He pull him down by da tail. (C. Mathews 1824, 25)

Later in the century, a famous British critic reported that all strata of society could be observed viewing a performance of *Hamlet* in which Shakespeare's tragic lines and the minstrel ballad "Jim Crow" were interwoven:

> Oh! 'tis consummation
> Devoutly to be wished
> To end your heart-ache by a sleep
> When likely to be dished
> Shuffle off your mortal coil
> Do just so,
> Wheel about and turn about
> And jump Jim Crow. (Hutton 1891/1968, 14)

These parodies brought the minstrel stereotype into sharp relief for the many social classes that ranked themselves above watching "Negroes" even as they watched blackfaced whites imitating "Negroes" as a human type. The blackface genre is immensely convoluted in its social significance, as it dramatized intricate and often contradictory attitudes among races, the laboring classes, and elites. It was, indeed, as Chamberlin and Gilman call it, a palette that could paint the most degenerate impulses of an entire society and color them black. Eric Lott describes the appropriation and distortion of black cultural practices, in one famous skit by the white blackface actor T. D. Rice, as "a neat allegory for the most prominent commercial collision of black and white cultures in the nineteenth century" (Lott 1993, 19). While minstrelsy has a hybrid lineage in the inextricable mix of cultures it drew upon, it chiefly depended on white attraction to black humor and self-deprecation and on racism's urge to vilify. Passed off as frivolous entertainment, it had ties to social criticism, off-limits manners, sexual innuendo, transvestism, and other vicarious forms of what were considered antisocial modes of behavior. Nonetheless, as I will outline, the whiteness of school songs emanates from the same sonic landscape in the northern cities. And just as minstrelsy could be heard in upper- and lower-class neighborhoods emanating from theaters, public spectacles such as circuses, advertisement venues, and taverns, school singing would model non-blackness, a kind of whiteface that was a fabrication of a persona hailed in school music instruction, as "merry," "bright," or "right," as in "right boyhood." Blackface portrayed Negro characters as in a permanent state of childishness that invited ridicule in such a way as to disclaim (at least overtly) any childish impulse on the part of spectators. All of this may seem to be a stark instance of racist invention; yet, if, as Eric Lott suggests in *Love and Theft*, we are to understand where we are and who we are with respect to race, it behooves us to trace the fault lines of black or white imagery so that their message is not felt as "natural" or "inevitable" and their surfaces begin to yield some record of the racial taboos that haunt institutions such as schools.

Specifically, with respect to the music curriculum, we would want to read a musical consciousness that appears, as if by necessity, to divide children by their entrainment and to catalogue one taste in music as more worthy, more citizenly than another. In school songbooks, childhood is innocence, but not to be mocked. The verbal and musical twists and turns that created such cultural gulfs between blackface minstrelsy, other popular entertainment, and school songbooks are my next subject. As Eric Lott writes in *Love and Theft*, where I first saw this illustration, the "thrill" of minstrelsy was the consciousness of imitation and ridicule, but also the vicarious liberation from whiteness. Figure 3.1 makes this point in the playbill that shows each actor in proper white comportment and the character, above, that they become in performance. In this sense, minstrelsy plays ambivalently with the very notion of whiteness. It is also possible to

Figure 3.1. Sheet music. The Harvard Theatre Collection. The Virginia Serenaders, 1844.

detect an underlying revelation: that whites too had, in them, what were understood to be black only attributes. School music, then, abided by these realizations and assiduously avoided putting songs that alluded to minstrelsy in the curriculum

Vicious Art

The vilification of "degenerate" verse and music was commonplace. Leading the crusade against these corruptions were figures such as John Ruskin, one of the most widely read art critics of the nineteenth century:

> Vicious art is . . . produced by a vicious soul; a just and strong person cannot produce it; every deed and force proceeding from him must be noble, effective and strong . . . music rightly so called is the expression of the joy or grief of noble minds for noble causes . . . [it] is the natural and necessary expression of a kingly, holy passion for a lofty cause . . . in proportion to the degree in which we become narrow in the cause and conception of our passions, incontinent in the utterance, feeble of perseverance in them, sullied or shameful in the indulgence of them, their expression by musical sound becomes broken, mean, fatuitous, and at last impossible. (Ruskin 1867, cited in Mark 1982, 100)

Ruskin elaborates that it is the wholeness of the lofty experience that makes the superior individual; it is not the single art experience that matters but the ongoing familiarity that comes with constant exposure. He warns that some are susceptible to intoxication; the most vulnerable would be children and women—the "weak"—or those who could be ruined by too much or too little exposure to art. "It is not a good thing for a weak . . . person to be momentarily touched or charmed by sacred art . . . only in the earnest disciplines of life can we enter into . . . the ordered perfect ness of sound . . . If we seek only the pleasure of the sense, then music searches for the dregs of good in our spiritual being and wrings them forth, and thus the modern opera with its painted smiles and feverous tears, is the modulated libation of the last drops of our debased blood into the dust" (Ruskin 1867, cited in Mark 1982, 101). Ruskin's tone is different from the democratic rhetoric of American educationists and critics such as John Sullivan Dwight, yet it sounds a similar note with regard to the moral power in music to influence society. An editorial written by Dwight in *Dwight's Journal of Music* in 1852 ties music, education, and political freedom together:

> Music enters into all our schemes of education . . . And whoever reflects on it must regard it as a most important saving influence in this rapid expansion of our democratic life. [It] is a true conservative element, in which Liberty and Order are both fully typed and made beautifully perfect in each other. A free people must be *rhythmically* educated in the whole tone and temper of their daily life . . . in order to be fit for freedom. And it is encouraging, amid so many dark and wild signs of the times that . . . [a feeling for music] . . . is beginning to ally itself with our progressive energies and make our homes too beautiful for ruthless change. (Dwight 1852, 4)

We can see that Dwight's editorial registered the turmoil of the pre–Civil War decades, turmoil that was reflected in the active abolitionist circles and publications in the Boston of that time. Music, referring to the genteel concerts and recitals of musical groups performing in the city, was to be a bulwark against degeneracy. Dwight's emphasis is on *rhythm* to counter "ruthless change" and the "dark and wild." Put in these terms, the rhythm of fine music would halt the degeneration of individual, family, and nation, all of which are essential to democracy. Dwight's remarks typify the Romantic[3] stance toward musical elaboration and the Romantic's frequent focus on rhythm. Dwight's view approximated those of the German philosopher, Friedrich Schelling, known primarily for his influence on Romantic literature. In his essay "The Special Part of the Philosophy of Art," Schelling depicts rhythm as making repetitious labor bearable through a periodic marking of time: "Man seeks therefore, driven by nature, to come to terms with multiplicity or diversity by means of rhythm . . . in everything that is meaningless in itself, for example in numbers, we do not long endure uniformity, we construct periods. Most mechanical workers lighten their work in this way; the inner enjoyment of the counting—even though not conscious, but outside of consciousness—permits them to forget their work, the individual falls into place with a kind of delight, because he would actually find it painful to see the rhythm interrupted" (Schelling, as cited in Lippman 1985, 74). In their agreement that music was an important element of civil society, Ruskin, Dwight, and Schelling provided important support and intellectual heft to the advocates of public music instruction. As members of the Romantic literary and philosophical movement, Ruskin, Dwight, and Schelling's notions of the beautiful in music included the polarities of human types as villains and heroes. This comes about because slightly beneath the surface of the conversation on music and rhythm lie constructions of peoples and nations. In the case of Boston's music curriculum, civic supporters were keen to mention its exactitude and the discipline it required of the student: "[The science of music] is a discipline of the highest order, a subordination of mind, eye, and ear, unitedly tending to one object; while any deviation from that object is at once made known" (Boston School Committee 1837, cited in Mark 1982, 138). In these early years, the "object" was, of course, the song, but the song foregrounded a system of thinking about musical beauty that would be common to a democratic citizenry and define that citizenry as cultured. Subordination to these values kept the child from becoming a mere amusement seeker: it focused the child's attention on edifying subjects. As Marc Redfield notes in *The Politics of Aesthetics: Nationalism, Gender, and Romanticism* (2003), our tastes are subject to improvement by group standards and by the labor of instruction. Through the discursive channels of literature and elite social institutions, taste becomes part of a national, cosmopolitan culture, yet it obtains its superior position through reference to an elite minority. The paradox lies in the assumption that certain values are universal, yet they must constantly refresh their universal relevance through pedagogy and, above all, through the posing of a temporal line of development from primitive society to the modern nation.

Many songbook prefaces[4] contained a statement epitomizing this Romantic, modern point of view on taste and musical practices. An example is the following statement from *The School Chimes*: "Care has been taken to exclude every piece against which the slightest objection was conceivable" (Baker and Southard 1852, "Preface"). School song collections featured praises of nature, friends, school, and God, to name the most prevalent themes, ensuring the enchantment of childhood, schooldays, and the national landscape. The preface to the school songbook *The Silver Bell* states that the songbook's contents will "lead pupils onward to the appreciation of a higher class of music that is now generally used in our Schools, and to form a correct taste and pure style. Most of the pieces are so arranged as to be played with piano ... a feature that will ... make the work especially welcome to the *social circle*" (C. Butler 1869, 3; emphasis added). Those who made up this "social circle" were families able to afford a piano.

The genteel musical home entailed a particular family life, imagined, along with the biography of the child, as the anchor of the model citizen.[5] Lowell Mason had written in his address to the Boston School Committee in 1837 that "vocal music tends to produce social order and happiness in a family. Those parents and children who sing together, have a stronger attachment for each other. The family circle is prized; for here can always be found amusement, and such as does not lead into temptation (words of Lord's prayer). They can truly sing, 'Home, sweet home,' . . . Nothing tends more to produce kindly feelings" (Mason 1837, cited in Mark 1982, 132). Music's governance over aesthetic values corresponds to the production of the normal home and the regulation of behavior in the home. Music instruction had a gendered role as well as a religious role to fill, according to the maternal provenance of wholesome music.[6] William Bradbury and Charles Sanders, music teachers and compilers of the school songbook, *The Young Choir*, wrote as follows: "The mission of music . . . ought to be that of the highest style of philanthropy . . . she should be the handmaid of religion, the teacher to truth and the inspirer of devotion . . . she should be the nurse of all gentle and pacific as well as patriotic Sentiments" (Bradbury and Sanders 1831, 111). The gendered, feminine role for music emphasizes the ideals of female servitude. These worked against the aggressive male registers of some rhythms, reasserting reason over action, the cultured over the rude, and a host of other binaries that Dwight had mentioned as wild impulses. In this period, among music critics and pedagogues alike, the balance of what is feminine and what is masculine swing wildly, at times veering toward racialist obsessions with the decadence and homoerotic fantasies acted out in minstrelsy[7] or toward challenges to cultured music as providing an effete pastime.

"Home" songs sung in music classes addressed different types of gendered experiences—from nursing the sick to pioneering on the western frontier—each revolving around a core memory of the lovely home presided over by maternal watchfulness and refined musical taste. The use of the woman as a rhetorical and metaphorical figure linked the character of the future citizen—that is, the school singer—to an image of domestic life that memorializes feminine virtue but, at the same time, incises as masculine the nation and its masculine protectors. In Asa Fitz's *The American School Songbook*, there is the following verse about a home situated in a nation as the loveliest and most memorable of places:

> Home, home, well I remember,
> Thee, thee loveliest home,
> Yes, yes though I may wander,
> Far o'er the nation to roam,
> Yes, Yes, I will remember my home. (Fitz 1846, 54)

A stark contrast to the school songbook images of home is described in Frederick Douglass's *Narrative*, where he recalls the haunting verses of

"I'm going to the Great House Farm," a song memorializing the plantation's main house as a manifest symbol of slave life, but most of all as a symbol of sorrow and pain.[8] "Great House Farm" was sung in both urban and rural settings during this period, in great variation to be sure, but it was never identified as music in genteel circles, although there were white champions of slave song.[9] Songs like "Great House Farm" had a noticeable presence in many settings such as the docks of the northern cities where blacks worked during this period. Their crossover into blackface entertainment reveals how much contact with (and adaptation of) slave song was available to white society, in the theater venues of New York, Boston, and Philadelphia, for example. The absence of most traces of minstrelsy in public school music instruction is testimony to the negative cultural burden these songs and dances bore in relation to the universal claims of beauty and discipline extended to songs such as "Home" from Bradbury and Sanders's 1831 collection.

One of the qualities that kept the child productively engaged was mental celerity but this also had to be continually watched over. Degenerative effects came out of an excess of mental activity such as studying. The report of the special committee to the Boston schools in 1837 recommended vocal instruction as, among other things, a form of mental refreshment to counter inattention and the degenerative effect of idleness that invited the Devil to take over the soul. Underlying the prescriptions for school music as a restorative was the notion that good music actively stimulated the mind, not the body. Such fine singing would not instigate dancing or set the body in motion. This was pointed out to allay fears that music in schools would lead to sexual activity or socializing that was morally degenerate, involving the pairing of male and female.[10] Unlike the waltzes, marches, and polkas that typified music marketed for social gatherings,[11] school vocal music would not stimulate large muscle movements.[12] "An alternation is needed in our schools, which without being idleness, shall yet give rest. Vocal Music seems exactly fit to afford that alternation. A recreation, yet not a dissipation of the mind—a respite, yet not a relaxation . . . music is not dancing—because Music has an intellectual character . . . and above all, because Music has its moral purposes, which dancing has not" (Boston School Committee 1837, cited in Mark 1982, 135). In one school songbook compiled by Lowell Mason in the 1850s, there is an articulation of how sensual urges, such as those symbolized in coupled dancing, could be stirred by music, represented as "Eve," but submit to the moderating authority of the poetry, represented as Adam. "In our 'Normal Singer' Poetry (the Adam) governs. Thought directs feeling: music (the charming Eve) the more gentle and lovely of the two; cheerfully submits; her nature inclines her so to do; modest and retiring she desires not to attract attention to herself, but only to express in her enchanting tones, and with deep felt emotion, the lyrics of her spouse" (Mason 1856, iii). As an important part of the trope of the madonna or seductress binary in Romantic poetry, the threat of the "charming Eve" made use of the common notion that women should be submissive and was directed at musical styles, such as Italian arias, that were thought of as overornamented, effete, and decadent.

The Adam and Eve story may also be interpreted as code for racial relations. Seen in this light, the submission of Eve (a racial other) assists the advancement of white civilization in fulfilling the destiny to which it was thought best suited.[13] Enveloping school music in a religious rhetoric, the use of the biblical Eve figure bound particular musical practices—an enlightened pedagogy, pastoral verse, and diction—to the task of making the future citizen, preventing the regression of the community, and building the nation. Submission to the authority called Adam heightened the governing faculty in music and rhythm that Dwight invoked to tame the "dark and wild" forces threatening society in the prewar decades. There was, for example, Beecher's *Plea for the West* warning against the specter of theocracy via the immigration of Irish Catholics. There was also widespread labor unrest during the early industrialization of former independent artisans and the threat of miscegenation, which was comically enacted in blackface minstrel skits. In the background, the abolitionist movement and constitutional crises created an atmosphere that many feared would (and eventually did) propel the Union into war with the South.

To summarize, the differentiation between musical practices that were degenerate and those that were appropriate for public music instruction emerged from several interrelated planes of social discourse. These were, generally speaking, notions of Protestant civic virtue, northern European musical aesthetics, and anxieties related to the new democracy's stability and the proper amount of stimulation for the brain and exercise of the gendered body. All of these had racialist registers that linked signature practices of civilization to whiteness against black otherness, including that of the Celtic race. Never explicitly racial, public school hearings and debates concerning music maintained the rhetoric of democratic ideals even as the Boston Primary School Board upheld the segregation of the public schools in 1846.[14] The social loop of the churches, music societies, and journals was well connected to the public school administration, as Michael Broyles notes in *"Music of the Highest Class": Elitism and Populism in Antebellum Boston.*

Racial Destinies and the Vocal Curriculum

The new fields of ethnology and anthropology fed racial divides through a poreoccupation with cranial volume as a measure of intellectual capacity, and the widely popular scale of values ranging from the "civilized" to the "barbaric," calibrating notions of mental capacity, sexual propensities, and moral inclinations. The categorization of music as savage or civilized formed points of convergence and disjuncture between craniometry and ethnological divisions of human types as an imaginary portrait of American culture riding on the will of white society took shape. While several noted critics of the pre–Civil War era mentioned black vernacular musical practices as more valid representations of what was uniquely "American," dismay at the racial implications of their remarks overrode consideration of the exclusively German tenor of concert and recital halls. One European observer commented that blackface was representative of American folksong, a sentiment that was met with dismay.[15]

The pre-Darwinian ideas of human inheritance, along with the broad tenets of colonial *noblesse oblige*, etched the classification of human types along successive axes of civilization as against sliding toward savagery; the latter connoted a biological breakdown of a species.[16] Several influential practitioners of craniometry marked the African as the lowest anthropoid type.[17] In 1799, Charles White, an English physician, wrote *An Account of the Regular Gradation in Man*, a treatise that had wide circulation on both sides of the Atlantic. White devised a fixed racial hierarchy with each race assigned to a point of maximum potential of intellect and bodily perfection. According to *An Account of the Regular Gradation in Man*, whites possessed superior abilities demonstrated by head shape, gait, sensitivity, skin color, and the physiology of Europeans as compared to non-Europeans, with fair skin color becoming the major dividing line between the civilized and the barbaric in the 1850s. Anthropological, ethnological, and physiological inquiry were not separate from the theory and philosophy of music in the early 1800s—ideas freely crossed the boundaries of literature, music theory, and science. Johann Goethe, as botanist, physiologist, poet, dramatist, and novelist, articulated his interests in human physiology, music, and education in his novels.[18] His investigations of plant and animal structures looked into how inner essence compelled growth as a way to formulate the influences of external circumstance and a morphological, inherent destiny. The search for a morphological archetype ran parallel to the projects of delineating human types that followed particular musical practices. Sketched largely in Goethe's fiction, the ideal and the malformed phenotypes brought the author to imagine an educational system that used music as a key pedagogical tool to optimize mental powers and foster a noble comportment in young schoolboys.[19] Goethe's *Wilhelm Meister* novels were popular in intellectual circles in the United States, forming part of the atmosphere or mentality in which the Boston School Committee operated. Caught between what were perceived as American national crises and the Romantic view of Protestant white males as the leading racial stock of democracy, public vocal instruction made its debut during a crucial time of restructuring of the racial hierarchy. The three-fifths rule for people of African American descent and disputes over the constitutional bona fides of immigrants were uppermost in the social anxieties of the civic actors who were to certify vocal instruction as purified of any sign of African

Americans' or immigrants' musical life. In addition to this tendency, the racial register of concerns mingled with the German tradition of public schooling, insofar as Lowell Mason's travels and acquaintances had put him in touch with Pestalozzian philosophy,[20] an outlook, in his mind, that focused on engaging the child's "nature" as delighting in music that was familiar to her: countryside, motherhood, and fatherland.[21]

However, the explicit meeting of ethnological, national, and musical traits was even more firmly established by the influential philosopher Johann Gottfried Herder. Herder combined music with a mythical history of the Teutonic race to delineate an ideal human type, or *Urtyp*. With enormous appeal to the Romantic Movement's proclivity to unify and thematize the cultural traits of an ethnic community or *Volkstum* that would save the West, Herder considered the relations between the *Urtyp* and the ethnic group to hold the keys to a lost power,[22] a form of living for the good of mankind. While Herder expounded a universal theory of music as thought, emotion, and rhythm, he also conceived of a hierarchy in which pantomimic movement to music, or what I earlier referred to as bodily entrainment, establishes the trait of primitiveness in highly active peoples. "The tones of music are temporal; they animate the body . . . hence also the gesturing bound up with music. Strongly moved, natural man cannot abstain from it; he expresses what he hears through appearances of his countenance, through swings of his hand, through posture and flexing. The dances of primitive and especially of warm, highly active peoples are all Pantomimic" (Herder, cited in Lippman 1988, 36). Herder's focus on a pantomimic response that he calls "primitive" echoes the sentiments of the 1837 Boston School Committee that vocal music "is not dancing . . . because Music has its moral purposes, which dancing has not."[23] Genteel Bostonians had a Puritan history that viewed dancing as the Devil's temptation; these strictures found a new secular home in nineteenth-century schooling along with ethnic and racial stereotyping, even though, as the historical literature shows, dancing had been popular among every class from the Puritan settlements through the nineteenth century.[24] Echoing the elite tones of northern European aesthetics, the puritanical streak in American elite society embraced this kind of polarized thinking in all spheres of life, especially music, as an enterprise that would benefit the white race and prevent its degeneration.

In Johann Gottlieb Fichte's work, Herder's notion of *Volkstum*, or ethnic community, became a vehicle of racialism in which all aspects of Teutonic corporeality and culture manifested racial superiority.[25] When admiration for the music of the German masters took hold of the middle class in the United States,[26] the notion of a racialized national prototype aligned public school instruction in vocal music mission with the civic music societies' ethos of cultivating the soul through music. This specifically northern European point of view was historically tied to the Protestant notion that worthiness, an inward quality, could be discerned externally from one's comportment, speech, and taste. Even as the Pestalozzian approach to schooling underwrote the singing method described in many, if not most, of the early school songbooks and this singing method was absent explicit racial categorization, its circulation within a musical world that shared outlooks derived from Fichte, Herder, and British notions of culture forced an overlapping of white or Teutonic destiny with a taste for cultivated music. In America, with its specific social conditions of slavery and the dispossession of Native Americans as well as mass immigration, the articulation of race was along white or black lines, different from, but exhibiting a hierarchy similar to, Caucasians and "others." A well-known Boston minister, Theodore Parker, gave an address titled "The Dangerous Classes," in which he described children as equals in one sense but unequal in another. Each baby starts at the beginning of the civilizing process but has a different potential depending on his race: "Some are inferior in nature, some perhaps only behind us in development; on a lower form in the great school of Providence—Negroes, Indians, Mexicans, Irish and the like" (Theodore Parker, cited in Schultz 1973, 243). Lowell Mason, a contemporary of Parker's, made a point of comparing the musical comportment of white operatic singers to other practices he observed in Europe. In the first of three comments in his travel letters, Mason reports attending a Jewish synagogue service in Paris. When he enters, he remarks on the oddity of hats worn at a religious service. His impression of the music is less the measure of comparison than the comportment of the congregation: "The men all sit or stand with heads covered ... there was very little reverence or solemnity; indeed none that could be observed ... the appearance of the assembly was somewhat like that of a New England town meeting, after having been called to order by the chairman ... All the service was chanted, in a responsive manner, with the exception of two airs or melodies" (Mason 1854/1967, 160). Commenting about the Catholic service he attended in Paris, Mason writes that it, too, fell short of being what he expected of a religious event: "The music was quite modern; as much so as if composed by the latest Donizetti or Verdi, and quite orchestral, antiecclesiastical style. It was indeed vocal words were sung; but as they could not be understood, the effect of the whole was such as is the musical effect of a grand pantomime. Indeed, the worship in the Roman Catholic cathedrals seems to be little else than a mute, gesticulatory action of bowings, crossings and kneelings, with grand processions, musical accompaniment, etc." (Mason 1854/1967, 161). Moments such as this recall the hierarchic scale of the "pantomime" Herder elaborated as the primitive musical signature. For Mason, it was the Jewish and Catholic services that had failed to qualify as true reverence. Neither chant nor "anti-ecclesiastical" sounds qualified in those settings as music. The Catholic service involved the body in "gesticulatory action," devoid of meaning except to leave the impression that this was a ceremonial and musical frivolity, a series of undignified acts that Mason and many other members of the musical elite in the Boston of his time also saw in Italian opera.

Given this set of values, it is hardly surprising that Mason's description of the body comportment of the soprano Henrietta Sontag meets the standard of statuesque composure and the Teutonic *Urtyp*.

> The man who, in describing her singing, said, "she is a statue with a music box in her throat," said well, so far as a perfect execution, touching or bowing, is concerned; but he should have added that is a statue of humanity, having a tone, look, gesticulation and movement. So easy, so graceful, so elegant, so chaste, so artistic and yet so simple and natural . . . one would think her to be a mere child of nature . . . for there is . . . the highest degree of cultivation . . . To listen to her is like looking at the most beautifully variegated bouquet . . . like the appearance of a thousand charming little girls of six years of age, wreathed with freshest roses and dressed in purist white. (Mason 1854/1967, 131–32)

With its superlatives tone, Mason's letter sounds all the notes in the feminine register of modesty in the chasteness and propriety of Sontag's gestures. The metaphors of the pastoral—flowers and innocent children—permeate the description, with no account of what Sontag sang. Since Mason's portrait of Sontag is about the archetypal northern European body and what it instigates in the way of praise and admiration, it is infused with what Charles White, in his famous *Account of the Gradations of Man*, referred to as a nobility of appearance and movement found only in the white European: "That nobly arched head, containing a quantity of brain . . . fullness of expression; those long-flowing graceful ringlets . . . rosy cheeks and coral lips . . . that noble gait . . . In what other quarter of the globe shall we find the blush that overspreads the soft features of the beautiful women of Europe, that emblem of modesty, of delicate feelings."[27] What is important in Mason's letters, though, is not the conventional and reflexive racism that exists across the literature on racial typology, but the fact that Mason's views on race correlated with the aesthetic underlying his singing methods and song collections. These are indications of what he, as the first public music teacher, and others after him would view as meritorious, to be encouraged, while other modes of singing were to be discouraged. In shadow in Mason's letters and pedagogical writing are the bodies measured against the soprano, Sontag. According to genteel customs, close physical inspection of lower types was not a proper topic of social intercourse. Scientists were freer to elaborate abjection, as this letter shows. Writing in the 1840s, Louis Agassiz, the famous Harvard naturalist, outlines the foil against which racial grace played: "They are not of the same blood of us. In seeing their black faces with their thick lips and grimacing teeth, the wool on their head, their bent knees, their elongated [and hideous] hands, their large curved nails and especially the livid color of the palm of their hands . . . I wished I was able to depart" (Agassiz 1846, cited in Gould 1996, 77). Agassiz portrayed black body carriage or comportment as ape-like: "their bent knees, their elongated hands." The use of these and similar images drove the narrative of bestiality and racism that was acted out in the temporal and spatial universe that Frederick Douglass described in relation to slave ownership, in his *Narrative* published in 1845. The formation of stereotypes in the intellectual circles foregrounded the long-standing strategies of control by slave owners who demanded reproductive exploitation of their slave as well as expecting self-denigration from them. Douglass brought the deliberate deformation of black character to the attention of his readers in several passages describing the psychological techniques used by slave owners: "[Blacks were encouraged to be] engaged in wrestling . . . fiddling, dancing and drinking whiskey. A slave who would work during the holidays was scarcely seen as deserving them. He was regarded as one who rejected the favor of his master. It was deemed a disgrace not to get drunk at Christmas . . . the holidays [were] among the most effective means in the hands of the slaveholder in keeping down the spirit of insurrection . . . these holidays were . . . safety-valves" (Douglass 1845/2001, 55). In relation to music, specifically, Charles White's and Thomas Jefferson's description of blacks saw having an ear for music as more a matter of "instinct" than cultivation. In White's *Account of the Regular Gradation in Man*, the Negro's ear was similar to the ape's, suitable for only certain kinds of expression, especially sexual seduction.[28] An avid violinist, Jefferson believed that black musical ability was capable of imagining a simple tune, while he doubted that anything more complex was possible.[29] It was his view that the Negro lacked the ability to go beyond the primitive stage of music making and that the Negro horizon was to emphasize sensation rather than thought.[30]

Jefferson's *Notes on the State of Virginia*, widely read by social elites in the early nineteenth century, promulgated notions about the negative effects of crossbreeding of the two races. In this work, Jefferson declares himself in favor of emancipation and education only if separate living areas were assigned to freed slaves. After slavery was banned in most northern states, in accordance with Jefferson's views, school segregation in Boston and other Northern cities was upheld in public challenges to separate schools for freedmen's children.[31]

Given the broad consensus among elites at this time in favor of the maintenance of a racial hierarchy, we can further contextualize the moral climate surrounding early vocal instruction as one in which notions of beauty were *necessarily* entangled with the display of white virtue and taste. The ornate and elevated tone of Mann's school board reports represent democracy as universal reason, but it was a universality that had racial boundaries. The moral ministry of musical salvation was understood as exclusively white, except for the Abolitionist cause. Quoting a medical authority of his day, Horace Mann's "Report for 1844" offers the following:

"Our best and highest music is that which is charged with loftiest principle, whether it breathes in orisons of sacredness or is employed to kindle the purposes and to animate the struggles of resolved patriotism . . . music may well be regarded as a most beauteous adaptation of external nature to the moral constitution of man . . . moral heroism most fitted to solemnize the devotion of the heart and prompt the aspirations and the resolves of exalted piety" (Anon., cited in Mann 1844, 150). The substance of Horace Mann's reports radiates into the channels of civic piety to establish "our best and highest music"—the music sung in public school—as the province of a genteel community. This imagined community is entrusted with an aesthetic hewn out of racial and religious comparisons. If there is an irony in the public schools' mission to make each child a reflection of these values, it is that the child is not as much formed by singing lessons as he is hailed and divided according to his physiological features and social comportment. By the 1860s, Darwinian evolutionary schemas in music (a dubious endeavor) were used to assemble a systematized compendium for the competition of human types.[32]

Minstrelsy Had a Plaintive Air

In the racially condemnatory atmosphere of the mid-nineteenth century, one might find it surprising that minstrelsy was popular among members of the northern music societies. However, this popularity embraced an ambivalent stance of attraction to, and revulsion toward, blacks, even while, as Eric Lott[33] and others point out, the players themselves were most often whites in blackface.

Audiences for blackface assumed the characters were white—yet, some reviewers frequently took these actors for actual Negroes, penetrating the color line in such a way as to belie a psychological fusion between acting "black" and being black. In sum, blackface appeared on the surface as entertainment, but it also represented the attempt to maintain stable racial categories in a society subconsciously aware of the tenuousness of doing so. Minstrel tunes were performances and, in a sense, feats of ventriloquism in which blackness had diminutive charm and nostalgic appeal. The elite *Dwight's Journal of Music* published an admiring commentary on the subject in 1852: "We confess a fondness for negro minstrelsy . . . The whole world is redolent of the sweet and plaintive air in which ['Dearest May's'] charms are chanted and the beauty of her shining form often comes over us . . . and makes Italian trills seem tame . . . as for poor 'Uncle Ned,' God bless that fine old colored gentleman, who, we have been so often assured has 'Gone where the good niggers go'" (*Albany State Register*, cited in *Dwight's Journal of Music* 1853, 124). These depictions of minstrelsy in the popular and literary press merged with the precept that colonial subjects were made to be subservient. A version of this tone rings through the mixture of affection and condescension given voice by Thomas Carlyle, one of the most widely read English essayists in the nineteenth century: "Do I then hate the Negro? No ... I decidedly like [him] and find him a pretty kind of man ... A swift supple fellow; a merry-hearted, grinning, dancing, singing, affectionate kind of creature, with a great deal of melody and amenability."[34] The strains of this type of double-sided racialism coursed through multiple hierarchical systems—the musical system in particular, infecting the later presentation of folk song and the Negro spiritual in school song collections and eventually, the gramophone industry. Early public school music instruction would not mix musical genres and song types, even while they were thought to be innocent entertainment for the genteel class. As John Ruskin's idea of "vicious art," minstrelsy had nothing to teach in the way of social manners, civic virtue, and reason. That it was present and prominent in the sonic landscape of Boston and other large cities made it a form of music making *against which* school music attempted to make its mark.

Rhyme, Rhythm, and National Memory

There is a long literary tradition of linking music with remembrance. Romantic philosopher-historians have remarked on the capacity of music to build memory: "Images abandon us and grow faint; tones come with us as our innermost friends, who from childhood on cheered and raised us up, delighted and strengthened us."[35] Herder had also observed that song has an important role in mapping national territory.[36]

Similar ideas percolated in intellectual circles in the United States and are reflected in the uptake of idealized, "fraternal"[37] images in American school songbooks. Constructing a national memory was a deliberate strategy of American songbook publishers in marketing their collections to educators who considered the formation of the patriotic brotherhood a school responsibility. In the preface to *The Young Choir* (1831), William Bradbury states that American songbooks should dedicate themselves to raising American taste by publishing songs that conveyed a specifically national feeling and devotion.[38] "School Song" in *School Chimes* offers an example of the imagistic and rhythmic cadences that marked the fellowship of American social life:

> Sparkling and bright
> In the morning light.
> The cheerful, smiling faces
> Of the merry throng,
> As they haste along ...
> So full of glee
> So free from care and sorrow. (Baker and Southard 1852, 112)

The verses above sweep the singer into a rhythm that restores the "tone and temper of daily life," as Dwight described it in his *Journal* in 1852.

The musical settings of school songs were, in general, not difficult to execute. Narrow in voice range, the harmonies were simple, either monodic or two- or three-voiced, with no ornamentation. Rhyme is the most characteristic feature of the song texts. In "School Song," the last word of each line, or the rhyme scheme of AA, B, CC—*ight, ight,* and *ong, ong*—frames the first five lines of each strophe. As mnemonics, or memory prompts, for the images of bright, happy schoolfellows, each of the paired lines contains a strong-weak-weak configuration, the equivalent of a 3/4, or waltz-like, meter: ONE two three, ONE two three.[39] The waltz pattern lifts the ordinary description of going to school onto a plane that makes walking dance-like. A slight complication is that "School Song" embeds the dance in 4/4 meter in verses three and four, when it turns to a rather sober comparison between the rewards of studying and the uncertainty of the pioneer's life on the western plains. Overall, the song has a didactic function, "an intellectual character giving it a moral purpose that dancing lacks," as the Boston School Committee report of 1837 (135) calls for. The combination of the song's gestures, the idyllic scene, and the lilting rhythm kinesthetically regulate a performance of liberty and self-discipline that produces the template for the recognition of the neither too boisterous (the oversexed) nor too melancholy (lacking the happiness of the free) student.

Romantic Diction

Through a close reading of the images, rhythm, and diction, one can discern that songs like "School Song" left the imprint of a northern European human type as a model for merit in school music. Linking up images from many songbooks selected at random from the late nineteenth century provides a way to consider how song texts and prefaces to songbooks sustained a pious shepherd or flock narrative within the secular public schools that is also relevant to today's music instruction programs. The following is a description of the music teacher as angel and protector of a flock of teachers taking direction from God: "The Good Shepherd has taught us that the angels who watch over little children . . . have reached the highest altitudes . . . an incentive to loving service for the lambs of the flock . . . We are co-workers with them who do always behold the face of the Father" (J. Murray 1887, 2). Songs such as "School Days" from *The Golden Wreath*[40] and "The Graduate's Farewell" from *The Silver Bell*[41] inscribe a sense of national life as a pastoral idyll. The following stanzas drawn from "Love of School" provide another example from *The School Chimes*:

> When softly summer zephyrs blow
> We read tread the schoolhouse path,
> And when old winter heaps the snow
> We little heed his wrath

...

> We love the pleasant tasks we learn,
> The pleasant songs we sing,
> We love the pleasant words we earn
> That faithful studies bring.
> But more we love the little band
> Who daily gather there
> The teachers and their little troop
> Who stand around in circles fair. (Emerson 1857, 119)

Another "School Song" describes another idyll:

> Sparkling and bright
> In the morning light
> Are the cheerful, smiling faces
> Of the merry throng,
> As they haste along...
> No life may be so full of glee. (Baker and Southard 1852, 112)

The phrases "merry throng" and "little troop" make the school synonymous with a group whose good fortune indicates God's grace. "As they haste along" is a phrase that reiterates the merry throng's purpose or duty, analogous to the duty referred to in church sermons.[42]

The Romantic style of diction added to the Puritan rhetoric would later surface in social Darwinism and in the Progressive Movement. Added to the atmosphere were the popular stereotypes that lent a racial cast to the description of even so seemingly neutral a phrase as "merry throng." Faces are "sparkling and bright"; the teachers and their little troop stands in circles *fair*. The word *fair* is from the old Saxon verb *fagr*, to clean. When used to reference hair and face, "fair" is the opposite of the dark or brown features of less favored folk—the "foul."[43] These ready descriptors are oblique but powerful lead-ins to recall of the dark and putrid imagery associated with blacks and other degenerate elements. In contrast, "fair" connotes auspicious, beautiful, blond, reasonable, uncolored, objective, and civil.[44] In the song, "Love of School," "fair" denotes the whiteness of the schoolchildren. When "fair" is coupled with "bright," from the Saxon *beracht*, or light, the term leaches into the dimensions of the higher human type.

The bright and merry were not merely the children, archetypes of those present in the school—they were the forerunners of the deserving citizens who stood apart from others seen as less so. The register of lightness conveys a biographical trajectory—the charm of appearance, overlapping the construction of whiteness in music and health regimens—for the esteemed citizen.

Recalling Lowell Mason's images of cultural nobility, the "merry and bright" persona is in striking contrast to Louis Agassiz's racial stereotypes and Carlyle's caricature of the Negro temperament as "merry-hearted, grinning." In this potent mixture or alchemy of the ideal types, the song portraits of the school cohort draw a sharp contrast between the physiognomy and affect of the dark versus that of the light. The lyrics distinguish the "darkie" as not good enough, as his life does not match the templated images. Without making assumptions about teachers' intentions, one can consider that the processes of self-selection and merit conferring merit enfolded in these songbook or curriculum outlines as metaphors of cultural nobility brought into confrontation with "others" *indirectly*. That is to say, the social recognition of particular dispositions is the "gold" outcome of the alchemic distillation of pedagogy and curriculum.[45] No single agent can effectively deliver the ordering of human types without the articulation of this ordering on nearly every scale of the school as a state, local, and interpersonal institution.

Judith Butler's notion of the hailing of the modern self provides a framework with which to interpret the effects of school songs at the level of the performance of individual identity. The illustration Butler gives of a hailing mechanism is a policeman calling to someone and that person recognizing the call as directed at him or her.[46] Butler's concept brings us to consider how self-recognition takes place in school where a student feels the call of membership in the school cohort. On the other hand, for others, there will be no such belonging.[47] With respect to school song lyrics, their rhetoric and diction either performs a familiar welcome or registers no mutual recognition between song and student. As discussed above, the performative qualities in the school songbooks run along dispositional, racial, ethnic, or class lines. In this respect, the teacher is an intermediary between the vocal curriculum and the self-classification of the child. "Love of School" hails the child who turns toward, figuratively speaking, the mood and image of the following: "But more we love the little band / Who daily gather there" (Baker and Southard 1852, 112).

Singing for Friendship

Friendship songs were object lessons on genteel sociability. The lessons enacted a formal protocol reminiscent of the minuet and its recall of aristocratic bearing and the concern for health:

> Come, sing the song we love so well,
> The song of Friendship dear,
> For while we sing
> We all do feel
> That we are doubly here.
>
> To Friendship then our voices raise,
> Be that our sacred word;
> As blend our voices in this song,
> So let our souls accord!
> Then here's a health to those we love,
> The health of friends afar!

The health of friends afar! (Baker and Southard 1852, 64)

The rhythmic notation, elaborated as repeated eighth notes, lends formal authority to the sense of the words, codifying them as a toasting of friends' presence, even in their absence, and wishing them good health: "To Friendship then our voices raise . . . As blend our voices in this song . . . here's a health to those we love." The formal tone also complements the overall theme of meeting and parting, involving little outward show of emotion or bodily gesture that would suggest a rough and ready companionship or romantic liaison. In this sense, friendship follows the etiquette of the genteel comportment associated with whiteness.

"Song of Friendship" spells out a harmony of relations, the according of soul to soul in a platonic mode that leans toward the nobility of manners expected of the rising class.[48] It also distinguishes the civilized countryman from barbaric peoples.[49] Both the regular, rhythmic treatment of the verse and its subject find resonance in the qualities associated with whiteness during the mid-nineteenth century.

John Sullivan Dwight's statement, "Free people must be *rhythmically* educated in the whole tone and temper of their daily life . . . in order to be fit for freedom" (Dwight 1852, 4), referred to the concert repertoire but mirrors the programmatic sensibility found in the school songbooks. Dwight may have just as well referenced Beethoven's Ninth Symphony, for example, with its text, Friedrich Schiller's "Ode to Joy," to convey the kind of "Liberty and Order . . . fully typed and made beautifully perfect in each other" (Dwight 1852, 4) that materializes in a song like "Song of Friendship." The school verses bear some relation to Schiller's famous poem, even if they are paler than his more widely known ode. If such sentiments resonated with the child's notion of ideal friendship, they would, as Dwight put it, help her to resist the dark and untamed temptations present in the unfinished republic. If she should fail to be hailed, the fellowship inscribed in "Song of Friendship" foresaw the benefits of liberty, for her, as unlikely.[50] Put another way, the performed normality in "Song of Friendship" paints a different picture of relationships that are less than the ideal, and, in starker terms, makes them abnormal.

School Songs and the Sublime

Nineteenth-century American Romantic verse, in general, with its portrayal of the sublime, is less formally rational than the European, more rooted in feeling and emotion. American school songbooks draw on the ruggedness and largeness of the American landscape, tending toward the awesome, fear-inspiring, or spiritually transcendent as the signature image of natural wonders and the vast wilderness that awaits civilization.[51] What became known as the American temperament in literature—as seen in Longfellow and William Cullen Bryant, for example—took much from the poetry of Wordsworth, Coleridge, and others with respect to nature, work, childhood memories, affection, loss, and other feeling states, but were often specific to America's vast landscapes.

School songs were modeled on the style of Bryant, whether it was the bright morning light or "Then Follow, follow, follow, Nature's praises sing" (Baker and Southard 1852, 66). Coleridge and William Cullen Bryant's poetry appeared in *Dwight's Journal of Music*, expressing the close connection between the world of music and the broader intellectual fabric of Boston that was instrumental in supporting public music instruction. Sketching the vast, uninhabited American landscape in a poem such as "The Prairies," Bryant was one of the first American poets to capture, in very few words, the sublime of American scenery—nature in its boundlessness and magnificence:

> These are the gardens of the desert,
> These the unshorn fields, boundless and
> beautiful
> For which the speech of England has no
> name. (Bryant 1832/1962, 260)

Although Bryant never achieved the international reputation of Coleridge, his work, along with the work of other nature poets, represented a sensibility that was considered native, but it was also part of a cosmopolitan interest in the transcendence of natural scenes that encompassed, for example, Beethoven's Pastoral Symphony and his Moonlight Sonata. An inclination to find God and truth in nature intensified through the nineteenth century. This was a pantheistic spirit that easily accommodated the ideal of secular governance.[52] The songbook *Flora's Festival*, designed to sell for many venues including schools, features many songs in the nature-as-sublime genre. "A Home that I Love" reads as follows:

> Give me a cot in the valley I love,
> A tent in the greenwood.
> A home in the grove. (Perkins and Perkins 1866, 82–83)

Pastoral scenes took up the space in schools, in proportion, one might say, to the growth of the increasingly industrial northern city, underscoring the difference in sensibilities between the educated classes and "others."[53] Laying emphasis on middle-class nostalgia, for example, a Boston press brought out *The Nightingale: Songs, Chants and Hymns Designed for the Use of Juvenile Classes, Public Schools and Seminaries* by W. O. Perkins and H.S. Perkins, which contained a large number of ballads celebrating the American landscape, with titles such as "My Native Land" and "My Native Hills." In this repertoire, the home in nature is the homeland of the nation, portrayed as a garden. The citizen is portrayed as comfortable in its midst; the land provides cot and tent—valley and grove are natural furnishings for the citizen's needs. Other songs in *Flora's Festival* describe the national expanse and fertility that were also replicated in American painting.

Musically, the sublime translates into the well-known melody and images, "O beautiful for spacious skies and amber waves of grain," from "America the Beautiful." In this sense, school songbooks featured a Romantic sensibility of middle- and upper-class provenance that guided the choice of lyrics and musical setting.

The latest vocal methods transported the singer into a feeling state that would balance the tendency for the middle and upper classes to overuse the rational faculty. Sublimity in "Our Daily Task," from *School Chimes*, echoes this anxiety as the "busy bee" conveys the idea that leisure is earned by study, a form of work for the young middle class:

> Our daily task is ended,
> The afternoon is splendid;
> Our pattern now shall be
> You careless zigzag rover,
> Amid the scented clover,
> Gay coated humblebee.
> . . .
> She opes her every treasure,
> To those whose hearts are true
> And poureth out full measure,
> The golden streams of pleasure
> If faithfully ye do
> The work that falls to you. (Baker and Southard 1852, 66)

The song's frequent rhymes convey a harmony and lyricism reminiscent of Dwight's statement about the rhythms of democracy. But if "Our Daily Task" fails to match the elegance of Longfellow's "Village Blacksmith," of Wordsworth, or of Coleridge, it succeeds in kinesthetically transmitting cheerfulness regarding work, portraying the child as a double for the busy bee's virtuous task. The significance of the song is not its subject per se but its ability to connect the singer, via the busy bee metaphor, to ways of reasoning about school and work that sanctify those values and hail the busy schoolboy as the future citizen—or not. This song and hundreds of others like it in this period present a template for a childhood corresponding to the images of whiteness that were fabricated through a range of practices related to sociability and work.

Another example, focusing on the meter and rhyme in "Our Pleasant School," is similar to the kinesthetic performance in "Our Daily Task," transporting the singer to a perfect world where simple pleasures beckon and schoolwork is calibrated to the rhythms of nature. The rhythmic iteration would have a singularizing effect in hailing the schoolchild who imagines a scene of warbling birds and finds for himself a set of familiar images and words in the following:

> Where do children love to go,
> When the storms of winter blow?

What is it attracts them so?
'Tis our pleasant school.

Where do children love to be
When the summer birds we see,
Warbling praise on every tree?
In our pleasant school. (J. Green 1852, 23)

Matching Dwight's notion of music as a "saving influence" through its rhythmic regularity, "Our Pleasant School" gives the classroom the feel of a domestic scene. Songs such as "Our Daily Task" and "Our Pleasant School" are, in some ways, performances of the Declaration of Independence in which "the pursuit of happiness" is paramount and the rhythms of the syntax provide an accentual grid that emphasizes meaning.[54] The school song repertoire, sung across thousands of public schools in these decades, accomplished, daily, the governing of new citizens in their performances of contentment and duty. At the same time, the songs functioned as material templates that compelled the singer to match her own sensibility to the characteristics with the citizen-images in the verse.[55]

Organic Form and the Magical Incantation of Nation, Child, and Citizen

The rhythmic aesthetic underlying the songs discussed above had a rich lineage in romantic poetry and philosophy, restoring a wholeness to life that threatened to come apart under constitutional and economic strain of nineteenth-century American life. For the German and British romantic with a cosmopolitan worldview, nature spoke as an organic whole; literature and music were to do the same.[56] "The Star Spangled Banner," found in most of the school songbooks examined, follows this organic understanding in several ways. "Oh say can you see" ascends a triad to the octave, sounding a unit of harmony that was theorized as one of several "natural" configurations of the musical scale.[57] This marks a musical space in which the survival of the flag, like a limb, keeps the body of the nation intact.[58]

The idea of an organic relation between singer, subject, and music spilled over into school verse as a life ethos in which the typical school song would inspire a serene demeanor and inscribe virtue as industry. Among other self-disciplines, vocal instruction was to combat indolence and a lack of productive purpose. The diction and images in "Our Daily Task," for example, hail the child of a complementary sensibility, one for whom the bee's tasks would recall a familiar world that bore on home and school. Verses fashioned "organically" would deliver meaning more effectually than the didactic verse that characterized songs about geographical places and the multiplication tables.[59] The songs were also to bring pleasure in and of themselves as an important aspect of the organic aspect of romantic verse. For today's reader, a poem such as Coleridge's "Kubla Khan" is conventional and dated, but for early nineteenth-century readers it was a new phenomenon. "Kubla Khan" represented the apex of incantatory, song-like style—it was a sensation in its day due to its musical qualities and the way in which imagination was given free reign, transporting the reader away from the mundane features of ordinary life. The opening of "Kubla Khan" is subtitled "A Vision in a Dream: A Fragment":

> In Xanadu did Kubla Khan
> A stately pleasure dome decree:
> Where Alph, the sacred river, ran
> Through caverns measureless to man
> Down to a sunless sea. (Coleridge 1797/1968, 1400–1401)

Coleridge's unique poetic sensibility converted words into music by devising rhythmic schemes that put demands on the reader as a voice, compelling her to become involved kinesthetically in meaning through the meter and rhyme employed in the verse.[60] Coleridge wrote that significant poetry was "that species of composition which is opposed to works of science, by proposing its immediate object [as] pleasure, not truth";[61] it also imagined music as the center of life itself: "What if all animated nature be but organic harps diversely fram'd that tremble into thought?"[62] For readers and singers in the early nineteenth century, the musical coordinates for feeling were revolutionary. The stuff of the school song may appear old hat in the twenty-first century, but it was electrifying in the nineteenth. Revisiting here, for a moment, Horace Mann's description of the effects of vocal instruction as "the energy and electric celerity of movement [in singing] … generated in a well-formed brain" locates him in the midst of a literary revolution that galvanized the intellectual circles he frequented.[63] A romantic, incantatory aesthetic in, for example, "Excursion Song," echoes the kind of in literary fashion that permeated school music:

> How pure the crystal fountain!
> How clear the purling rills,
> How sweet the tufted flowerets that
> blossom on the hills. (Emerson 1857, 44)

The iambic meter puts the singer kinesthetically in touch with romantic imagery. It makes the child, as future citizen, an appreciator of nature and brings her closer to the poetic conventions that characterized the art songs or *lieder* performed in genteel recitals. Through the (en)chantment of the nation, the singer is taking up a performance of the disinterested persona that compels her to recognize her own social locus within this elite frame or find herself missing from these performances.[64]

Romantic metrics, sentiments of ennobling nationality, the sublime, nature, and the exercise of reason come together in the school songbooks as they are given a dramatic presence through the idealization of school and friends. While the mark of poetic distinction is absent from school ballads, as they lack the "ethereal trembling into thought" that Coleridge and others had in mind, they nonetheless provide a fabrication of the singer's identity as exalted citizen.[65] In funneling the values of romantic philosophy and poetry into music instruction, songs sung in school became the vehicle for the enchantment of the white citizen's qualities.

In this chapter, I gave detailed attention to the organic performances provided by the diction of racial whiteness, verse forms, and rhythm in the school ballads. These pointed to the ideals of an innocent childhood, the Romantic aesthetic embodied in the typical school song, and the hailing of the genteel, civilized future citizen. The next chapter attempts to elucidate how an amorphous language of degeneration constructed blackness through oblique referents to a human type who would, left unchecked, precipitate a national decline.

4

Bacchanalian Chaos, Degenerate Hymns, Public Music Instruction, and the Discursive Fabrication of Whiteness

When, in 1844, Horace Mann recognized the potential of song as a vehicle for instructing the child in the proper attitudes toward nature, his fellow human beings, and the sick, he did so within the political terms of the secularism that replaced the churches as moral institutions. According to the role of instructional ballads such as "Love of School," the schoolhouse becomes the biblical refuge from storms, where storms are, metaphorically, life's troubles:

> When old winter heaps the snow,
> We little heed his wrath:
> But boldly face the roaring blast . . .
> Till safe we take our seats at last
> Beside the schoolhouse fire. (Emerson 1857, 184)

The school song also intensified the domestic cottage scene to make it a "natural" site for counteracting a mechanical, dreary life. Casting a spell of delight over the routines of work, the fermentation of church, nature, home, and music also portrayed the degenerate as a danger to those institutions. Alcoholism was associated with "Negroes" and Irish Catholics, familiar characters in the popular blackface minstrel shows where the "smoked" Irish often played the role of blacks.[1]

For school songs, temperance was a fervent issue that remained in the air throughout the century. It was a social movement connected to racial notions of degenerate behavior that were often projections onto population groups with contradictory attributions of blame and prevention. To exercise citizenship with regard to social behavior, one joined with others in the cause of temperance. In the school context, this meant saving families and neighbors from ruin. Freedom for oneself carried an obligation to free others. School songs fought alcoholism with temperance slogans.

> Friends of Freedom! Swell the song:
> Young and old, the strain prolong.
> Make the Temperance army strong,
> Shrink not when the Foe appears;
> Spurn the cowards guilty fears,
> Hear the shrieks,
> Behold the tears of ruined families! (Bradbury and Sanders 1831, 136–37)

The common narrative of temperance scenarios depicted the loss of childhood to the worry and cares of adulthood. Wordsworth writes in "Resolution and Independence" from 1803,[2]

> I was a Traveler then upon the moor;
> I saw the hare that raced about with joy;
> I heard the woods and distant waters
>   roar:
>   Or heard them not, as happy as a boy;
> The pleasant season did my heart
>   employ.
> My old remembrances went from me,
>   wholly
> And all the ways of men, so vain
>   and melancholy. (Wordsworth 1803/1968, 1309–10)

Wordsworth's vision of childhood was not representative of the social classes ranking below his own but, rather, it was a construction of that idyllic childhood in the lake region of England.

The loss of innocence and its attendant heartache were themes that resonated with the fallen and lost Promised Land that American Protestant educationists such as Horace Mann encountered in the miseries of city life. This language, both spoken and written, circulated through the mentality of the literary publications, music journals, and the genteel music societies in the eastern United States. Literary ideas flowed to and from these intellectual circles and the educational leadership in New England through numerous religious institutions and music periodicals, such as *Dwight's Journal of Music*, and the popular press, such as *The Christian Science Examiner*.[3] These carried exemplars of the new poetic style that had made its way into school song. Their significance is not only the transformation of genteel ideals into a school curriculum, but also the governance of conduct that flows from the comparisons of children who match or fail to fulfill these ideals. In this sense, the songs are a part of a paradoxical technology that unites the nation while dividing it. The songs urge freedom in order to further the social good when, at the same time, that social good was located in particular social classes and manners.

As noted in connection with the Romantic aesthetic of school song lyrics, there is an obvious impulse to rescue man from the perils of the industrial age. The ethical purpose alluded to in almost every school song cited here, as Ian Hunter writes in relation to the history of the teaching of English in *Culture and Government* (1988), was underwritten by the belief that the society would either march forward or sink into a degenerate state. For this, there had to be degenerative personae, a continuance of the presence of "fallen" brethren from the previously religious era. In early Republican terms, the degenerative label fell most heavily on the poor, the black, and the Irish, perceived, in musical terms, as makers of "noise," "obscenity," and "deviltry." While the performances of blackface made fun in the Negro mode that made it a vicarious moral transgression for the audience, school singing reassured the genteel of their separation from the acts on the minstrel stage. Eric Lott points out that minstrelsy created an imaginary folk culture and vernacular, splintering what he calls "racial publics." This splintering was to blend into a national symbol of folk culture. Being in the spotlight of public entertainment for nearly a century, this also raised racial anxieties exactly because of its popularity. Its gestures of self-deprecation in clever rhyme schemes were the direct opposite of artificial, moralizing style of school songbooks.

> Music now is all the rage,
> De minstrel bands am all engaged,
> Both far and near de people talk.
> 'Bout nigger singing in New York. (Lott 1993, 251n3)

And to point out the interchange of audiences who found both minstrel and "high" culture to be part of their cultural world, the following:

> De Astor Opera is anoder nice place;
> If you go thar, jest wash your face!
> Put on your "kids," an fix up neat,
> For dis am de spot of de eliteet! (Lott 1993, 251n3)

For the instruction of children, school songbooks continued to be at the center of the curricular structure along with the set of musical techniques outlined in many of these songster albums. Its role was to mediate the sonic fields outside the classroom in order to produce a persona of a whole population either needing rescue or taking up the role of moral supervisor. *The New Favorite* songbook offers a collection of laments and warnings against the seductions and consequent "fall" of virtue symbolized by darkness.

> Don't stay late, don't stay late,
> Oh don't stay late tonight,
> This world so cold, inhuman,
> Will spurn thee if thou fall.(Giffe 1875, 74)

Carrying with it the connotations of darkness associated with the street and sinister bodies lurking to ensnare new sinners, the song (below) alludes to neighborhoods of the Irish Catholic and freedmen where taverns proliferated. If the night was dangerous, those who worked in the night, whether as dockworkers, entertainers, prostitutes or tavernkeepers, were the source of evil. As one school ballad put it,

> From many a poisonous rill,
> God calls us to deliver,
> The victim of the still!
> The victim of the still! (Bascom 1855, 51)

Right Boyhood

The element of rescue expressed in these temperance ballads performed several functions of citizen-making—they inscribed childhood as a memory that could be mapped onto the pastoral images of the nation and the child as the citizen rescuer who, through the exercise of a kind of his own ministry, steered others to the path of right living. The disposition highlighted for rescue was the "other" among the singers whose family and surroundings were vilified in song. On the other hand, ballads that described how to be virtuous in vivid detail, as the uptake of duty in the following verses from "The Happy Schoolboy" in *The Young Choir*, show the alternation of leisure and duty that characterized the striving middle class.

> When I play, I will play, like a pleasant boy,
> And my play shall be cheerful and free;
> When I work, I will work, like a Yankee boy,
> With a right good will it shall be:
> At work or play, endeavor still
> To do it all with right good will;
> Then away, then away, O Yankee boy
> With a smile and a pull, all so free. (Bradbury and Sanders 1831, 111)

The mood in the song above could also be found in popular songs, but it carries a special weight in school songbooks through its reiteration of the "Adam" or masculine archetype whose assertions about winning prevail as a sign of the right reason about school, manhood, leisure, and work. The hailing of an Adam type conveys that *he*, the Yankee, is the deserving boy, the cheerful laborer. Idle play squanders the freedom that comes from a calculus of happiness, duty, and urgency:

> Let's away with a cheer and with glad horrah!
> Like a man, I will toe to the mark;
> Leave my play—all my play at the school-room door

With a heart like a cheerful lark:
And I will work all the time I'm there,
I'll keep each rule and I'll work with care,
Come away, haste away, there's the school-bell,
Hark, I will try to be first on the floor. (Bradbury and Sanders 1831, 111)

Recalling the comparison of Adam and Eve in Mason's preface to *The Normal Singer*, the projection of masculinity as a governing principle for action operates as a foil for the feminine: "Modest and retiring she desires not to attract attention to herself" (Mason 1856, iii). Gail Bederman and Eric Lott, two Americanists who have written on this subject in different historical periods, discuss this maintenance of masculinity as specifically attached to a hierarchy of racial types.[4]

School songs were almost unremittingly humorless—their tone drew upon a comportment that was synonymous with proper church behavior as ideals amongst the white middle class. This code separated the child's world of play from the high purpose of schooling: "Like a man, I will toe to the mark . . . And I will work all the time I'm there . . . I'll keep each rule and I'll work with care" (Bradbury and Sanders's "The Happy Schoolboy"). This split is articulated on the philosophical level as the mind-body dichotomy, running through all of the social hierarchy and portrayed with special vehemence in the black or white dramas of minstrelsy in which the privileging of body over mind, attributed to every aspect of black culture, formed the basis of uproarious humor and a dubious, at best, denial of white middle class foibles.

Devotions, Noise, Musical Vulgarity, and the Bodies of Others

The template of a contemplative and reverent soul divided children along racial lines, doubling for the ideal citizen of the new nation. In one song, the busy bees and joyous songsters etch the comportment of the ideal inhabitants of the sublime landscape.[5] Excluded from the charmed circle, by virtue of their circumstances, children of lower class backgrounds were differentiated from children of promise—the "fair" and the "bright"—separate from the stream of immigrants or even native-born Catholics, Jews, and African Americans that filled the cities' cheaper housing and jostled one another on the crowded, narrow thoroughfares of mid-nineteenth-century Boston, New York, Baltimore, and Philadelphia. A society sliding toward degeneration, but heroically making the effort to commit all its energy to overcoming corruption, found the school and school music to be plausible receptacles for social evangelism. The work of the late eighteenth-century American composer, William Billings, provides the occasion to consider a different treatment of this cultural theme and how it became the keystone narrative of public music education. Specifically, the "degenerate" hymnody of singers and some composers trespassed on the imagery of goodness touted in the conventional hymn and child's songbook. Billings's work announced the intent of inviting the common, unfinished singer to star in church.

> Let horrid jargon split the ear
> And rive the nerves asunder;
> Let hateful discord greet the ear
> As terrible as thunder. (Billings, as cited in Perkins and Dwight 1883, 24)

This kind of verse shocked the pompous but delighted rebellious churchgoers. Billings's hymn recalls unharmonious Sundays and celebrates them in a satirical take on the stiffer forms Protestant piety. In ways that are similar to minstrelsy's mocking "high" culture, purposively degenerate hymns portrayed the mix of manners scattered throughout the social hierarchy. And like minstrelsy, the hymns that were seized upon as "degenerate" became the most popular and, therefore, the most threatening. According to the conventional historical narratives on the origins of public music instruction, the school subject was the "saving" device that would reverse rebellious tendencies in church singing. But we must also consider the threats of racial and class mixing in these hymns that range alarms for the standards of comportment required of bona fide citizens. For civic elites such as Samuel Eliot,[6] Lowell Mason, Horace Mann, and John S. Dwight in the mid-1800s, public music instruction would raise the level of national taste that had fallen off through inattention to proper singing technique. Histories of music education published between 1928 and 1999 describe hymn-singing schools as forerunners of publicly funded music instruction and as the original institutions out of which school music programs grew.[7] Yet, I think it is necessary to unsettle this foundation by reinserting the story of a "declining" hymnody and musical taste into the tension-filled society from which it has been set apart. In the pages that follow, I attempt to bring clarity to the religious, linguistic, and racial anxieties that have been obscured by the notion of cleaning America's musical house.

The central theme in foundational narratives has been that the lackluster musical life of the country was the product of inferior beings and fallen hymn singing.[8] What attempts to reform the church hymn covered over were that public music instruction became a principal site for the differentiation of bodies *qua* citizens. In the early 1800s, Samuel Eliot, president of the Board of the Boston of Academy of Music, wrote, "As for our country, music cannot be said to have any history, scarcely an existence here" (Eliot, 1836: 83). Michael Broyles described the churches' role in teaching hymn singing: "The clergy's position was to teach them from the psalm books, sticking with the tunes as presented by the authorities, and there would be a stop to the chaos, anarchy, indeed almost Bacchanalian license with which parishioners sang . . . Lyric poetry [hymn verse] should indeed be impassioned but it should be passion within the bounds of Moderation"

(Broyles 1992, 42). The effort to improve the nation's sorry musical state propelled the dream of elites that American music teachers would eventually be trained in the European traditions taught in the music conservatories of the German states. Lowell Mason expressed this vision in the 1830s: "Even now there are but few who have a just conception of the previous preparation, time and labor necessary to thorough knowledge . . . The subject is better understood this [the European] side of the Atlantic . . . It were to be wished most heartily, by all lovers of music, that such an Institution could be founded in every large city of our own country. The rapidly growing taste of our good people seems to demand some such effort" (Mason 1967, 71–76). Mason's outlook had been codified in *Dissertation on Musical Taste*, a treatise by the early nineteenth-century American composer Thomas Hastings. As a general critique of vocal art, Hastings's text typified the dispositions of music educators from the early years singing instruction to the music appreciation curriculum in public schools in the early 1900s.[9] Hastings argued that the standard of singing would continue to decline if left exclusively to a young generation he described as "light-minded and vicious."[10] In the *Dissertation*, he took special pains with the facial expression of singers, outlining a set of prescriptions for pronunciation and rules for hymn writing. For example, he provides a critique of Handel's musical setting of the line, "far from all resort of mirth/save the cricket on the hearth,"[11] claiming that Handel had rhythmically distorted the syntax, making a cricket the supplicant in place of man. However, reading this line differently provides an opportunity to hear Handel's lapse of taste as a musical joke in which the cricket's rasp represents the rustic, untutored sounds commonly heard in church. Hastings's view was that those having "the gift of a musical ear"(Hastings 1825, 13) were charged with regulating the singing of ordinary folk. According to Hastings, the only singers to be encouraged were those trained for the choir: "He who speaks or sings too loud needs to be reminded of his fault . . . harshness, as well as a guttural, dental and labial or nasal quality of tone is [accompanied by] wrong confirmation of the mouth" (ibid., 24–25). He also wrote that an undisciplined congregation would become silent when hearing a well-disciplined choir that knew how to execute sibilants, vowels, and produce accurate pitch. Vocal performances with less rigorous standards were "disgusting and ridiculous" (ibid., 70).

Particular human types were incapable of musical refinement: "There is a degree of refinement delicacy and invention which . . . the Asiatic [non-white is incapable of]" (ibid., 135). Untoward hymnody, with its more than occasional dissonance, whether intentional or not, provoked the musical anxiety of those concerned with maintaining pious comportment in church congregations, concert audiences, and schools. These were the songs that came to occupy the sonic place of "noise" and slave music, as noise made the good ear a symbol for particular racial, national, and religious distinctions.[12]

Another target for the charge of musical degeneracy was the crude imagery of popular hymns. By bringing the degenerate hymn and its defiling sounds within earshot in the present account, the censured hymn verses and harmonies provide acoustical clues to the life of the spurned singer, composer, and listener who broke with church etiquette and performance standards of the European vocal aesthetic. A look at one example of degenerate hymnody[13] cites the genteel as a group of monsters whose extroverted singing was purposively rebellious:

> Ye monsters of the bubbling deep,
> Your Maker's praises shout;
> Up from the sands, ye codlings, peep
> And wag your tails about. (Christopher, as cited in Perkins and Dwight 1883, 21)

These images play against the traditional expectations for reverent comportment. "Ye" points to many miscreant singers and calls for a show of them from the "bubbling deep" (a primordial substance) in order to reclaim their places in the congregation.

The most famous hymnodist from this period, William Billings, had enjoyed sufficient popularity as a hymn writer to allow him to propagate something called the "fuguing tune," a musical form that gave greater independence to voices than traditional hymns. These popular tunes sounded intentional musical dissonances and their verses often called for musical defiance.[14] "Let an Ass bray the Bass, let the filing of a Saw carry the tenor, let a Hog who is extremely hungry squeal the Counter and let a Cartwheel which is heavy loaded ... squeak the treble" (Billings, as cited in Perkins and Dwight 1883, 26). Meanwhile, the views of Hastings, Mason, and others offered resistance to the inflections of colloquial speech and mannerisms that were being given more play in the social fabric (through minstrelsy, most prominently) and hymnody of the early 1800s' Jacksonian populism.[15] For example, prefaces to school songbooks such as *The Young Choir* described their contents as omitting the childish, vulgar, and showy forms of European vocal music. Claiming that their songbooks would promote purity and happiness in contrast to other song collections, rhetoric such as the following commonly appeared as prompts in advertisements and flyleaves. "Hitherto, as a nation, we have given musical utterance chiefly to our lower feelings, while in the nobler workings of the soul, we have done little to express, and still less to excite and develop the more worthy moral and patriotic emotions" (Bradbury and Sanders 1831, flyleaf). "[Well-composed ballads and hymns] can and ought to be made the handmaid of virtue and piety ... truely [sic] heavenly and delightful ... It would be a sure and excellent means of national improvement" (Mason 1834/1982, 130).

A similar preoccupation with this mission suffuses the prefaces to several songbooks from the 1850s to the last decades of the nineteenth century.[16]

As a contrast to ill-mannered gatherings, Lowell Mason's translations of music into the balm of family solidarity sketched a picture of the ideal white Protestant family united in song—the epitome of tranquility and reserve: "When singing is employed in the family devotions, it tends to produce a proper frame of mind, and to calm the feelings. It throws a delight and interest into the exercises, which calls up and fixes the attention. In the pious families of the Scotch, singing is a necessary part of the devotions ... and in no families in the world, do all the members more heartily unite in these exercises" (Mason 1834/1982, 132). If civic music enthusiasts and educators' efforts to bring early musical training to the public were successful, civic leaders such as Samuel Eliot thought it more likely that America would produce brilliant composers at the level of the German greats: "How does it happen, that a nation of kindred origin with our own parent stock, and with a language almost as unmusical as ours, should have produced such an uncommon proportion of musical genius of the highest order ... Is it not manifest that the national practice of giving the rudiments of a musical education at school must have contributed largely to the development of the national talent for the art?" (Eliot 1836, 85). In 1839, Adolph Marx wrote a beginner's guide for understanding music. Popular with music enthusiasts and teachers in the United States in an English translation, Marx's instructions (written for a German-speaking public) correspond to similar ideas in the writings of Eliot, Mason, and music teachers and composers, but it went further in its equivalence of music and nation: "We conclude that music cannot be destroyed and lost but with the nation itself ... [It requires] an elevated contemplation of what our nation is made of and what music requires to elevate the capacities" (Marx 1839, 4). Later in the century, John Sullivan Dwight tirelessly wrote of efforts to consecrate classical music performances and to eliminate frivolity from the concert stage.[17] "There is a public here for just this class and character of concerts—concerts pledged to nothing but to standard music of the highest order ... our musical culture is blown this way and that by the caprices of fashion and the tricks of advertisement, we want ... opportunities for refreshing our knowledge and our feeling of the great masterworks of men of Genius" (Dwight 1873, 102). Similarly, The Handel and Haydn Society's major purpose, reaffirmed by a historical account of the society's founding, was to provide the kind of music that would inspire the nation.[18] "Too long have those to whom heaven has given a voice to perform and an ear that can enjoy music neglected ... subduing the ferocious passions of men ... so absolute has been their neglect, that most of the works of the greatest composers of sacred music have never found those in our land who have even attempted their performance" (Perkins and Dwight 1883, 39). John Fillmore, a music teacher and music historian who promoted school concerts in the 1870s and 1880s, testified to the low taste in repertoire by touring artists: "Mr. Weiner's flute solo was a medley containing 'Home Sweet Home,' 'Yankee Doodle' and 'O, Susanna!' I was curious to know how this would impress the thoughtful part of the public, some of whom had complained that artists would not play simple things which they could understand . . . I believe I speak the exact truth when I say that . . . the audience regard this performance [with] mingled disgust and contempt. They had become familiar with the notion that artists were above that sort of thing" (Fillmore 1877, 10). Even more in need than the eastern cities, however, was the Western frontier where degenerate socializing and tavern-going[19] encouraged singing the lowest kind of popular ballad. Fillmore, in Wisconsin, which was part of the Western frontier in the mid-1800s, noted that the effort to bring European musical taste and concert decorum to the West was spearheaded by a relationship between schools and traveling musicians: "There are teachers all over the West who do their best . . . and look to the traveling artist to meet their [students'] needs . . . to study the great composers" (Fillmore 1877, 10). To take these statements at face value, as similar testimonials are interpreted in historical accounts of formal music education in the schools, is to miss the exclusion of the rich vernacular traditions with which those students that Fillmore mentions were familiar. It was not just that cultivated musical taste was central to the education of the citizen, but that the bona fide citizen should not be confused with former slaves who lived in the these territories or the many thousands of Eastern and Southern European immigrants who would settle in places like Michigan, Illinois, Iowa, and beyond. As one prominent educator stated at a music teachers' conference in the early twentieth century, "Everything in education should unfold from music. We have not been a musical nation . . . We have been a pioneer people, drifting away from the influences of music in conquering a new world" (Claxton 1915, 48). These aspirations set the tone against which public music education and the civic music societies were to labor, eliminating from possibility nearly every kind of music making that was not of European, classical vintage.[20] In that spirit, training courses in music for the classroom, a feature of the teacher training that would later be offered in colleges such the Normal Musical Institute in Binghamton, New York, and the Normal Institute in North Reading, Massachusetts, conformed to the aesthetic views of Dwight, Mason, and Fillmore in overlooking any American or European vernacularmusic's as appropriate for classroom use.[21]

As Horace Mann had written in 1844, the aim of public music education was civility in the traditions defined by civic, cosmopolitan elites. This quality was seen as rare, but essential to national progress, submerged, as many thought, beneath a mass of sheet music and public performances of "the insipid, vulgar and frivolous" (Mason 1847, ii). Given the rhetoric of more than a hundred years duration, twentieth-century music education continued to cast the curriculum as rowing upstream against a corrupting onslaught of musical vernaculars and popular musical styles. Maintaining the boundaries between degenerate and elevated tastes, every new text and curriculum change reinscribed the demographic contours of musical "habitus" and dispositions that were linked to a way of thinking about music that mirrored the historical patterns of racial and ethnic segregation.[22]

Looking Ahead to the Foundation of the Good Listener

Early nineteenth-century views on uncouth singing and vulgar diction found their double in the comportment that signaled bad listening habits in the early twentieth century. The strictures imposed on vocal composition and performance continued to be of concern in music teacher manuals,[23] yet it was not only musical matters that were referenced in those curricular guides, but any aspect of social life that imbricated the body of the singer and listener to set him apart from his schoolmates. For example, Surette and Mason's manual for music teachers made a special point of visually recognizing the "lazy" listener: "It is safe to say that out of any score of persons gathered to hear music, whether it be hymn, song oratorio, opera or symphony, then are not listening at all, but are looking at the others, or at the performers or at the scenery ... Five more are basking in the sound as a dog basks in the sun ... in a sleepy, languid way" (Surette and Mason 1907, 1). With music appreciation's formal entrance into public school classrooms, the education of families and future generations was concerned with the body's demeanor while listening to music. Instructions on social decorum and listening occupied music appreciation texts, school board documents, professional journals, and manuals for teachers from the 1880s well into the heyday of music appreciation when radio and phonograph became commonly available, circa 1920. Concert etiquette now demanded a new seriousness as a standard for music lovers similar to the more severe attention called for at church—silence.[24] John S. Dwight argued for higher standards of behavior in concerts and for more sophisticated programming; he went so far as to leave town during the National Peace Jubilee in 1869 in Boston rather than take part in what he called musical vulgarity. This extravaganza included a performance of Verdi's "Anvil Chorus" from *Il Trovatore* in which one hundred hammer-wielding firemen struck one hundred anvils at the appropriate moments in the score.[25] Since revulsion typified the reaction of music educators, critics, and connoisseurs of the genteel class to anything associated with the overly ornate style of Italian opera, as it was thought to be vulgar in comparison to the sublime in instrumental works, the hundred anvils struck at once was directed at the new classes of moderate means. The spectacle aimed at instilling the feeling of sublime awe,[26] but it also appeared as a species of "lowbrow" mentality, somewhat grotesque or gothic.[27] When differentiated from more "tasteful" performances, the thrill of synchrony in that show created a sonic space where extroverted expression served as a foil for the cultivated listener.

The interplay of the undignified and the low brow was also associated with a style of extroverted, religious behavior of the revivalist movement that periodically swept the country. Revivalist traditions encouraged believers to allow the spirit of Jesus to seize the body and display what were conventionally thought of as "undignified" postures. Prayers and hymns were shouted spontaneously and often accompanied by body movements that others perceived as akin to sexual ecstasy. Even worse in the minds of conventional Christians, evangelism seemed to override mores that separated the sexes and races, provoking fears that revivalism promoted miscegenation.[28] The images of raced and gendered bodies as coequal in religious rites signaled the unreason of religious excess and surrender of self-governance that civic pietists and educators were eager to leave the behind.[29] "The loosening of passions threatened the structure of society itself; taste and decorum were integral to the Congregational outlook" (Broyles 1992, 88).

Revivalism and Catholicism posed a threat to the civic authority of the genteel class.[30] For public music instruction, this resulted in an increase in the number of "religious" songs with a cool-tempered, secular ethos. Between 1830 and 1900, the proportion of conventional Protestant hymns in relation to other types of songs outnumbered nature songs by two to one.[31] This increase corresponded to the diminishment of the Protestant tradition in cities where mass immigration had begun to penetrate. The largest single jump in immigration, from 956,000 to 3,201,000, occurred in the years between 1880 and 1884 when Italians, Poles, Irishmen, Jews, and others brought with them religious and musical practices that were sharply different from the established Protestant churches.[32] In the northern states, those of evangelical faith, newly arrived Irish Catholics, freed slaves, slaves (in New Jersey), and "others" were well over half of the population.[33] Changing demographics exerted pressure on songbook authors to feature a secular worldview consonant with the Unitarian or Congregationalist ethos.[34] While the genteel sentiment against "monsters of the bubbling deep" kept the bodily practices and allusions to revivalism and Catholicism out of school songbooks, the effort to promote a secular and largely white, European vocal tradition became more urgent.[35]

Large-scale demographic change had a role in dividing schoolchildren as listeners.[36] Music appreciation lessons were to raise the child not only above the level of street life, but to level of the good ear purveyed by the early music societies, music pedagogues, and symphony conductors. Agnes Moore Fryberger wrote in *Listening Lessons in Music: Graded for Schools*:

> The most important mission of technical musicians today is to teach the rest of the people to listen . . . Early in life one should learn that music is something to think about, something more than entertainment. I cannot

resist referring to an experience some years ago with an eighth grade class ... [These children were] from the poorest district and their acquaintance with the phonograph was gained from cheap playhouses, restaurants, and from open doors of saloons where it was a feature of entertainment ... [When I played Handel's *Hallelujah Chorus*] imagine my surprise when the children began to giggle and laugh ... It is a strange fact that we must learn to listen, and it is regrettable that so few have acquired the habit. (Fryberger 1916, 213–14)

With values close to those in Hastings's *Dissertation on Musical Taste*, Agnes Fryberger's cosmopolitan musical aesthetic had discursive connections to early theorists of the sublime: Edmund Burke, the Earl of Shaftesbury, and Kant. Traveling in music society programming, periodicals, newspapers, and listening guides, a transnational literary coterie supported the idea that inward cultivation could be read through a charmed appearance. This was the famous "je ne sais quoi,"[37] or quality, of carriage characteristic of the noble mind.[38] The emphasis on modes of listening to music concurred with the ethos of improving one's social status, and both would converge in the production of charm that marked the superior "mind."

To appear to reflect on music as an experience of the "mysterious depths" of music was the privilege of a new musical identity in the late eighteenth and early nineteenth centuries.[39] Edward Hanslick had summarized the interior of the soul and ear as marking the detachment from material concerns. The superior musical mind was one who could "[give] an objective existence to his musical ideal and [cast] it into a pure form."[40] Used as benchmarks for the comparison of compositions, the "pure" musical ideas of true artists moved Hanslick to place particular composers, such as Beethoven, at the apex of disinterested accomplishment; at the opposite end were those composers, such as Berlioz, who merely imitated emotion and pictorial images.[41] Hanslick ranked listeners according to these polarities, concluding, "an interpretation of music based on the feelings cannot be acceptable either to art of science" (Hanslick 1854/1857, 74). These views were shared by a broad cross section of the educated elite. Writing about the phonograph and radio's ability to abate the effects of demographic change in the early twentieth century, Thomas Briggs, Professor of Education remarked, "[These technical wonders] make extrinsic interests intrinsic ... to enable pupils to share the inheritance of the race ... [and] to encourage the [discarding] of sentimentality ... for the silent, liquid and mysterious depths" (Briggs 1925, 5–7). A high school music teacher, writing in the *Journal for the Music Supervisors' National Conference*, quotes a line of poetry to capture the sense of delicacy and hushed anticipation that the appreciative listener must "bring to music worthily interpreted" (C. Adams 1929, 89). Such a cultivated listener would be able to make judgments between the properly reserved vocalist and one that "projects temperament into the atmosphere as a hose would squirt water" (Giddings 1910, 10). These descriptions overlapped with the characteristics that blackface entertainment stereotyped as Negro and that other observers associated with the lowbrow and sentimental entertainment of dance halls, taverns, and theaters.

W. J. Henderson, a music critic for the *New York Times* and author of several books on music listening, characterized the untutored listener as the proverbial tempted sinner—he was one who, by tasting the forbidden fruit, exposed his naked ignorance to the world. In books on listening that circulated on both sides of the Atlantic—such as *What Is Good Music?*—Henderson described how an individual who has been exposed to bad music, or "unstable, incoherent and pernicious sound" (Henderson 1929, 218) can fall ill from its effects.[42] Another musical tastemaker stated that the highest effect of music was its ability to bring the body into correct relation to the group and to sublimate raw energy into virtuous feeling; however, this speaker added, "The energizing influence of music may conceivably work harm by exciting the listener to action [an agitated, nervous state] rather than to thought" (Birchard 1923, 74). As one article in *The Etude* put it, the challenge of acquiring good taste demanded that the listener separate himself from the crowd and the mass of music printed which was vulgar.[43] Musical vulgarity meant extroversion, either through the promotion of emotion at the expense of intellect or through the instigation of undignified body movement. It contained offensive diction or idioms that bespoke a lack of character or puerility—descriptors that merged with popular cultural, racial and ethnic notions of worthiness.[44]

Music as Language: Songs for Home and School

In the early 1900s, as Eastern and Southern European immigrants changed the tone of cultural life of the large cities, music appreciation lessons in public schools were seen as a way to teach civility in school and at home.[45] "Musical progress of the nation is not concerts, it is home that counts the most . . . Before we can boast of being a musical nation, we ought to have a far more varied and versatile performance in our own houses" (Goepp 1910, 33). Underpinning this way of thinking was the discourse of music as language, and language of all kinds as the reflections of a "racial" heritage.[46] For example, American popular songs made use of the notion of inferior language through linguistic stereotyping and mental deficiency in the Asiatic. A verse from the song "The Chinese Laundryman" reads,

> Me comee from Hong Kong Chinee
> To workee for the Mellican man . . .
> Me no talkee much English. (Levy 1971, 158)

With regard to what was called "Negro song," as distinguished from minstrel tunes, Elias Nason, a composer and critic, remarks in his monogram, *Our National Song*, that well known ballads such as "Columbia," "The Star Spangled Banner," and "Yankee Doodle" were superior to Negro song or

"Spirituals," as they were called: "[These were] . . . much to be preferred over Negro song for the truly national ballad . . . [in which] attention to the flag proclaimed freedom . . . [This] consolidate[s] us into one vast free people . . . harmonious, high-minded, hopeful, grateful and aspiring" (Nason 1869, 14). Blackface minstrel tunes, sometimes referred to as "coon songs," promoted the satire on dialect or versions of what, today, is called African American Vernacular English.[47] The model of whiteness purveyed in school songs as the "fair" and "merry" in the idyllic garden is not only a visual model, but is also a mode of diction that played against dialect that was satirized in the blackface performances. One version of the song, "Jim Crow," reflects a consciousness that would see humor in couching high art in a caricature of Negro speech. But it also pokes fun at politicians in Washington bringing an imagined black point of view into alignment with the common white person's sentiments about the high and powerful. Being neither white nor black, but an inextricable blend of working class, laborer, and petit bourgeois resentments, minstrelsy used black stereotype to express the "common sense" of the classes of people who were interested in "Sal" or low pastimes as opposed to "Pagganninny."

> I'm a rorer on de fiddle,
> And down in ole Virginny,
> Dey say I play de scientific,
> Like Massa Pagganninny
>
> O den I go to Washinton
> Wid bank memorial;
> But find dey tork sich nonsense,
> I spen' my time wid Sal. (T. D. Rice, 1828, as cited in Marrocco and Gleason 1964, 262)

The language and sentiments in and about minstrel ballads were unrecognizable to W. E. B. Du Bois. For him, Negro music consisted of "sorrow songs" that evoked the harmonies of a people and their message from across the color line:[48] "A-dancing and a-singing . . . raised but confusion and doubt in the soul of the black artist; for the beauty revealed to him [in Negro music] was the soul-beauty of a race which his larger audience despised" (Du Bois 1903/1989, 3–6). Frederick Douglass denounced blackface minstrelsy as racist counterfeit, relating what he remembered of slave music that, during the first half of his lifetime, did not qualify as music: "[Slaves] would sometimes sing the most pathetic sentiments . . . I have sometimes thought that the mere hearing of those songs would do more to impress some minds with the horrible character of slavery, than the reading of whole volumes of philosophy" (Douglass 1845/2001, 20). The theme of the music making of degenerates, whether slaves, immigrants, or religious "others," reflected the discursive atmosphere of bodily comparisons much larger than the downward slide of hymnody was able to accommodate.

As a saving-from-degeneracy narrative, its premise distanced public music instruction from the richness of musical differences heard on the American sonic landscape during the long nineteenth century. This only served to confirm "the color line" that Du Bois observed on each scale of existence and imprint its limitations on the public school music curriculum.[49]

The Negro Spiritual, Native American Primitiveness, and Musical Wildflowers

Pursuing the notable absence of African American musical idioms from the curriculum of the nineteenth century along with many other forms of music that were heard as uncultivated, we are well advised to look at the way in which they were eventually caricatured as they began to appear in the pedagogical materials available to teachers in the early twentieth century. Readily available to school songbook writers, neither Negro songs nor Native American music appear in a survey of two hundred American school songbooks published before 1900.[50] This was also the case with several texts selected at random from a large collection of school songbooks published between 1900 and 1912;[51] however, in the 1920s, there was, for various reasons, some having to do with recording technology and profits to be made on folk music sales, a sudden inclusion of folk forms in curriculum manuals sold to teachers and school districts. "A deeply religious fervor, child-like [sic] conception of heaven and marvelous gift of natural harmony of the Negroes are immortalized in their weirdly beautiful spirituals" (Victor Talking Machine Company 1923, 167). This manual's inclusion of the Negro spiritual in a nationally popular music appreciation series packed a double description for black otherness—the noncorporeal spirituality of the gospel—in its representation of this "primitive" musical form. The historical precedent for this were the post–Civil War reports and notated versions of Negro songs that were reputedly a genre of the sublime that would garner the exaltation of sophisticated Americans.[52] Separating the notion of the spiritual from slave song facilitated a transformation of the spiritual into music suitable for the classroom as a sample of sublime, spontaneous creation.

Early accounts of the black oral tradition stress the level of emotion that went into their performances, something altogether strange to white observers, and, as such, was heard as undisciplined in comparison to the process of musical composition associated with the white, genteel world.[53]

By the early 1900s, when the genre of the Negro spiritual first came to have a place in Frances Elliott Clark's music history course, it was an exemplar of the musical sublime, a way of sequestering it as a kind of art that was not crafted by human intelligence but inspired by religious belief.[54] In Edith Rhetts's text, written for the Victor Talking Machine series with accompanying recordings for classroom use, the "Negro spiritual" was classified as a separate nationality, along with Swedish, English, French, and other "national" genres. Similarly, a Native American exemplar was categorized as "primitive." Rhetts writes of some folk genres: "Folk song and folk dance are the wild flowers in music—the spontaneous expression of primitive peoples" (Rhetts 1923, 20–21). In another collection of songs for the general public, the "Indian" ballad represented as "Alknomook, The Death Song of the Cherokee Indians from Tammany," features a piano accompaniment.[55] This genre often characterized Indians as bloodthirsty and unfeeling. In one of the Victor Talking Machine manuals for teachers, a lesson on a Blackfoot song serves as an example of primitiveness of the beginning point of musical evolution. It represented the monotonal "ancestor" of Western classical forms that, along with Negro music, exhibited "childlike" ideas of the afterlife.[56] In distinction to the mourning songs of the genteel, American Indian laments were exemplars of a foreign sensibility, forming a part of the colonial discourse referred to as "orientalism."[57]

Peopling the child's world with black or Native American song took the form of providing a separate niche for an untutored, "natural," and naïve approach to music.[58] Agnes Moore Fryberger's *Listening Lessons in Music*, describes the black choral group, the Jubilee Singers, as a chorus whose singing "had a peculiar fascination for all who heard them . . . That these uncultured people could bring all Europe to their feet by the inherent beauty of their song, demands for the negro a distinct place in the musical world" (Fryberger 1925, 240). Early twentieth-century music educators formed a consensus around the notion that the genius of European musical culture had no relation to the sonic surround of other languages, dialects, and musical expressions, other than its authorization to represent many of them as primitive artifacts or as decadent.[59] However, the alchemical process of making the school subject of music a vital part of the mission of public education compelled civic elites to use comparative objects of study and methods to underline population differences and the moral, aesthetic, and hygienic dispositions that would hail the bona fide citizen. The color line, as W. E. B. Du Bois called it, with its projected savage moods and childlike innocence onto black singers, instrumentalists, and composers, is quite evident in the paternalistic tone in music appreciation manuals.

However, this tone was also ubiquitous in almost every cultural practice having to do with race and nation. If language, and music, as a type of language, was one of the key markers of national pride and identity, dialect was as an important means of distinguishing "degenerate" from "fine" singing. Notions of backward development and stages of childhood are evident in the "songs of childhood" genre in old school songbooks in which an interlocutor speaks, ostensibly, as a child. On close examination, the child's speech has characteristics of black dialect that appears, like its minstrel cousin, to ambivalently address the nostalgic memory of childhood as black infantilism and represented in the songs as quasi-vernacular remnants, As Toni Morrison writes about the novel, the shadow of racialism[60] in American writing "[offers] no escape from racially inflected language... A racially inflected language [hid]... the thunderous theatrical presence of black surrogacy" (T. Morrison 1993, 26–27).

An example of black surrogacy, a ventriloquism of black voice and a racially inflected language makes its appearance in, for example, "Yeep! Yeep!" from the 1870s. The song's intimations of an African American Vernacular English[61] imbricate the child who has yet to master English:

> My chickie's name is Cuddle:
> Jus' see him blink his eyes;
> He's Brownie little orphan,
> And listens dreffle wise.
> My but I think he's awful Funny little peep;
> The way he says he loves me is "Yeep, yeep, yeep!
>
> My chickie's coat's the softest
> It looks like puffy gold;
> I wis' it wouldn't turn to
> Big fedders when he's old...
> ...
> I think it's most r'dic'lous
> But granma says it's true,
> To think the eggs the hens lay
> Can turn to chicks like you.
> ...
> Jus' see him pick the crumbs up
> An' drink and lift his head
> > That is the way he thanks God
> > For givin' daily bread. (Stewart 1873, 79)

One can hear the echo of blackface imitations of black dialect in this song. These inflections are cued by their phonological departure from Standard English—deleted syllables, consonants, and grammatical forms, such as "an'" for "and" and "fedder" for feather. The dialect corresponds to some aspects of African American Vernacular English, for instance, the voiced "d" for the Standard English "th" in the word "fedders" and the performed "ff" for the middle d in the standard "dreadful." Similar events in the song occur in the realization of "jus'" deleting the "t" and the elision of syllables in "r'dic'lous."[62] "Yeep" serves as foil for linguistic gentility and like the popular genre of dialect parodies exemplified in "The Chinese Laundryman."

The most important aspect of the linguistic features of "Yeep" is that it links the childlike persona to the not fully present African American subject.[63] In this sense, "Yeep! Yeep!"—a common type of song in school collections—enters into proximal relation with the minstrel-like verses and sets this child or black dialect apart from the perfected diction looked after in school song collections.[64] Its language represents the persona whose reason is stunted—frozen at a beginning point of evolution, closer to the savage than the civilized.

The indistinct traces of the black voice and body played against the backdrop of debates for and against school segregation that were framed in the register of degeneration versus progress. The Primary School Board of Boston[65] made claims for the physical, moral, and mental inferiority of "colored" children: "The distinction [of race] is one which ... is founded deep in the physical, mental and moral natures of the two races ... no legislation, no social customs, can efface this distinction ... Teachers of schools in which they [colored children] are intermingled remark that in those parts of study and instruction in which progress depends on memory or on the imitative faculties, chiefly, the colored children will often keep pace with the white children; but, when progress comes to depend chiefly on the faculties of invention, comparison and reasoning, they quickly fall behind" (*The Liberator*, 1846, as cited in Ment 1975, 45). Those supporting segregation relied on arguments about the evils of miscegenation in the interest of staving off degeneration that was an inherited propensity, yet could also infect the superior human type.[66] Civil rights groups challenged laws against intermarriage and they were repealed in 1843 after a long period of tense agitation.[67]

The internment of the Negro spiritual, except for its later inclusion as a pale reflection and disembodied innocence, for example, in the song described above, reflects the impossibility of merit for black-embodied presence. New research on black music documents that the sonic space was or is neither black nor white but an amalgam of linguistic and musical forms.[68] Yet, this concept has not meaningfully penetrated music education, as I argue, due to its history and a foundational philosophy that set the cultivated tradition in an embattled position against dangerous, popular elements.[69]

In this chapter. I broke from the traditional read of the foundation narrative as a rescue of the hymn and civilization. I did this in order to demonstrate how the theme of musical breakdown shored up divisions of the population by race, musical disposition, comportment, and language. In place of the "normal" musical hierarchy, I posed a hetero soundscape in which the déclassé hymn could be heard as a reversal of the ordering of bodies that gave the white or black body its concrete cultural and racial classification. To borrow Peter Stallybrass and Allon White's general thesis about low and high culture in *The Politics and Poetics of Transgression* (1986), William Billings's hymns exemplify ritual strategies of social reversal (Rabelais being another example) that matched repugnance with attraction, not only acquiring a following in music but portraying abject singers in churches in a celebratory way. Such reversals, ventriloquistic tactics, and parodies often contained the germ of sexual attraction as Lott discusses in

Love and Theft, sounding the alarm of contagion and degeneration that remained thematic in the school music curriculum.

The nineteenth-century genteel music societies' condescension toward folk song and their explicit rejection of Billings's work established the boundaries of what could and could not be included in public music instruction. The primordial base of degenerate music, literally *baseness*, referring to the darker forces of unreason, represents, I have argued, those left out of the partitioning of the beautiful in music in public schools. Such a division of beauty defers democracy and has left the vernacular traditions in inferior relation to the mind versus body ethos.[70]

In the next chapter, I examine a traditional system of distinctions in the turn-of-the-century music appreciation curriculum. Enfolded in the discipline over the ear, music appreciation curriculum in the early 1900s mapped the mind-body divide as cultural progress through music contests and the reiteration of music masterpieces on a national scale through the radio and gramophone.

5
Ranking the Listener, Disciplining the Audience

> Is it the twilight of nightfall or the flush of some faint-dawning day? Thus, sadly musing, I rode to Nashville in the Jim Crow car.
>
> —W. E. B. Du Bois, The Souls of Black Folk

In the early 1900s, Chicago's music in the parks programs witnessed a series of skirmishes with police that took place as some in the informal audience danced to the live music being performed. The police moved into the crowd to rough up and arrest several individuals. In this era, dancing in public places represented a lower form of socializing than many civic music organizers wanted to discourage. On the other hand, the labor movement and musical "progressives" wanted to encourage attendance by the working class and insure that park performances were broadly inclusive, an outlook that accepted popular dancing at public concerts.[1]

Newspaper headlines aggravated the popular dance issue in reporting that Negro "coon" songs, the Bunny Hug, the Grizzly Bear, the Cakewalk, and the Turkey Trot took place at what were, according to some, "quiet" cultural events.[2] This public fracas was symbolic of the larger debates between musical progressives who wanted the park concerts to be what today would be called multicultural events and those who envisioned offering classical music as an edifying education for the masses. In neither camp was the presence of African Americans given attention, except through the dancing associated with ragtime and the negative image that music had among critics.[3] This was also the case with public school listening "concerts," whether from phonograph, live performance, or radio when, before 1928, an official rule banning African Americans from Chicago radio studios was lifted.

This chapter describes public school music's position in relation to those issues and cultural elites in the first decades of the twentieth century when the nation was preoccupied with building its white European image. It focuses on the overwhelming tendency in school music programs to authorize curriculum, even what was called the "progressive method," in music education that treated cultivated music as an object of reverent study, limiting physical gestures and motion. At the same time, leading proponents of public music education were keen to organize community events according to their own interpretation of John Dewey's progressive, democratic principles of action.

As overlapping social groups, music teachers and cultural elites were especially sensitive to the European critical literature and journalism with respect to the judgment of musical quality. With few exceptions, they were quick to deny that American culture included so-called African reverberations. However, that was not the view of detached observers. According to one prominent European intellectual, "Negro" influences were everywhere in evidence in America in the early 1900s, in raucous laughter and loose movements of the lower body with the consequence of sexual promiscuity.[4] This point of view was typical of a system of reason about the cultivated tradition that tied it to Northern European notions of Nordic or Teutonic superiority.

American musical institutions shared in this musical mythology, yet attempted to combine it with the democratic idealism of using education as a means of social progress. Underlining the importance of developing a habit of reverent listening in the population, the President of the Julliard School of Music wrote, in 1930, "We ought to prepare ourselves for hearing music, even though we have heard it often before . . . We should make a fresh study . . . to find something new" (Erskine 1930, 648). While Erskine's attitude represented the European aesthetic that had become the trademark of the musically cultured class, his inclination was to exchange casual attitudes for a more reflective, immobile form of study.[5] Race was not an overt concern. However, it was widely understood as an important factor with respect to "serious" involvement in music.

In his recent book, *Racial Paranoia: The Unintended Consequences of Political Correctness* (2008), John L. Jackson, Jr., argues that society's turn away from overt forms of bias (as is the case with, for example, present state and national standards for music) only perpetuates its material effects through indirect means. This chapter follows this point of view in early twentieth-century music instruction by describing the racial reasoning that codified the listener through a hierarchy of embodied forms of musical response. In the music curriculum, there are no declarations of racism, but the symbolic extremities of high vs. low culture, the classical body image vs. the abject body of unreason, have, I argue, made race an indissociable element of the traditional music appreciation curriculum.

My method of unraveling the inscriptions of racialism in the music appreciation curriculum of the first half of the twentieth century plumbs some of the depth of cultural classifications found in the teacher journals and manuals of the time. These documents are rich in emblematic motifs of the base and the exalted that favor the classical pose and the kind of reasoning about music that made it, in Erskine's (1930) words, an exalted mental activity. It is reasoning that erases the visible and audible entrainment of the body and its heritage historically overlaps the lofty prescriptions for comportment coupled to whiteness.

Erskine's idea is typical of many music educators whose prescriptions for making a study of one's own and others' listening contrasted sharply with the embodied musical cultures of the actual public school population. Many children entrained their bodies vigorously in religious singing and danced *as* they listen to music with family and friends, hence the myriad references in music teacher conferences and journals for the need to reform the popular musical taste. Keeping this domain of the low-other in mind, we can trace its shadowy presence as it shaped the structure and values of music teaching.

Music Appreciation, Listening Methods, and Technology

Distinctions between ethnic and socioeconomic groups in within music classrooms, as described earlier in this book, activated the racial and ethnic coding of music evident in song texts, popular beliefs, and historical preferences for images of white, Northern European traditions of self-cultivation. Just as school songs featured prescriptions for proper behavior, hygiene, and comportment, they also established vocal music as enlightenment and cure for social ills. School singing's hailing of genteel comportment was a way to reward a musical taste that would also teach others. The same dynamic applied to music listening. Modeling oneself on the comportment of *The Thinker* made one a reasoning individual and marked one apart as qualified for an exalted form of citizenship. Music appreciation extended the reach of school song texts to discipline over good ear and qualities of whiteness.

This comparative arc ran through most cultural activities, but it was most firmly anchored in the philosophical, anthropological, and psychological literature, especially the child study movement of the nineteenth century.[6] In order to fully articulate how the template for the musically cultivated listener operated, I will explore the social context that made it possible to open the "hearing" of the private citizen to regulation and put it into direct relation with the cultural theses that compared dispositions and abilities.

Radio and phonograph in the early 1900s enabled music educators to simultaneously reach large audiences. Armed with a set of standard procedures and rubrics from evolutionary frameworks in anthropology, hierarchical rubrics encouraged the teacher and student to keep watch over repertoire, kinesthetic responses to rhythm, and the kind of attention they summoned when listening to music. Music appreciation was the name given to a range of methods that taught the student how to listen to classical music and how to take note of its structural and thematic characteristics. Instruction in the rudiments of music listening typically began in kindergarten, progressing in level of complexity from basic rhythm exercises through high school elective courses that taught the style and form of the "greats" of music to students in many school districts. Protocols included teaching students how to distinguish whether a particular piece of music was in sonata or "theme and variation" form, for example, and note its national origin, major rhythmic characteristics, and melodic material.

In the United States, music appreciation became an industry and a byword of individual and national progress.[7] It marked the musically knowledgeable person as one who was cosmopolitan—a citizen set apart in her ability to represent the high level of culture of the United States to the world. By the 1920s, music appreciation was a prominent feature of many public school's general music programs all over the country. Curriculum guides and detailed lesson plans of that period were presented to nationwide audiences of music teachers through professional music teacher publications, phonograph manufacturing companies, teacher education programs, and music publishing houses, creating an image of the nation, en masse, as intent on making its level of cultivation an equal of Germany, France, and Britain. The untutored masses were considered a prime target of public music appreciation for national radio broadcasting, school instruction in the German greats, and settlement house music programming. "The big public, the mass of the people, should be music-lovers. To this end the schools must work, so that our public may come to that point of intelligent listening where in forming an audience they "assist" as the French say instead of merely attending" (Cundiff and Dykema 1927, 6).

Throughout the nineteenth century and most of the twentieth century, the perception of blackness was enough to register *any* musical behavior and performance as substandard. Blackness registered doubt about the subject's abilities, rationality, and loyalties. An example of this treatment was the way the African American singer, Paul Robeson, encountered racialist comparisons in performance reviews. One drama critic[8] praised aspects of his performance in Shakespeare's *Othello* only to state that Robeson's chief handicap, astonishingly, was "his uncompromising African cast of countenance and bearing" (Mantle 1930, 14). On the other hand, Robeson had been lauded for his voice's Africanist qualities in his delivery of "Ol' Man River." These two examples typify the narrow band of opportunity afforded African American concert and stage performers, including even blackface minstrelsy, for which blacks were considered "too real" as Negroes and therefore not fit to be viewed by a white audience.[9] They also represent the direct references to notions of race that were absent from curriculum documents because, among other reasons, the racial hierarchy was structured within musical values that were or are never entirely inseparable from gradations of whiteness and blackness. In this sense, racialism's broad sweep is important to the formation of the music curriculum that sometimes consciously, often unconsciously, drew the limits of musical merit in demographic terms.

Racial limitations infused music appreciation lessons in which entrainment was measured with reference to evolutionary milestones of child development (rudimentary to complex), the productive use of leisure time, psychological testing, and concepts of body culture that instructed her in how to model dignity and "racial" nobility. Here again, race and religion were rarely explicit disqualifiers because whiteness was widely taken for granted as the grounding characteristic of social fitness and merit.

Historically proximal to the beginning of music appreciation in public schools, the turn of the century series of world exhibitions and national centennials provide a picture of the value placed on racial comparisons in American music audiences. Occurring between 1876 and 1920, these spectacular fairs drew on the intellectual, artistic, and commercial aspects of American society to show the comparative advantage white Americans had over other cultures, chiefly, African and Asian. The effect was to align America with the empires of Europe and to isolate the nation from American indigenous, as well as from specific voluntary and involuntary immigrant, cultures. The Centennial Exhibition of 1876 featured minstrelsy as a crowd pleaser. Memorabilia and black commemorative statues referred to blackface as a "stray piece of midnight."[10]

The Chicago World's Columbian Exposition of 1893 positioned blackface performances in the most crowded venues. The exposition itself was built around the idea of the White City on the lake and was themed as a celebration of industrial progress. One of the chief organizers, G. Browne Goode, was a hereditist and an adherent to the idea that racial or family biology was destiny. He had set out to make the Chicago World's Columbian Exposition a definition of the modern by way of comparison to "primitive" civilizations. As part of this theme, there was a concert series that replayed some of the Chicago park controversies around what etiquette was required for serious music listening, whether there would be free admission for the working class, and what kind of repertoire would be featured. W. S. B. Mathews, a well-known music pedagogue, had envisioned the art music exhibits and concerts as events that would expose the audience to the concept of national edification and awaken the world to the higher forms of music.[11] So-called non-Western music, drawing huge crowds, included Scott Joplin's "Maple Leaf Rag."[12] At other exhibits, for example, in the demonstration of pancake making, an Aunt Jemima figure (also the name of a catchy popular tune) served as an emblem of loyalty to the white social order.[13] Within this highly and deliberately bifurcated, racial setting, the concerts and lectures on cultivated music took place, but they did not meet with the enthusiasm they would garner from music educators just one decade later. There were serious disagreements about the concert hall and programming, which led to a falling between planners, commercial sponsors, performers, and the conductor. For some, following prevailing ideas of high and low forms of music, American musical taste was represented by European composers to be treated as exemplars of the cultivated "national" character of Americans. For others, the concerts appeared to be exclusive events that ignored the desires of many to hear "lighter" music within a concert hall setting.

Building Character and the Nation Through Listening

The exposition in Chicago and its controversies over who made up the concert audience took place at the same time as a high school music appreciation curriculum class in Massachusetts began to attract the attention of American music teachers.[14] At the center of what would become a kindergarten through twelfth grade formal music appreciation curriculum were ideas that had percolated for over a century in the form of listening guides by European music theorists and more recent texts by American writers.[15] Building on listening guides that were popular since the eighteenth century, American guides were similar to the recommendations in Charles Burney's *A General History of Music from the Earliest Ages to the Present Period* in urging listeners to join the ranks of a cosmopolitan crowd through knowledge of the universal laws of music. Knowing these laws would make them the judge of a range of musical material that they would encounter: "There have been many treatises published on the art of music composition and performance but none to instruct ignorant lovers of music how to listen or to judge for themselves" (Burney 1776/1957, 2). Burney's (and others') cultivation of the listening art percolated for a century until a new genre that directly bears on the music appreciation curriculum, and schoolteachers' preparation to teach it, appeared in W. S. B. Mathews's *How to Understand Music*.[16] Mathews was a frequent lecturer in normal school courses in music for public school teachers, facilitating music appreciation's welcome as part of the general music curriculum for public kindergarten through the higher grades in the early 1900s. One prominent educator summed up Mathews's influence: "It is now ... thirty years since Mr. Mathews published his book, 'How to Understand Music' ... Within the past five years there have appeared in ever increasing numbers, in the curricula of secondary school and collegiate institutions, courses designated Music Appreciation" (Cady 1910, 49). The protocols Mathews recommended featured classical musical selections to be played on the piano by lecturers, followed by the analysis of thematic development and formal structure. This approach remains largely intact in general music programs to the present day, but with recorded musical samples. "The aim [of this course] is . . . training them to appreciate rather than to produce. This is a feature of high school work not found elsewhere in the country, and the best classical music is played. A part of the time is spent ... to aid the listener in grasping the form and meaning of the works heard . . . and the mastery of the elements of musical structure" (Balliet 1897, cited in R. Dunham 1961, 18–19). While the elite music societies in the big cities had sponsored lectures or recitals along music appreciation lines for twenty years, teachers like Regal and Mathews are widely credited with establishing music appreciation, as Balliet described it, as a fully developed curriculum for the public schools.

European Self-Cultivation, a Blueprint for American Music Appreciation

Insofar as music educators would come to consider music appreciation a discipline of the listener for reasons of spiritual and practical necessity (relief from industrial labor and boredom), the flowering of public music appreciation, in venues other than schools, such as settlement houses and civic exhibitions, follows the logic of the religious quest, but also placed personal dispositions and expressions under scrutiny and governance through those public institutions. Public schools were a part of a larger movement to govern musical behavior and sonic space of the cities. Particular musical accomplishments, such as reverent attendance at formal concerts were prophetic of success and converged with demographic, cultural differences. These overlaps templated the preferred persona and marked the lesser social and psychological life course of the uncultivated individual or group.[17] The higher social regions made use of uncultivated behavior and tastes as a symbolic domain against which genteel institutions compared themselves as "cultured" citizens.[18]

In this respect, never before was the vehicle for mass distribution of culture as promising as the public school. Pursuing musical cultivation through mass music instruction provided tremendous momentum for the dissemination of anaesthetic philosophy that claimed sublime beauty in art music as a universal principle that flowed from human reason and divine (godlike, but not strictly speaking religious) inspiration. This state of reason verged on the sublime "Astonishment is that state of the soul in which all motions are suspended . . . In this case the mind is so entirely filled with its object that it cannot entertain any other" (Burke 1957, 5). Possession of this kind of judgment required both sensitivity and the ability to contemplate qualities in works of art in which reasonable minds found greatness.[19]

The pursuit of the sublime in school music reached a new peak during the early decades of the twentieth century when technological advances such as the radio and the phonograph promised the delivery of music's soul and genius for practically everyone. Settlement houses and music clubs sponsored recitals and lectures on music appreciation as well. This would not only raise feelings of astonishment, as Burke called them, but would wholly win over the attention of the listener, an experience that matched American strategies of political speech and patriotic song to establish loyalty to the concept of a democratic melting pot. It was an odd combination, as many European visitors would note: populism with elite ideals, but it had its own "American" story in the sense that some of the semiaristocratic forefathers denounced the old monarchical structures of governance for Republican ideals that also reinscribed limits and partitions for democracy. Music as feeling without words was a kind of sublime speech that could move nations without the trappings of syllogism and rhetoric.[20]

However, the quest for the sublime in music ironically led back to the business of making distinctions between music(s) that were and were not exalted. Aristocratic standards of taste reevoked the colonial and anthropological categories that would find themselves in the forefront of music appreciation protocols, leaving the sublime to settle on the biographies of a few, select German composers. There was also a kind of sublimity alternately extended and withdrawn with respect to musical genres such as Negro spirituals and folk dances that bestowed a kind of bodiless sublimity to the spiritual. Well into the twentieth century, devotion to the sublime delight of other kinds of music associated with black bodies, such as the enormously popular (among all social classes) minstrel, ragtime, and jazz genres, propelled music educators to position them beyond the limits of music itself.[21] One way to interpret these shifting parameters of sublimity and baseness with regard to racialism in music is to view the transgression of the color line as active element of that racial formation. As notions of sublimity moved across social and artistic boundaries, their appearance gave further cause for alarm and deeper entrenchment of genteel values. The formal cementing of racial or musical boundaries in school music, where it was possible to codify restrictions in educational language, followed, to a great extent, the vacillations of the political climate, the buttressing new forms of social work in the industrial cities of the North, and ambivalent attitudes toward black migration northward. In this sense, the value placed on music listening by civic elites and the music curriculum was intensified by greater fears of black music's popularity.

A perfect scenario for templating the worthy listener emerged in the form of national Music Memory Contests sponsored by the National Bureau for the Advancement of Music, a.k.a. the Committee on Trade Advancement appointed by the National Piano Manufacturers' Association of America.[22] Coincidental with these developments was the rise of new social tensions and anxieties about race,[23] housing, and city administration of street life. Just as the uncertain categorization of the Negro spiritual through the nineteenth and twentieth centuries forms a case in point of the white fear of slave resentment in the form of a disembodied voice that transcended bondage in nonmaterial ways, ragtime and jazz were excluded as forms of music that excited the lower social body legible in the form of sexual myths, fears of miscegenation, and anxieties over the fate of eugenic projects tied to patriotism and progress.[24] As Jim Crow legislation took hold, musical progressives who were making a mark on the national consciousness, in terms of the acceptance of socializing and dancing of different classes and ethnic groups, remained ineffectual with regard to blacks. In the 1920s and 1930s, it was the city center, not its parks, that became the hub of musical entertainment. Commercial availability of recordings and the growth of the tavern into the nightclub were phenomena that accompanied urban growth, but these clubs were also locales where the fear-fascination and racial "transgression" with blacks and black musical culture took place without actual integration.[25] Simultaneously, the availability of radio performances and recording of classical music gave public music instruction its best weapon for ruling out jazz and restricting the teaching of indigenous forms of music to oriental (including Native American) and Negro exemplars.

From the early decades of the twentieth century to the present, the polarization of good and bad music was left to schools (and agencies connected to education of youth, some workers, and immigrants) where the structures of authority and material provision for enforcement were firmly in place.[26] That composers such as Gershwin borrowed and adapted Negro motifs was not fully registered by the national and state music teachers' organizations until the 1960s, although some music educators, like Anne Faulkner Overndorfer, included Gershwin in her text, *What We Hear in Music*, of 1943. Some of this belated uptake may be due to the fact that music educators and civic leaders had positioned themselves so well within the instrument, sheet music, broadcasting, and recording industries, that they were convinced that the music appreciation movement was a mass phenomenon and not the minority musical disposition that it actually was. "Modern science, the Records and the Radio have brought the great orchestras, the finest artists, opera and oratorio into practically every home . . . Music has taken its place in social science . . . the greatest cultural force in the daily life of every individual, home, and school" (Clark 1923, 6). Central to what one might call a myopia about the size of this audience were the links to orchestra conductors and record companies that provided a special series of concerts, recordings, and teacher manuals for sale to school districts. These materials defined the embodied difference between true musical comprehension and the more misguided musical dispositions.[27]

Longstanding Traditions from Europe

The American music appreciation movement was not a thorough-going double of the Northern European tradition of self-cultivation or *Bildung*. The latter is an affair of the individual whose knowledge and status are measured by the dimensions of his *inward* nature. American musical self-cultivation was explicitly aligned, as I discussed in connection with republican governance and the ideal of contributing to the progress of the democratic community. However, some aspects of the theory of *Bildung* did imprint the models for visual or linguistic recognition of the uncultivated, unworthy soul in order to compare him to the singular and awesome superiority of the elevated individual.

For school instruction, this meant that the *Bildung* tradition incited the music curriculum in order to emphasize the appearance of an inward contemplation in each individual's demeanor. Teacher manuals on music appreciation in the early 1900s disseminated images of the person who was interested in both an education in classical music and a practical route to achieve it. Numerous listening guides selling in America, including Mathews's text, presented the sublime as a pragmatic program for the cultivation of taste; that is, listening guides took beauty apart and reconstructed it through a set of protocols that would enable the keen listener to apprehend the architecture of music. For Northern Europeans, however, musical learnedness turned inward was a constant and unfinished examination of the soul. In the German territories, this individual aspired to true *Bildung* or inner illumination that was more than could be gleaned from one text. Wilhelm von Humboldt, one of the most important figures in German education in the early nineteenth century, defined *Bildung* in the following way: "When we say *Bildung* in German we mean something at once higher and more inward (than civilization), namely, the disposition which harmoniously imparts itself to feelings and character and which stems from insight into and feeling for man's whole spiritual and moral striving" (Humboldt 1963, 266). Other German writers described *Bildung*(self-cultivation) as a political institution in its own right, "the great affair of any man."[28] According to Johann Fichte, an early theorist of *Bildung* in the late 1700s, "*Bildung* cannot be directly acquired from nature . . . reason is in continual battle with nature" (Fichte, as cited in La Vopa 1988, 363). Most central to the difference between American and European traditions, genuine *Bildung* could not be attained through instruction.[29] Its predominant meaning is a life that is "always moving on the path of self-discovery" (Koselleck 2002, 184). In this sense, the path of inwardness is assisted by an image of the sublime in music, although its American application bore little resemblance to the soulful *Bildung* tradition. The American cosmopolite was more aptly described as one who followed contemporary developments in government, the arts, and the new scientific fields.[30]

Early nineteenth-century American uptake of self-cultivation brought it into contact with Calvinist values that translated individual betterment as a step toward national salvation and progress. In the late 1800s and early 1900s, it was the kind of step-by-step cultivation through music appreciation texts and manuals that imbued the would-be cosmopolite with an aura of civilization. Although transmuted, the importance of *Bildung*'s imprint on American listeners was its recognition of the appearance of cultural nobility in a republican outlook associated with the ambition to reform society and bring it into alignment with the values of whiteness.

The pragmatic outlook of American music educators made music listening in the United States significantly different from the ideal of self-cultivation in Northern Europe. "No other country in the world has public school music supervision. Is there any other country where such a democratic nation-wide organization could exist?" (Keith 1929, 141).

In America, a mass audience could consecrate experiences with classical music and receive a benediction through sheer numbers, although the "mass" quality and superficiality of some means to acquire culture were suspect. "The 576 selections were the most commonly used in secondary school contests. [They] have also been used, most of them with great frequency, in contests by younger pupils . . . [this] suggests a strong influence by the [National Bureau for the Advancement of Music] rather than independent judgment . . . The success of the music memory contest is to raise . . . the standard of music taste in the majority of people" (Briggs 1925, 6–7).

The memory contests gave emphasis to mnemonic techniques as opposed to the more spiritual and learned approach of the European tradition. In the American context, mass procedures distributed the necessary discipline over the aural environment that would assure the well-being of society during the "crisis" of the popular dance crazes of the 1900s through the 1920s.[31] "Modern social conditions required the keen sensitivity of hearing that awakens gentler emotions and higher aspirations" (McConathy 1910, 70). In a similar manner, the adaptation of "Dalcroze Eurhythmics" is also an example of a pragmatic translation of a European regimen for cultivation of the soul into pedagogical mnemonics. Jacques-Dalcroze's work stressed that the "true" musical ear is a "faculty of the soul."[32] Interwoven with music appreciation, the Dalcroze art movement was interpreted as a pragmatic means for teaching rhythmic and metrical concepts.[33] What Jacques-Dalcroze understood as the continuation of self-cultivation of the soul came to mark the progress of American music education among the masses and would make the Untied States a member of the international community of nations.[34]

According to European observers, the American pedigree would lack humility and inward piety. Yet, these were not qualities that struck American anxious to incorporate that coveted aura—the "Je ne sais quoi" or "that certain something"—marking him apart from his less-civilized countrymen.[35] As I will discuss later in this chapter, "that certain something" was represented in the body culture movement with a large following in the United States. Modeled after a revival of classical art in the Victorian era, books and lessons on body carriage, gesture, elocution, and etiquette in the U.S. were designed to give the appearance of the self-knowledge and taste that one attributes to an "inner nobility."[36]

At the turn of the nineteenth century, the body type that could be ennobled by instruction included immigrants from Southern, Eastern, and Central Europe, and white Southerners from Appalachia. Many settlement house boards turned blacks away, some seeing them as beyond social acceptance and the redemptive courses on life improvement offered by many houses. At the Christamore institution in Indianapolis, for example, when increased migration of blacks from the South began to outnumber poor whites, the board decided to relocate the settlement to another part of the city.[37] This trend added to the de facto segregation that was in place in most of the Northern industrial cities, in effect, segregating public schools and music instruction in other local institutions. Notions of self-cultivation and who was prepared to undertake the instruction to realize its American configuration were about nationality, class, and race. Cultivated negroes, especially in the Northern cities were interpreted as "Dandies" or imposters in blackface performance. The measure and limit of Negro status, as W. E. B. Du Bois reminded his contemporaries, was not lack of self-cultivation in high culture among the many he could cite with regard to high achievement, but lack of the right to vote and the means for most black people to obtain more than a basic education.[38]

Mapping the Cultivated Nation

The idea that American culture was modeled on the European culture was firmly entrenched in music education for much of the twentieth century. Commercial interests in nationwide sales of phonographs and records began to increase exponentially around 1910 due to their felicitous intersection with the pastoral obligation to convert the mass of listeners to good music. The responsibility for this task fell most prominently on the educational directors of the recording companies. "We must proceed orderly [sic] in a given direction, thinking clearly, planning progressively, sifting, sorting, segregating sequentially the material needed in every phase of the work, building toward a definite goal, that of a nation of intelligent listeners to good music with ever-increasing percentage of performers and creators" (Clark 1924, 273). For Frances Elliott Clark and for others, the Music Memory Contests, even with reservations about their repetitious nature, provided a pragmatic measure of success. The size of radio audiences and the orders filled for phonographs and records, principally the Victor Talking Machine Company, as the record player was then called, were also closely monitored and calibrated to the spread of Music Memory Contests to the public schools. "Time and the advent of the talking machine have made possible greater things, so that today there are few school systems whose course in music does not include something known as music appreciation, attained through intelligent listening . . . the movement of the music memory contests . . . is spreading rapidly and bids fair to mark an epoch in the development of music in our schools and communities" (Williams 1921, 147). Radio symphony broadcasts for children, conducted by Walter Damrosch, and the sales of Victor phonographs served as gauges of the growth in audiences nationwide. A graph published by the Victor Company notes that in 1911, very few school districts had purchased phonographs and records. By 1924, that number had enlarged to the thousands.[39] A more independent source for tracking this trend was an article in *School Music Monthly*, stating that the Chicago School District had purchased eighty-seven talking machines in the school year 1911 through 1912.[40] Although difficult to obtain information on the use and distribution of this material, de facto school segregation in Chicago at this time probably affected the participation of black students in the Music Memory Contests and other events connected with music appreciation. As a measure of educational inequities during this time, there was a parallel effort on the part of white social clubs, such as the one in Hyde Park in Chicago, to exclude black renters and homeowners in spite of efforts such as those of John Dewey and Jane Addams, who opposed social segregation of all kinds.[41]

With respect to public schools, there was continued strife, sometimes violent, around the hotly debated issue of racial integration. The upshot is that informal and formal practices, physical threats, and housing's de facto divisions magnified the concentration of energy on black churches to provide music instruction for black children. The precise dimensions of musical segregation along educational lines are difficult to discern from the data related to employment, entertainment venues, and schools,[42] but it appears that the separation of musical events, against many efforts to bring black and white social groups into contact with one another, further incited the projections of inferiority on to black musical performance, music instruction, schooling, religion, and embodied practices associated with gospel music and social dancing.

The differential proportions of cultivated music appreciation in schools and in the normal school courses for teachers also manifested itself in the elite attitudes of Protestant college academics and administrators.[43] Quoting President Henry King of Oberlin College, Professor Edward Dickinson wrote, "They [the general public] do not realize how large a part the faculties of aesthetic appreciation ... play in human welfare ... Hence, the almost exclusive weight thrown upon observational reasoning and memory ... art and literature make an appeal that no abstract principle can make ... we have no way of expressing a general principle but by putting it into some define concrete, individual action" (King, as cited in Dickinson 1915, 32). These views on the arts and culture were part of a debate on the direction of formerly religious colleges as they made their way from being Christian institutions to secular bastions of scientific research. In the discourse that emerged from this shift in higher education from religion to the liberal arts, various forms of action—scientific and artistic—were related to progress and community building.[44] Music Memory Contests were also a sign of this transition, as they drew on rote methods—that is, repetition of pieces of music to make them permanent artifacts of knowledge in ways that scripture had been revered and memorized. On the progressive side, Music Memory Contests promoted the idea of a broad community by listening to cultivated music. Community action, scientific research, and art were the modern, secular rituals that made music listening a part of the social evangelism of the times.[45]

Music appreciation taught in public school music was marshaled to confront political and material challenges through democratic interchange. Addressing an audience of public school music supervisors, Louis Mohler, Professor at Columbia's Teachers' College stated, "[Music appreciation] brings the learner into an understanding of relationships as they are expressed by one another. Such a common understanding may lead to cooperation with fellow men in groups; this is one of the fundamental principles of democratic living" (Mohler 1924, 262). The appearance of the success of the Music Memory Contest, nationwide, was that public music instruction ministered to every child's musical experience. This was to situate school music programs in their most progressive light while disconfirming differences in the cultural endowment and interests of the mass of children in public schools. Community cooperation rarely meant crossing racial lines. Black church-centered neighborhoods sequestered musical genres associated with blackness further from what was imagined as an educational or national mainstream.

The sector of white middle-class life that represented the "normal" of schooling adopted a Romantic care for the soul in the same era as Jim Crow legislation segregated almost all private and public institutions. Sometimes referred to as an enlarged humanist sensibility underscoring feeling, the irony of Romantic humanism, when juxtaposed next to the struggle for existence of blacks and laborers all of the country, traveled in the classical composers' works and in the writings of Schiller,[46] Goethe, and Humboldt, entering music criticism by way of aesthetic philosophers[47] and given a special Teutonized form with Herder and Richard Wagner.[48] Each, in their own way, of these august figures addressed what was, for some, the worrisome transformation of the spiritual realm into a mere technical instruction. For the cultivate elite, the danger was the loss of an organic wholeness of society and, with it, culture. "We see not merely individuals, but whole classes of men, developing but one of their potentialities, while the rest, as in stunted plants, only vestigial traces remain . . . Enjoyment has become divorced from habit, the means form the end, the effort from the reward. Everlastingly chained to a single little fragment of the whole, man himself develops into nothing but a fragment . . . thus little by little the concrete life of the individual is destroyed in order that the abstract idea of the whole may drag out its sorry existence" (Schiller 1795/1954, 15). This discursive tradition appeared in the teacher manuals of the early twentieth century, making music appreciation lessons a means of recapturing some of these Romantic ideals. Acting as the Victor Company's education director, Frances Elliott Clark expressed the always-present fear of degeneration that would give way to music appreciation: "The joyousness of life is being lost in the mad scramble to earn or get money, and the resultant fevered and unnatural madness for amusement of some sort of an intoxicant, is bringing temporary but dangerous oblivion. [For the music festival] Victor records seemed the solution . . . furnishing a basis for and means to achieve a nation-wide movement for the uplift of healthful play" (Clark 1913, 14–15). Such psychological needs became the province of music education. In this discursive vein, school music drew parallels to musical progressivism in its appeal to the mass of people suffering from poor use of leisure and over work. Frequently, music appreciation focused on particular icons and works of genius as ways to bring out the "inner" life of the ordinary person and facilitate ministering to the soul. Calvin Cady, a prominent figure in college and public school music education, championed music appreciation as a method of bringing out the inherent expression and interest in the child through melodic composition and rhythmic movement. He explicitly stood against using music for what he described as political ends, yet his own point of view introduces the political dimension through a singular example of musical merit:

> [You hear it sometimes said that] Beethoven's music is good to the degree that it forbids thought and compels feeling. An idea is expressible in words, and words may be set to music, but the words are not part of the music . . . This echoes a very prevalent notion and you hear on every side the argument for a widespread pursuit of music as one of the humanizing agencies because of its social function or its supposed power of seeing 'only the heart' . . . It is urged that it [music] should be extensively used to normalize and regenerate the student's emotional life . . . to sanctify and idealize the affections, and even war. But the attainment of [true appreciation for country and religion] is utterly impossible apart from the virile, broad, pure, truthful, sincere, reverent intellectual grasp of the idea embodying all this supposed potency. A just recognition of the function of music, especially its interpretative study is therefore vital to the realization of its humanizing influence. The two can no more be separated in art than in religion. To seek for permanent moral stability . . . apart from a mental grasp of the underlying principles of moral or religious life is futile. (Cady 1908, 150–51)

Reading a social purpose or agenda into music, he argues, is to misapprehend its central status as an intellectual achievement and to miss its potential as a humanizing agency.

Like Calvin Cady, Frances Elliott Clark's curriculum guides and material provisions are aligned with this point of view, carrying with them the subtext that an unreflective musical entrainment engaging the body, not the mind, is dangerous. Their thinking reveals social anxieties that gesture toward racial and socioeconomic conflict similar to those that played out in conflicts over park concerts and racial integration. The priority for the intellect to replace the body comes hand in hand with the sense that high musical culture would make it possible to heal the fractures of the modern age. Works such as Beethoven's Fifth Symphony and the setting of Schiller's "Ode to Joy" in the Ninth Symphony carried this message.[49] A music supervisor in the Chicago schools in 1908 mentions the "moral direction" that Beethoven's work adds to studies of language, science, and literature.[50] Insofar as his music conveyed a conjunction of feeling and reason, it would deliver values associated with *the* self-governance of the American citizen. She was not an individual that danced and listened to popular music, but one who showed that she was "climbing the mountain trails with thoughtful effort and real study" (Cady 1923, 9).

The differences between the group represented by Cady and Clark and musical progressives in the larger cities came down to the kind of music that would be improve the society. Progressive programming for parks, fairs, and other public venues was perceived as catering to low tastes in may instances[51] and the split between "high" and "low" even went so far as to divide the regular audience for Chicago symphony concerts in the 1890s. Theodore Thomas, the conductor, disliked "requests" because they were often sent in bulk through one interested musical camp versus another. The larger issue between them amounted to "light" versus the heavier classical works.[52] As several studies of social institutions of this period attest to, audience attendance at different musical events took the absence of African Americans for granted, reflecting the segregation of most of society's institutions, their programming, and the reputation of the artists to perform.[53] Public music education mirrored this macrocosm in dividing the populace into camps of listeners. This difference was fateful in separating school music from the vernacular and popular forms of entertainment and listeners who, for the most part, did not identify their interests with cultural or moral crusades. However, musical progressivism did inform some educators' views. Anne Shaw Faulkner's series, *What We Hear in Music* (some published under the name Anne Oberndorfer), appeared in different editions from 1921 through 1943 as a high school teachers' manual. Those texts and her articles published in *Ladies Home Journal* carry recommendations for the Victor recordings of "Creole Negro" tunes, folk songs from South America, and "Smoky Mountain Melodies."

Taken as a whole, though, the professional literature of music teachers reflects a very narrow range of musical debate on what repertoire would best suit a democratic society. There was an overwhelming tendency to prescribe the German greats as models of intellect and social consciousness. In contrast, and all but forgotten until the 1960s, the activities of musical progressives to present popular and classical fare were more manifest in institutions such as Hull House, the Cleveland and Gary settlement houses, with the Chicago Symphony Orchestra wage-workers' free concerts of mixed programs of classical and lighter fare. It was these Midwestern cultural venues that were models for other orchestras, music schools, and opera houses in the following decades.[54] For social progressives, this was a sign of positive change in an age when labor was beginning to agitate for broader participation in the benefits of democracy, a living wage, and the leisure to enjoy it. Yet, the image of the wageworker who was also a concertgoer was overwhelmingly white, as most black laborers were personal service and domestic hires who did not qualify as "workers."[55]

Freedom and Leisure

There was more than one dimension to broadening concert audiences and these were not necessarily progressive in philosophy. Bringing larger numbers of people into the fold of music appreciation reinforced the concept of prescribing leisure activities and teaching music as a way to direct youth's leisure. Writing in *School Music Monthly*, a professional journal for music teachers, John Franklin Bobbitt, Professor at the University of Chicago in the early 1900s, renown for his "efficiency" models for schools, stated, "Where workmen are engaged eight hours a day or forty eight hours per week, after allowing twelve hours a day for sleep and meals, there yet remains a surplus of thirty-six hours per week to be devoted to leisure occupations . . . It is also probably the portion of his time for which he needs the greatest amount of education . . . among leisure occupations of a healthy sort there is certainly none of a wider appeal than music" (Bobbitt 1912, 27). A music supervisor for a district near Boston wrote, "Music appreciation . . . is becoming more and more necessary to modern social conditions . . . It is when he [the worker] leaves his desk or bench that the call to his better or worse self determines his good or evil influence on his community. Let the community see to it that he has the means for elevating diversions . . . for the very safety of the community" (McConathy 1910, 73). The director of the U.S. Bureau of Education, Louis Mohler, wrote that music appreciation courses offered structure for an otherwise aimless stretch of leisure time that distinguished the citizen from the ne'er-do-well. George Upton, a music critic for a Chicago newspaper, wrote, "The subdivisions of labor which are continually increasing tend to make the workingman's life more monotonous. The shortening of hours is likely to give him more leisure. How shall a portion of that leisure be occupied? Why should not music, with its elevating influences, assert its rights to it?" (Upton, quoted in K. Miller 2003, 149). All of these comments about the relation between the worker and the proper use of leisure accorded well with the social evangelism of the Progressive movement and with more conservative moves to supervise time off work.[56] It saw music as an outlet for human development and cure for idleness, but it also made music listening into a technology to regulate behavior and the interior consciousness—what used to be called "soul." If listening habits were unmonitored, some educators reasoned, the youth or worker was in danger of losing purpose and aim.

Music educators commented on the mental benefits of good listening. Calvin Cady described how a child recovered from a severe mental illness. The cure, Cady concluded, was the result of her classroom music appreciation lessons. "Music appreciation results [in] the child's appreciation of melody as the free and spontaneous expression of his poetic fancy and experiences. This value is expressed in terms of general and specific mental awakening and development" (Cady 1910, 53). These themes—mental health and leisure time spent with inferior music—intensified the racial and class divides between prospective citizens. The social group that identified with cultivated, classical tastes called attention to what they saw as the "destructive" tendency of an unsatisfied mass. Music appreciation was a beneficial use of time insofar as it produced that "happy" state that fought the monotony of industrial labor and combated the sexual licentiousness instilled through lower musical pastimes. It also prevented the boredom that some perceived as the driving force behind labor unrest. The idea of a productive use of leisure accorded well with a progressivism that saw music as an outlet for human development, but it did not fundamentally change the range of distinctions over the listener that would order her (whether laborer or domestic) as a person who spent leisure wisely or, in Mohler's words (1929), was the "unproductive" human type.

In his appearances at music educators' conferences, Louis Mohler had taken a leading role in representing those aspects of the progressive movement in education that music teachers could see as commensurate with child development and the stepwise phases of listening skills. The benefits were opportunities for the exercise of freedom on the part of the child through his decisions about his own activity. At the same time, group instruction ensured that even freely chosen music was monitored by the teacher. Speaking about the application of a "project method" of education, Mohler wrote, "His [the child's] activities are being continually directed but his development is entirely unfettered . . . The result will be self-activity" (Dewey, cited in Mohler 1924, 262). Self-activity, as a child's preparation for liberty, would be directed outward and carried over to community participation in the form of the moral responsibility of democratic life. Participation in social activity led to happiness. Quoting John Dewey, Mohler linked the child's exercise of initiative in music to the attainment of happiness and the habits of productive adulthood: "Happiness is an emotional accompaniment of the progressive growth of a course of action, a continual movement of expansion and achievement" (attributed to Dewey, in Mohler 1924, 261). Previous to the twentieth century, rescue of the "unhappy" child centered on performing songs that tied his persona to a tranquil domesticity and pastoral landscape. In the twentieth century, happiness emerged from shared musical activity vital to the coordination of action in society. For educators who shared the views of Cady, Clark, and Mohler, social prescriptions were entwined with the music appreciation curriculum, making it not only music learning but a citizen-certifying system of governance over the listener.[57] Underpinned by the universal aesthetic of the sublime and the ethos of self-cultivation, music appreciation enabled a system of reasoning about the citizen, as listener, that was consistent with a racialist order of human types equipped to participate in those projects.

A New Role for Rhythm

Rhythm played a large role in deciphering the citizen of promise from inferiors. This was partly due to rhythm's ability to accumulate meaning form one context to another. First, it appeared as a unifying pulse in the rhetoric of the American Enlightenment. During this time, rhythm also distinguished slave song from other kinds of music—the trademark an African primitiveness—later called "hot rhythm,"[58] that was reported to have been heard in performances of slave drumming, singing, and dancing forbidden by slaveholders.[59] Later, rhythm exercises would deliver knowledge of rhythmic primitiveness as a comparison of the individual's development. These different ideological uses of rhythm were facilitated by the way rhythm informed racialist theories of music and affect, prompting interests in psychological experiment beginning in the late 1800s.[60] From roughly 1880 to the early decades of the twentieth century, intellectual, artistic, and scientific interests in Europe and North America formed a discursive web around the question of how industrialization and urbanization affected attention, primarily from the visual point of view.[61] Another line of interest was centered around aural perception and its propensity to engage or distract the listener or the casual passerby. One of the music listening (appreciation) activities thought to create a sense of order and engagement in the child was rhythmic marking of music heard. Exercises in rhythm in the early grades would produce happiness because rhythm in and of itself was the essence of life activities. Since biological rhythms laid a foundation of regularity and pace for the rhythm of work and play, thus organizing time into periods of work and relaxation, it followed that the "right" kind of rhythm would remove the bad effects of city and industrial noises. Psychological experiments concerned with the rhythm of schoolwork, for example, noted the interspaced rhythm of fatigue and concentration in schoolchildren doing arithmetic calculations.[62]

The broad project of investigating rhythm was concerned with its relation to the periodicity of attention and how to capture it. Many experiments oriented investigations toward classifying different kinds of attention, but also toward taking note of which rhythms in what kind of music (jazz or classical, for example) enhanced a steady physiological state and mood.[63] Enlisting the language of psychological experiment, attention became a reinvigorated issue in music listening since it signaled a student's level of maturity and ability to control impulses as well as to sustain interests.

Auditory testing also focused on rhythm and attention, spilling over into the benefits offered by the classical repertoire as distinct from other types of music. Appreciation lessons formed life-lesson principles linking biological rhythm to the regularity necessary for sustaining common interests and heal communities. According to many who spoke and wrote for national school music organizations, music appreciation lessons and specifically their rhythmic exercises were prescriptions for an ailing population of industrial workers susceptible to socialist propaganda. Speaking to high school music supervisors, Earl Barnes warned, "new conceptions of politics cry aloud at every street corner ... 1, 200,000 youths and maidens in our high schools are being trained largely as individualists, and we are content to do little more than police the rising social forces" (Barnes 1915, 35). Barnes continues that the emergency created by the democratic license to exercise individual taste, "willy-nilly," called for serious music to channel the mind to healthful rhythms and beautiful music. Reason and the capacity to turn away from stimulants depended on a sense of order that, in this period, among educationists, was related to the use and perception of time.

Anthropology, musicology, and experimental psychology flowed together in the literature, theorizing that the hold that rhythm had over the child, the adult, and the community would order their sense of time and purpose. Public school music reflected this preoccupation with rhythm as a defining element of culture but also as a means to expose cultures that were "uncivilized." The uncivilized in the population were too often attuned to the irregular rhythms of popular music associated with low life and impulses. Syncopation, the characteristic that was singled out by music educators as the most corrupting influence, would contribute to drifting away from home and joining gangs.[64] Rhythm was to be taught through supervised experiences calibrating age and development of the child to the type of rhythmic activity appropriate for her stage of growth. Rhythmic protocols would model "democratic" impulses. Thaddeus Bolton, a psychologist whose work is closely associated with the child study movement wrote, "a highly civilized people is not affected by mere rhythm. A simple tone is not so expressive as it is to the lower classes of people. The Negro preacher often resorts to recitative speaking to produce the desired emotional state in his hearers which is generally known as the 'power'" (Bolton 1894, 164). In *The Musical Education of the Child*, used widely by music educators in the United States, Stewart MacPherson stated, "[Music listening and study have] an invaluable agency on corporate action ... [and] in the formation of character ... while personality counts, it demands at the same time subordination ... to a common purpose and a common ideal" (MacPherson 1915, 10). Another European with ties to American music education is Jacques Dalcroze.

The links between the study of classical music, rhythm, time, community, and nation leaned heavily on the new experimental insights into the rhythms of human attention. Symphony concerts, opera, and recitals were also to preserve democracy by healing the destructive effects of hard labor and class divisions. For example, a community of listeners ranged from charwomen to the ordinary middle-class crowd standing in line for tickets in the rain. Reports of this type of scene testified to the ability of cultured music to forge a community of listeners representative of national character and ambition.[65]

Wagner was quite the sensation in America. His operas attracted a wide audience in the United States, intensifying attention to the mythic, Teutonic themes that resonated with the racial logic of whiteness and national destiny. Although the Wagner phenomenon was circumscribed within an elite musical coterie, the mentality it came to represent challenged musical progressives and the educators among them to found music clubs in schools. The following provides a sketch of the students that were drawn to and recognized by the music appreciation club or society: "The special group [whether for appreciation or performance] . . . furnishes a natural outlet for the kindly, altruistic spirit which music study should produce. It has valuable social possibilities . . . in the establishing of fine relations . . . membership stresses ability and initiative and produces high individual qualifications" (Cundiff and Dykema 1927, 153). Stressing the "altruistic spirit," there is something of the social reformer in her image, as musical interests were elements of character formation. The fusion of the music appreciation activity with a civic purpose in its "valuable social possibilities" presented music in such a way that the focus was, peripherally—especially in the elementary school grades—on the music the students would listen to. Social effects were the central concern.[66] In later grades, social morality aligned with compositional technique insofar as the composers' struggles became models of moral triumph over disabilities and mediocrity. In this sense, public school music set itself apart from the activities of elite music clubs, the training and apprenticeship of musicians, and the performance world. Although the field maintained its prerogative to teach the rudiments of music, in the early twentieth century, it focused on classical music as a mode of recognizing the meritorious listener as future citizen.

Bred in the interstices of the new psychological literature on the concertgoer, public music instruction and its leisure time recommendations fabricated a listener who fit particular specifications. He was not to be the average child but the exception—a persona whose biography would make him an exemplar of civic-mindedness. The body of this listener was attuned to the regular rhythms and pulse of "normal" physiological function—for example, regular beats of the heart, moderation in physical exertion, and the alternation of activity and rest.[67]

The early twentieth century was the time when the music curriculum was given an expanded role in the "civilizing process," as many recognized the potential of music to be read as a map of the mind that was formerly the soul. Different from the earlier vocal curriculum, music appreciation focused on how one acted while listening or the quality of the attention of the individual. This focus on psychology made it congruent with the growing interest in scientific theory about society and the individual's failure or success in adjusting to modern conditions. "Science had a dual quality: to reform the city in a complex industrial society, and to study and order social and individual lives in order to make them more efficient for a conditional life associated with democracy . . . The laws of psychology were to bring efficiency to the techniques of teaching by enabling students to develop and increase the desirable qualities in their nature" (Popkewitz, *Cosmopolitanism and the Age of School Reform*, 87). Even opera could be enlisted as a method that brought psychology, aesthetics, action, and democracy together. This is apparent in a lesson using Humperdinck's opera *Hansel and Gretel*:

> The children have the story of the two children and the witch in their readers . . . When they hear the music they . . . build a dramatized situation around the music . . . From this will come the desire to express themselves . . . an appreciation of the masterpiece of the composer . . . [It meets] the demand of modern education in its claim for the development of the individual *to adjust himself to social conditions*. We readily see that the entire procedure is similar to a social group activity, either in school or . . . in business or in a profession for to arrive at a satisfying success there must be co-operation, congeniality and deference one to another in a group. (Mohler 1924, 263; emphasis added)

The lesson dramatizes a set of interrelated ideas about music listening that would imprint the value of the music with principles of civic conviviality and cooperation. On its face, the lesson's psychological, social, and aesthetic precepts have nothing to do with race. Yet, read at another level, they activate a template for musical activity that elides music with a particular of kind of reason about the identity of enemies of the community and state. The lesson is exemplary of the civilizing process discussed in connection with school songs in the previous century, a process in which the symbolic exclusion of a grotesque calls up the historical otherness of abject races, creeds, and social classes. This is not to say that civic cooperation is always a question of symbolic projection or that evil does not exist, but to draw attention to how repulsion that churns up *a stereotype* initiates wider forms of exclusion: reason (the mind) versus unreason (the body), darkness versus light, and black versus white.

Hansel and Gretel presents many possibilities for interpretation. The one given elaboration by Mohler is not the only one possible. Its logic and rhetoric normalize a simple repulsion as civic duty: the rescue of the good from the bad. If these dichotomies appear as inevitable and "natural" to us, we deny their delimitation of what it is possible to think and do in relation to music.

In the next chapter, I will consider this issue of partitioning social worthiness further in relation to musical publics and notions of the primitive.

6
Goodbye, Darwin
Music Appreciation and Musical Publics

> The people of the east and of the Old World can have no comprehension of the eagerness and sincerity with which the West [in the United States] is pursuing, under many difficulties, the study of better culture.
>
> —Van Brunt, as cited in Kiri Miller, "Americanism Musically"

A famous anecdote about Beethoven quotes him as complaining about the behavior of a Berlin audience after one of his performances there. He was not applauded, he reports, but people crowded around him and wept, which was not what he wanted, preferring that they show their appreciation through applause.[1] According to Beethoven scholars, sometime after this incident, clapping became the norm. Beethoven's career coincides with the beginning of a middle-class classical music audience and the initial formation of the rules of concert etiquette in which listening, replacing more active forms of participation, became the norm.[2]

An image of the unemotional, reflective listener has been used, with variation, as a model for the American concert audiences in school music instruction in the twentieth century. As an imaginary for a national enthusiasm for reverent attendance at concerts, its scope and significance promised rescue from the crudeness of vernacular entertainment. Music educators active in promoting a new music appreciation curriculum reiterated the European ideal, hewn in the middle of Beethoven's late career. Calvin Cady, college professor and Normal School music lecturer, wrote, "In the mode [of music appreciation one is] elevating the listener into *an active mental attitude* . . . he is awakened to that attitude of mind that marks the *real student* . . . of reflection, of inference, of judgment" (Cady 1910, 52). More recently, a similar incitement: "Let us consider some differences [between casual hearing and] listening that require focusing attention on its unique qualities. This takes energy and self discipline; if our minds wander, we lose the meaning of what we are hearing" (Anderson and Lawrence 2001, 266). Whether we are looking at the older music appreciation texts or present day advice to teachers, the cultivated audience is set apart from what is a tremendous variety of musical *publics*. The word *publics* is used here to acknowledge different and overlapping audiences that appear to be eclipsed in the curriculum texts by an imaginary of a normalized public at large, exclusively interested in classical music. In this sense, as David C. Goodman notes, the musical hierarchy created its own cultural supremacy by dislodging something else.[3] This norm, as I have argued, emerges from the racialism of musical representations for whiteness and "others," whether in relation to popular ballads, minstrel songs, and dance music of the earlier era or the pop genres of today's electronic mass communications.[4] Given the fabricated "normality" that is traceable to an "origin" of superior mentalities, public school music appreciation manuals often made a point of citing famous conductors and authors as the source of their mission: "The study of music appreciation emanates from the effort to prepare audiences at symphony concerts to listen more intelligently... First used... in [the conductor, Theodore Thomas' program notes]... W. S. B. Mathews' 'How to Understand Music' set forth the nature of this study" (Cady 1923, 10). In spite of these references, it is important to unearth the racialist components of this ethos that lay just under the surface.

The present chapter sketches the tensions in the social atmosphere surrounding the construction of this normative, cultivated audience for music education during the period from 1890 to 1940. During this period, economic pressures to revise concert programming and matters pertaining to pedagogy's categories of degenerate and primitive were reworked to fit a progressive agenda in the education of the young child, the adolescents, and the adult. Music appreciation also reworked notions of pre-Darwinian and Darwinian selection in teacher manuals, heavily weighting the importance of music to the child's "evolution" into an adult. Publishers such as Silver Burdett, Victor, and Ginn put out scores of series manuals that interwove stages of maturation with ideas on the benefit to attention span in the study of music. These concepts were backgrounded by fears of contamination by musical fare associated with the growing black, urban populations in the Northern cities.

The version of Darwinism that appears in Sir Hubert Parry's *The Evolution of the Art of Music* (1884) outlined a ladder of ascent for complexity from primitive genres to Western musical forms. Since it was the prototype for American music listening guides up until 1920, it had an important influence on the ideas about race and nation as the "origin" and endpoint of musical expression. However, when stages of musical complexity became linked to stages of growth of the human being, a certain ambiguity entered the picture. This was the idea that the musical primitive could mature, through systematized exposure, to appreciate "higher" musical accomplishment. Development that was formerly frozen for more primitive types became the more flexible progress one *might* expect from the individual in a democracy. Recapitulation of the primitive stage narrative for every child became a byword in the music appreciation manuals: "It is true that we are always dealing with children who are recapitulating racial experiences and trying to catch up with civilization" (Barnes 1915, 33). The new child-centered consensus recorded in these conference proceedings centered on a belief in the programs of education of the "unfinished" listener available on radio and through books and records.

Radio played an important role in the movement to democratize access to the cultivated tradition and it also further strengthened the redemptive, civilizing role for public music instruction. As Goodman points out in his forthcoming book, *Radio's Public: The Civic Ambitions of 1930s American Radio*, the sacralization of classical music and the fervor to distribute its benefits to the masses went hand in hand. There was nothing contradictory in an elitism that had populist aims. As a prominent music supervisor in the 1920s wrote, "Within the last few years, music study in the public schools has acquired a deeper significance. After many years of development along rather restricted lines, there is now a well-defined movement toward liberalizing and broadening the study so that it may more fully attain its real purpose" (McConathy 1925, xiii). Even with the generous conditions for redemption and rescue provided by records and radio, however, school music instruction met up with (and used as a foil) the "uneducated" musical public, always just steps away from engulfing the imaginary audience exclusively committed to classical fare.

Music Appreciation as a Great Awakening

Music appreciation was commonly compared to a religious revival in its spiritual renewal: "Aesthetic interests have a vital relation to essential elements in individual and collective progress . . . The great awakening has come—the advance in the appreciation of [music] . . . its application to the promotion of civic and industrial progress, is proceeding in America with a rapidity unparalleled in history" (Dickinson 1915, 29). This awakening, wrapped in Calvinist rhetoric, would purge the nation of obstacles impeding its progress and disclose the order necessary to liberty. Such deliverance was a frequent theme at music teacher conferences. "In music, above all other arts or manifestations of intellect and imagination, freedom is of paramount importance. Music, like such elemental phenomena as fire and air, is and must be free, and yet this freedom should be controlled and applied by a guiding intelligence" (Claxton 1915, 50).

The exercise of tracing the "guiding intelligence" in great music would uncover the clear articulation of themes, their harmony, and their development. Taking apart and reassembling the sonata, lied, or symphony, for example, made manifest the clear relation of part to whole. For music critics in the German states, part and whole had a symbolic significance in the task of identifying the spirit of a people through culture,[5] while political unity was ambiguous and troubled. For Americans, there was a long, established tradition of using the rhetorical register of music to symbolize national unity and a discourse about the harmony of blended voices and points of view. William Billings's rebelliousness stood for the freedom of the individual in the context of the whole nation.[6] The rhythms of the text of the Declaration of Independence, in Jefferson's reading copy, marked out duration in order to "divide the piece into units comparable to musical bars or poetic lines" (Fliegelman 1993, 10).

These early American rhetorical traditions carried over to the social progressives of the late 1800s as well. Most familiar to us today as the image of the melting pot, the nation would harmonize parts through music—that is, by encouraging the a reciprocal cultural exchange between social classes (with the absence of black participation).[7] However, what was *rhetorically* the case was not the actual social and political situation. The country still had to deal with the contradiction of racial inequality at every level of life. Rationing morality to those who counted, the appeal to the recognition of different "dispositions" between the races had, a priori, *made* men socially unequal.[8] Black music, according to Jefferson, was limited to monotonously sounding four chords.[9] Discounting black sensibility followed not just from race, but also from the constitution of the *public* as those who could exercise liberty as the "nurse of genius."[10]

The theme of a cultivated, white nation made its twentieth-century appearance in school music manuals, radio symphony broadcasts, series editions of classical records, and educational events such as the national Music Memory Contests. The latter elaborated an aural, national community from the technological reach of the new media.[11] Using national maps as visual representations of the aural landscape, the memory contests

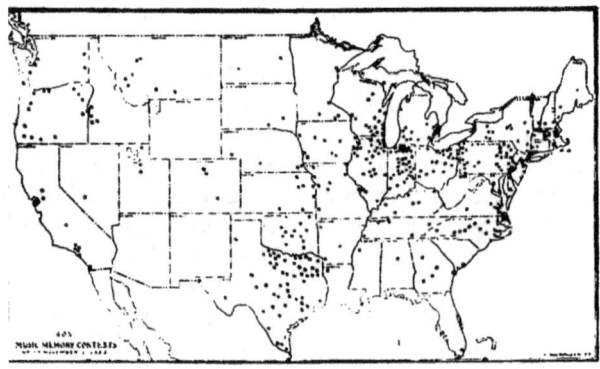

MAP SHOWING THE DISTRIBUTION OF MUSIC MEMORY CONTESTS
IN THE UNITED STATES PRIOR TO NOVEMBER, 1922

Figure 6.1. National Bureau for the Advancement of Music, "Fifty Newspaper Editorials on Music Memory Contests" (New York, 1922).

appeared as coast-to-coast events. They required children to become familiar with a list of musical selections so that they would recognize a small portion of a selection when it was played on the Victrola. "Much music is built on themes and the recurrences, the development and the contrasts [are] the design of the music. A quick and retentive memory is required if the listener is to hear and remember the themes with sufficient clearness to recognize them when they appear again" (Giddings et al. 1926, 33).

Teacher conferences mentioned the scope of the Music Memory Contests, often correlating the growth of public school music contest winners and piano purchases. Figure 6.1 is one of the maps published by a private organization of piano manufacturers whose role was to promote piano sales.

The ability to recognize themes and subthemes was the central principle of structural analysis offered in music appreciation courses. It restored the organic whole to the perception of the fractured nature of modern life, normalizing the atomization of the sonic environment, but reaching beyond that state to fabricate an aurally unified nation. The growth of musical organizations, along with memory contests, appeared predictive. A table published by the piano manufacturers shows the entire continental United States with dots indicated the overlap of piano sales and contest winners. A similar map plotted the relation of piano sales and music supervisors. Thomas Briggs, Professor of Education at Columbia University, commented, "Initiated by Mr. C. M. Tremaine, Director of the National Bureau for the Advancement of Music, the contest has spread rapidly all over the country ... Contests have been held in some three hundred or more cities and in numerous counties, with several statewide final contests directed by departments of public instruction" (Briggs 1925, 5). The national bureau that sponsored the memory contests published figures that showed 1,193 cities, towns, and counties participating overall:[12] a sublime, awesome quantity of national significance.[13]

Orchestras, Teachers, and Audiences

In a manner not much different from today, survival of newly organized musical ensembles depended on wealthy patrons, but also on ticket sales to a broader public that forced a revision of the exclusive programming many elites favored. For that reason, and to some extent for the sake of civic unity desired by progressives,[14] an amalgamation of sensibilities resulted in programs that reflected broader tastes. However, the school music appreciation curriculum placed greater emphasis on the traditional European repertoire that would develop the more elite listener. How and why this entrenchment occurred is related to the lecture circuit of professional musicians, journals, advertisers, and teacher training institutions. With auditory technology reaching millions of schoolchildren and families, the disciplinary nexus formed by the professional organizations and the commercial sphere in particular made it possible to believe that there was a ready-made, *national, coast-to-coast* audience for classical music.[15]

Yet, despite this view shared by many music pedagogues and conductors,[16] some historians doubt the existence of a large scale, exclusively genteel listening audience in the United States outside of the relatively small music societies. They point to the overlap of popular sheet music sales with classical music editions, radio offerings of all kinds, and other forms of musical entertainment that blended "high" and "low" cultural forms.[17] And in the shadows was the omnipresent threat of jazz.

> There is no doubt that a significant audience for classical music existed and that it was growing in size, though not at the epoch-defining rate the reformers would have liked ... NBC's Samuel Chotzinoff pointed out, quite reasonably, 'how astonished Beethoven would be to learn that his symphonies are now heard simultaneously by millions ... ' Classical music optimists extrapolated forward form the demonstrable increase in the audience for serious music and imagined a future in which there was no cultural divide ... It was also important to know that the existing audience for classical music was a hold out against the tides of mass culture ... Jazz was understood as invasive and addictive—something which might take over all musical life if the upholders of true standards were not vigilant. (Goodman, "Radio's Public")

An indication that vigilance was in order was the substantive decline in piano sales due, in part, to the radio and the popularity of the phonograph, correlated with a drop-off of interest in music literacy.[18] If people could hear music without investing long study in mastering its rudiments, why buy a piano? While piano manufacturers initially believed they only lacked sufficient advertising to reverse falling sales,[19] the scope of available venues for music told a different story.[20] A rise in concert ticket sales, radio audiences for classical concerts, and recital venues spoke to a broader spectrum of mixed musical publics in attendance that also enlarged audiences for concert and opera performances. As John Kasson points out in his book, *Rudeness and Civility*,[21] concertgoers were drawn from all walks of life and all tastes since classical music had become a mark of high social status for the new middle class. For many music educators who observed these trends, there was an expression of the troubled visions of a national culture they hoped would match European standards. Music appreciation, they wrote, would rescue the nation from the "bad" music to fulfill this promise.[22] For others, as made plain in the guidelines to music teachers, musical rescue needed the new tools of child psychology and findings from laboratories about the effects of music on the mind and body. All of these enterprises took on the cast of the racial struggles going on within and outside of sites for musical performance and instruction.

One of the more dramatic cultural confrontations concerning the national musical public took place in the concert debacle at the Chicago Columbian Exposition of 1893.[23] As discussed in Chapter 5, the overall theme was superior progress of Western civilization. Racial destiny figured strongly in its music exhibits, especially on the Midway, where minstrel performances dramatized the idea that "American" Africans ("American" came before "African" in those decades) had made great progress in the United States compared to other aboriginal groups. These performances were seen, by some, as part of a strategy to draw big audiences for the other educational, musical, industrial, and trade exhibits, including symphony and classical music concerts.[24] If this was the case, it posed problems as the cross-flow of audiences disrupted the reverent atmosphere some musical elites held sacred for the classical venues.

Some music planners favored programming that would exclusively attract a genteel sector of visitors, while social progressives argued that orchestra programs should combine cultivated or popular tastes. The hope of broad participation was raised in both camps with elites attempting to educate audiences through rules about talking and lateness, others working for the music industry to garner a cross-section of the exhibition's attendees as consumers of sheet music and instruments. The majority of planners apparently acknowledged that concert audiences would come from an undifferentiated mass whose tastes were not as important as the means to capture their attention.[25] In this sense, the exposition offered a microcosm of the issues and controversies public school music supervisors would speak and write about as they adapted W. S. B. Mathews's text, *How to Understand Music*, for music classrooms.

As the largest undertaking among these, music appreciation enjoyed the national attention of various industries[26] and the government agency, the U.S. Bureau of Education. Looking for model content for high school music appreciation, the bureau's music committee of 1912 recommended the analysis of master works of the German repertoire. This request, representing a common view among music teachers across many institutions, might have prompted the Bureau chief, P. P. Claxton, to comment that he favored a more inclusive program that would include varied tastes. In 1916, Claxton addressed a national conference of music educators, proposing that the bureau help music educators devise methods to attract mass interest by avoiding emphasis on the intellectual aspects of music. Sharing the progressive view of many others, Claxton envisioned music as a rescue from the burdens of industrial labor, He felt that public music education should connect the musical interests of the worker to music and, in a sense, recognize the need to use music, much against elite views, as a palliative for the fragmented experience of city life.[27]

For the most part, music educators and the public school curriculum took a different route, as developments in music teacher training institutions across the country were requiring more teachers to become licensed specialists who could teach music history and theory. The trend was decidedly to the side of an academic interest in music. Added to this was of the proliferation of public high schools, greater requirements for teachers in general education and the demand for music appreciation courses as a domain of study in colleges. Teacher training programs were primarily technical. Syllabi on singing and ear training carried over into the courses on music history and theory, offering skills that were perceived as vital not only to teachers of choral and instrumental ensembles but to those teaching structural and analytical concepts in lessons on music appreciation.[28] "The Special Teacher's Course [ear training and music theory] of the Iowa State Normal School [later the Iowa State Teacher's College] is an example of the evolution that took place in music curricula [c. 1910] . . . A special teacher's diploma was award to students completing the program" (James 1968, 98–99).

Training offered by the Normal Schools observed the customs of racial segregation with little or no little intermixing between African American musicians, educators, and the chief architects of music appreciation. Yet, formal segregation was no impediment to the percolation and cross-borrowing of musical cultures. The famous Fisk Jubilee Singers had created an audible and visible hybrid of so-called black and white music, a combination that assured welcome for their recordings in some schools.[29] Radano notes that racialized parameters for stage demeanor were carefully observed in their performances: "What becomes known as the Fisk sound fulfilled a curious contradiction: challenging the conventions of European choral practice, it also accommodated the norms of white concert music. In their blackness, the singers had already upset the standards of 'good music.' But in adopting the posture, behavior, and performance practices of the concert stage, they also affirmed the racial sense of place that pleased white audiences" (Radano 2003, 259). The principle of accommodating norms of whiteness in entertainment venues was also evident in objections to the "real" Negro performers even in blackface(!).[30] This complication in the lines of exclusion observed in entertainment reiterate the irony that so-called "black" musical gestures and comportment were forms of musical delight but their actual embodiment by blacks was out of bounds. Here, we also see the odd paradox that blackness in imitation of so-called white practices had a very narrow window of acceptance in the Jubilee Singers, but that, in general, even this form of singing could not pass as prerequisite for conservatory or university training. Black "origin" marks voice and gesture as inherently unacceptable.[31] This was, and continues to be, axiomatic in most postsecondary music teacher training institutions.[32] In effect, even while the rhetoric about accommodation to the working classes was everywhere (blacks were not, by and large, considered working class),[33] the cultivated tradition remained closely aligned to the social hierarchy and the values of the "civilizing process" that nineteenth-century educationists such as Horace Mann had so compellingly interleafed with vocal instruction methods.

Evolution and "High" or "Low" Culture

Well into the nineteenth century, these attitudes reflected the power invested in conductors who demanded genteel etiquette wherever their orchestras performed.[34] With respect to the recording industry of the early twentieth century, divisions between Red Label (classical), and Black Label (popular, blues, ragtime, and other types of folk music),[35] propelled the curriculum toward a larger list of classics, under the Red Label, for school instruction.

Some teachers had reservations about the influence of the flood of materials on music appreciation on prioritizing listening over instruction in technical matters; however, in less than ten years after the initial music appreciation course appeared in the late 1890s, pedagogical journals formed what appeared to be a consensus in support of teaching a classical repertoire.[36] Yet, even if the emphasis in professional venues for which we have records was on music appreciation, this does not mean that it actually displaced vocal music as a core practice. Likewise, it is difficult to assess the use of popular tunes. There are allusions to these practices, but with no specifics.[37] There is much evidence that, for example, John Philip Sousa's marches, already part of band instruction by the late 1800s, and the "light" music at park concerts, would have blurred the lines of the popular or cultured split that many wished to perpetuate.[38] What the record shows is that those involved in teacher training and music journals felt popular forms such as band marches to be crude and simplistic.

Lurking around the edges of the entrenchment of music appreciation instruction in many institutions was a point of view that was cause for alarm, namely whether or not there was an indigenous music, here before the importation of European genres, that was "truly" American? If so, could it be that it was, in large measure, African?[39]

Evolutionist thought in music made this proposition unlikely. Histories of the music societies and, in particular, the history of music instruction, traced only the white, religious tradition as relevant to a native "American" musical culture. In this regard, one of the most well known music supervisors in the public schools—Thomas Surette—was in great demand at music teacher training institutions, public forums, and teacher conferences.[40] Forming the core of a discursive matrix from which teachers adapted evolutionary theories for their own classrooms,[41] Surette's work embodied the assumption that American music would emerge as a cultivated tradition since the nation was approaching the height of human development.

Surette and Mason's book, *The Appreciation of Music* (1907), outlined a comprehensive framework for Western music's "origins" in what Herbert Spencer, Hubert Parry, and other intellectuals saw as progress. Surette's listening guide provides a sample of aboriginal Australian music to demonstrate "a single motive endlessly repeated without relief." "All savage races," the authors assert, "are musically like children; they cannot keep more than one or two short bits of tune in mind" (Surette and Mason 1907, 10–12). *The Appreciation of Music* was to provide guidance for the listener by directing her attention to what defined primitiveness and to use these characteristics as a yardstick for comparison to the great masters. This text received attention in the national publications of the music teachers' associations as a guide for advanced listeners. "The Surette and Mason book on Music Appreciation is used and followed quite closely. [It is] to give the student sufficient knowledge of the formal growth of music to enable him to listen to the subject matter of a composition ... Enough history and biography is combined with the work to give the pupil definite knowledge of the great masters" (Bowen 1908, 176). The book also crystallized a prominent theme of teacher conferences—namely, that music appreciation covered material on school subjects such as geography and history. "Racial characteristics are more or less dependent upon climatic influences ... and mode of life of the people ... If we wish to understand any particular people, we must consider the national life [as it is reflected], perhaps more clearly in its music" (Giddings et al. 1926, 44). Hopes for a less intellectual approach to music instruction, voiced by Claxton and others, was not part of the evolutionist agenda. In 1930, well after the national music teachers' conventions backed the nationwide adoption of the Victor Company materials on music appreciation, the national Music Supervisors Conference of 1930 featured a speaker whose concerns were that public schools would have the classics "poured into" (Coffin 1930, 239) the child.[42] Some of the objection came from commercial interests such as sheet music and piano manufacturers who felt competition from the recording industry.[43] Some objections came from within regarding the danger of overanalysis and "passive hearing" (Clark 1924, 271). Nevertheless, the early twentieth-century intellectual framework of Surette, Mason, and others provided models for teachers who envisioned raising the tastes of the public school audience as the singular goal for music listeners across the country.[44] To what degree actual classroom practices in general music changed is unknown. One indication of its smaller dispersal is that the presentation of music appreciation papers in this period is less than the choral directors' entries. Still, as a cluster of aesthetic values representing cultivated tastes, music appreciation and listening in general had large implications for the way music educators would identify the meritorious student.

Stages of Musical Development

Early twentieth-century music appreciation was pulled from a different direction in that evolutionism and the hierarchical point of view about musical culture began to allow for ambivalent inclusion of Negro song as music. This involved the introduction of a new aesthetic outlook and psychological concepts of the child that would stress the integration of the primitive and advanced stages of human development into the pedagogical ordering of grade levels. Music educators linked their approach to the psychologist G. Stanley Hall's notions of age-based plateaus.[45] Relating rhythmic exercises for young children as access to a "primordial" impulse, dances and songs of Native Americans exemplified simple emotions often referenced as "lost" to industrial societies. As a response to these pictures of modern, urban life, music teachers urged listening to the series of Victor recordings of primitive peoples, including Negro spirituals, as a way to disrupt the attention to popularmusic.[46] As the educational director of the Victor Talking Machine Company asked, "What must be done? . . . Return to the simple, original, joyous expressions of the folk, which are safe, sane, and beautiful" (Clark 1913, 14). For these educators, classical composers were the inheritors of analogous Northern European folk music. Analogy was key insofar as representation of a pure past would incorporate a romantically imagined Negro and Native American song that were on the level of the young child. Exposure to the disembodied "native" music of America would regenerate the "pure" state of feelings and rhythms that had been interrupted by other distracting and disconcerting forms of music. Regarding the status of Native American and Negro song, Frederick Ritter, a musicologist well known on the Normal School circuit, represented the elite side of the search for American musical roots when he said that "Indian" music was an unfruitful direction for composition.[47] His comment underlines how the double of musical purity in the "primitive" ethnic context both elevated and limited its horizon in the curriculum. The blending of developmental psychology, musical elitism, and evolutionism[48] was an alchemy that produced the curricular gold.

Outside of schools, the music played in dance halls, in particular, held fascination as places for individuals to move across racial and class lines.[49] And while dance halls became part of urban experience that simultaneously defied genteel codes, these continued to underpin denial of employment, educational, recreation, and housing rights to populations and to African Americans the most severely.[50] Discrimination met constant challenges in the courts and on the streets. News of change coming with regard to the allocation of radio time and audience space for black attendees at concerts entered the consciousness of music educators and elites along with reviews of African American singers abroad. [51]In Europe, Negro song received wholehearted acclaim and it played a role in rethinking the place of the spiritual within public school teaching. Music educators crossed paths with wealthy patrons[52] in the concert hall and in after gatherings on musical evenings, reinforcing the web of alliances of music publishers, concert benefactors, record manufacturers, and teacher conferences.[53] From the interconnections of these enterprises and the reconsideration of primitive art in European cultural capitals, music education forged new links between classical music and Negro song.

Gestures to the primitive as worthy musical objects, noted in early guides such as "Outlines of for a High School Course in Music History,"[54] began to appear regularly in teacher manuals after 1920. Folk dance, Negro spirituals, and Native American motifs on Victor records[55] included theories of children's grasp of rhythm and feeling. In these registers, appreciation tames the primitive and romanticizes it through a supervised exposure to musical knowledge appropriate to age level. Here, oversight is intertwined with an early experience with freedom. For the moment, levels of musical development are a pragmatic means to teach larger principles. "If children are to be sent to playgrounds there to live out in miniature the occupations and primitive life of our ancestors . . . such play must be organized. Directed, encouraged, and made attractive by every attribute and embellishment . . . [through] . . . the singing game and the folk dance . . . the widest opportunity [for musical expression] must be afforded to the children to exercise their rights" (Clark 1913, 13). In this kind of literature, one can trace the pedagogical tension between inclusion and estrangement of particular kinds of folk music, especially Negro spirituals. At one level, there is an aesthetic appreciation for folk genres and their use for young children. At another, Negro song occupies a separate nicheas the childlike other. Some manuals explicitly addressed strangeness. "Review characteristics of Indian songs. [These are] dominance of rhythm, melodies ending on 'sol' instead of on 'do,' weird minor effects, use of tom-tom and flutes, etc." (Victor Talking Machine Company 1923, 134). Historically, this represents the ambivalent treatment of Native American music and the Negro spiritual that emerged from the fascination with primitiveness. Twentieth-century music educators continued the reformist spirit from the Abolitionist movement that had first championed "Negro" expression as a form of singing worthy of white audiences and emulation. Abolitionist use of the Negro spiritual made a mark against the absolute hierarchy of musical value by sponsoring the transcriptions of many Negro spirituals and songs. The movement also contributed to a reworking of whiteness in musical terms as lacking qualities "heard" in the exotic spiritualism of black music. Moreover, even as Negro songs were memorialized in print, they were also captioned as strange, impossible-to-categorize forms of expression. With these statements of fabricated racial difference came the comparison to an overbred lack of expressiveness in white society and a yearning for the innocence of the Negro spiritual.[56]

In one sense, instruction in the Negro spiritual exposed the doublet of expressiveness versus white restriction of emotion through its focus on the young child's musical feeling as an object of deliberation. Implying something of a confrontation with white social mores as limitations on freedom, the music instruction of the young child should "recognize the emotional nature of the child," which is "something deeper than the purely intellectual understanding of music" (Giddings et al. 1926, 25). "Are we teaching the child or are we teaching music? Are we remembering that we are teaching the precocious, enlightened, untrammeled child of today and not his grandmother?" (Clark 1929, 307).

Embedded in this sentiment are the developmental principles that organize stages of childhood. In line with the Romantic ethos observed in the vocal curriculum of the nineteenth century, more recent manuals gesture toward the "lost" freedom of the overly civilized adult. These themes permeated literary culture in the transcendentalists, Emerson and Thoreau, and in a different kind of narrative in, for example, in Mark Twain's portrait of Huck and Jim, the runaway slave. They would also surface in the indigenous idioms of Dvorak's *New World Symphony* and in the cultivated musical tradition's rediscovery and incorporation of jazz, for example, in Gershwin's *Rhapsody in Blue*. The change was gradually registered in public school teaching as well.[57] In all of these renditions of what were called sometimes called "African" idioms, the body of the African American, as Toni Morrison notes, is more absent than present with regard to the representation of her social and material condition.[58]

The question of how to represent the musical spirit of the unspoiled innocent and the shadowy, victimized, Africanist figure was complemented by awareness that civilization had produced its own forms of brutality. Entwined with the progressive rejection of social Darwinism were Freudian psychology and a new primitivism in music (e.g., Stravinsky's *Rite of Spring*), painting, and literature. With the interplay of high and low impulses, it was possible for the cultured persona to be conversant with untamed forces in the form of body regimens, art exhibits, and music methods as long as these were contained in educative movements and programs.[59] New approaches to displaying rhythmic response interposed race and nation as rhythmic languages.[60] "The characteristic musical tendencies of race come to light in rhythm . . . It is thus important that each be given the means by special training, of externalizing the rhythm peculiar to the race" (Jacques-Dalcroze 1921, xiii). Characteristics of race as rhythmic entrainment spread through teacher manuals of the 1920s.[61] Imported to the United States from Switzerland and used in the New York City Public Schools and in American private educational institutions,[62] Dalcroze "eurhythmics" harmonized with child study[63] through graded rhythmic exercises such as percussive toy orchestras and similar exercises.[64] The shift in models for comportment from the old cosmopolitan gentility and cultivation to a different set of pastoral idylls in Native American and Classical Greek themes linked music appreciation to a lost, simple past. A new pastoral spirit positioned the universal "humanity" in Beethoven and folk myth in Wagner's operas as sources of "pure" expression.[65] In this context, the inclusion of Negro and Native American song brought music instruction closer to plebian ideals and distanced an elitism that had disparaged Dvorak's New World Symphony for its use of American folk themes.[66]

But what was also at play in the repositioning of evolutionary notions of cultural rank was the grid of practices and ideas that linked to Dewey's trademark philosophy of pragmatism and schooling to the globalization of education even as early as the period under discussion here.[67] This was, in brief, the idea that knowledge and action were inseparable and that the greater social good would come from education founded on the trials of experience. For music, this meant involvement in the very rhythms that were the "primitive" to sustain a comparison between what "was" and what is to be or what the child had to pass through in order to become the modern, adult citizen.

Still, the cultured tradition overshadowed the relatively small dignity allotted to indigenous music and completely ruled popular music out. Agnes Fryberger's 1925 guide for teachers, *Listening Lessons in Music Graded for Schools*, provides an action lesson to dramatize the lack of positive qualities of ragtime. First, she asks her pupils to raise their hands if they like that kind of music. Most did. Then she guides a comparison between ragtime melodies and "good" music.[68] After the listing of many aspects of the rhythm and lyrics, she states, "The fact is that a person whose voice is really good" will disdain popular forms of music (Fryberger 1925, 79). The point of citing her lesson is to show how Fryberger's book formed part of a well-organized network that fabricated the good ear, the meritorious listener, and a correspondingly judgmental musical public.[69] On the other hand, Anne Shaw Faulkner Oberndorfer's series, published from 1913 to 1943, is more progressive[70] but also retains racialist dichotomies. In this sense, her texts, as well as others by Frances Clark and Edith Rhetts, rehabilitated formerly "primitive" music and recruited exemplars to train the listener who would be the fully disciplined future citizen.

Democracy and the Fragmentation of Music

In the above, I have attempted to outline some of the issues and controversies that connected the twentieth-century music curriculum with the fabrication of a select listener and musical public. In the following, I consider the intersection of music, notions of race, democratic principles, and childhood that made music appreciation a formative system of reason about music and an enduring school subject. This power to govern the selection of the meritorious listener becomes more transparent as we explore the world of radio, Dewey's pragmatics, psychological experiment, psychoanalysis, and a new musical language.

The fabrication of particular citizen types in school music coheres across practices that made their way into teacher manuals and conference papers in the early twentieth century. For example, there is the role of the Wagner societies, comprised of both wealthy cultural patrons and educators, which offered lectures on Wagner's operas. This provided a social field for what some saw as a cult formed around his personality, related to the mystical and erotic themes of his operas.[71] Yet, some music educators[72] found a common strand between the heroic motifs in Beethoven and Wagner's work,[73] seeing in those legendary operas the reverence one should emulate in listening lessons. The ethos forwarded earlier German traditions but it also reinstated the importance of austerity in music audiences at home.[74] There is a thoroughgoing presentation of this ethos in the graded series *The Music Hour* (1929, 1930) and, later, in *Our Singing World*, by Pitts, Glenn, and Watters (1949), for example, that made listening an exercise in silence and reflection.

Silence was not just etiquette. It was also a yearning for restoration of the community identified with the German greats, democratic ideals, and, at the turn of the century, American social reform and progressivism. Frequent references to John Dewey in music teacher conference papers inscribed the ideals of democratic community that would be achieved through the union of art and science, specifically that one could use the fine arts as sources of pragmatic action and social reform. For example, as developmental psychology made it possible to thematize the life course of the individual as a growth in musical complexity, these procedures would lead to valuation and appreciation through teaching listening as an ability to discriminate in general whether in music or other aspects of life. Aesthetic education would also insure the progress of the nation and community.[75] The principle of action, as thought, served as an often-cited platform for the protocols of music appreciation that would produce discriminating taste. It is not easy to decipher the sources that music educators quote since the teacher conference papers of the early twentieth century did not use full citations, but one can locate what traveled as Dewey's thought in music and art as moral habits instilled through action (not verbal instruction). "[The nature of valuation is] important as standards of judging the worth of new experiences that parents and instructors are always tending to teach them directly to the young. They overlook the danger that standards so taught will be *merely* symbolic, that is largely conventional and verbal" (Dewey, in Mark 1982, 167). In a different vein, Dewey's musical social values also accommodated the aesthetic principle that music had no literal meaning. In *Art as Experience*, he supports the idea that music cannot be, for example, a picture of things or actions. "If all meanings could be adequately expressed by words, the arts of painting and music would not exist. There are values and meaning that can be expressed only by immediately visible and audible qualities, and to ask what they mean in the sense of something that can be put into words is to deny their distinctive existence" (Dewey 1934, 74). Such seemingly irresolvable propositions about music's capacity were made resolvable through the pragmatic attitude of American music education in which Dewey's notion of fine music as sound for its own sake (Dewey 1933, 104) assured the "harmony" of community and nationhood, reminiscent of Jefferson's idea of musical phrasing in the *Declaration* of

Independence as a rhetorical device to entrain his citizen audience. Where entrainment serves harmony in political matters, the converse of this idea is that the "wrong" music tears society apart.

Educators' use of musical metaphors for unity and harmony were common (see, for example, Birchard's (1923, 77) paper on music teaching). Such statements sustained agreement among educators, functioning as calls to develop active methods to manage, to regulate, and to build the moral community.[76] This was the "intelligent action" associated with Dewey's thought cited throughout the conference literature of this period. If references to Dewey seemed to be a merely reflexive bow to his reputation, paraphrasing him was also important in securing stature for public school music as a site for overcoming the crisis in the listening habits of the nation. In the context of immigration and seeming urban chaos, this "crisis" was shaped as a situation of embattled whiteness. "The changing racial composition of urban neighborhoods in northern cities produced a crisis in the settlement movement . . . When blacks first began to move into Gary . . . Campbell house excluded them, fearing that the settlement would not be able to serve these newcomers while continuing its programs with immigrants" (Crocker 1992, 130–31). For radio, whiteness displacing "black" sound took place through blackface minstrelsy, white studio bands playing the syncopate beats of black jazz musicians barred from studios (in Chicago, for example, up to 1928), and comedies such as *Amos and Andy*, with its stereotyped impressions of black bungling. With the exception of the "Negro Hour" in Chicago, many cities and towns provided little or no access for blacks to the airwaves until after the Second World War.[77] From the first decade of music appreciation work in the schools, controlling the musical environment and screening out "black" sound would be important in the "saving" of urban communities.

> A just recognition of the function of music, especially its interpretative study, in its intellectual import, is therefore vital to the realization of its humanizing influence. The two can no more be separated in art than in religion. To seek for permanent moral stability through the stirring up of emotional activity apart from a mental grasp of the underlying principles of moral or religious life, is futile . . . This has been very clearly and strongly voiced by Professor John Dewey. He says that music and art are as "valuable and important as any other work in the schools." (Cady 1908, 151)

Distinctions between different kinds of musical experiences overlapping the urban demographics of race were to play a large role in constituting a rehabilitated musical public. "But if his own experience is what he has been most accustomed to and has most enjoyed is ragtime, his active or working measures of valuation are fixed on the ragtime level" (Dewey, cited in Mark 1982, 167). Dewey's reasoning illustrates how popular dance music could identify the unworthy mind and his reference to ragtime indicates how limited reasoning was attached to particular segments of the population. In this period, racial limitations also surfaced in reference to rhythmic practices of one race or nationality (as in the Dalcroze method) and through experimental studies on the effects of "primitive" rhythm.[78] In this connection, psychoanalysis served to call attention to the mix of primitive and civilized impulses in the individual unconscious.

Interest in character and the unconscious forces associated with mythic, racial archetypes became a signature of the change in musical taste in the American cultivated audience that brought Wagner to the attention of music educators. *The Ride of the Valkyries*, for example, was used as a lesson in repeated, dynamic rhythm.[79] Participating in this broad discourse of rhythm's importance to the psyche and perception, some educational psychologists felt that there were significant overlaps between musical activity, a "natural" pulsation of attention, and the optimum functioning of mind. "Rhythm [in music] adjusts the strain of attention ... Genetically, the ordinary measure in poetry and music is determined by what is known as the attention wave. Our attention is periodic. All our mental life works rhythmically, that is, by periodic pulsation of effort or achievement with unnoticed intermittent blanks ... The rhythmic [musical measure], then, simply [takes] advantage of nature's supply of pulsating attention" (Seashore 1938, 140–41). Since significant pieces of music could be represented in fragments to stand for a greater whole, repetition of these fragments would minimize distraction and loss of valuable attention. "Repetition, repetition, again and again is an essential in teaching" (Clark 1926, 217). Clark goes on to say that the impression of the whole (when the listener hears it in its entirety) will be superimposed upon the fragments so that memory will be permanent. Audible bits sampled on the phonograph in the routine of the listening lesson fit the periodicity of human attention where the teacher's role was to make thematic fragments "stick" as prompts to memory. For this task, rhythmic patterns were important, mimetic tools for remembering which fragment fit which tune, such as the triple eighth note figure of the "fate" motive in Beethoven's Fifth. Protocols for identifying thematic material thus had the backing of psychological theories about fragmented attention and their role in the recognition of a mastery of classical themes and motifs—a musical disposition that was foreign to the musical experience of most schoolchildren of this period.[80]

There were, in effect, two sides to the coin of using fragmentation and rhythmic pulsation to conceptualize the listener. On the one hand, recognizing rhythmic pulsation of image and sound as part of the general human condition underpinned a reclaiming of the primitive in music in general and in public school instruction as well. Primal instincts of the collective psyche were seen as cruder human emotions.[81] As one observer put it, "When I go to the theatre, I feel as though I were spending a few hours with my ancestors, who conceived life as something that was primitive, arid, and brutal" (Maeterlinck, as cited in Nichols and Smiths 1989, 21).

Similar remarks were made about Stravinsky's *Rite of Spring* and Richard Strauss's *Till Eulenspiegel*, both musical works that dramatized the "irrational" side of humanity.[82]

On the other hand, atomized musical material assisted as shorthand for human comparison, creating a language for physiological, psychic, and auditory phenomena as indicative of race and human type.[83] Carl Seashore, a psychologist well known to music educators through his musical ability tests, noted that one of the characteristics of Negro singing is laughter. Using a graph connected to a recording device, Seashore writes, "Figure 4 is a Negro laugh . . . [it] plays a very important role in the jovial Negro song, and here we have, perhaps for the first time, the preservation of a hearty laugh in Negro style" (Seashore 1938, 380). Both perspectives—the acknowledgment of the primitive within civilization and a notion of primitive music as a product of a particular human type—stocked music appreciation with a comparative language to describe the uncultured or cultured voice, melody, and rhythm.

Musico-psychic theories were also put to use in partitioning compositional ability along racial lines. Wagner's "Jews in Music" claims, as Lowell Mason had intimated earlier, that Jews lacked the innate ability to articulate the genius of the German language and music.[84] Likewise, racialism's enclosures of rhythm with respect to what were perceived as Negro forms of entertainment, i.e., popular dance, was a major concern among American public school teachers. Music instruction, according to the bulk of conference papers dealing with jazz and ragtime, was to be instrumental in orienting the child's nervous system to its proper external, musical object that was not too stimulating or too mesmerizing. In this, stated one educator, "She [the music teacher] must regulate and standardize dancing" (Barnes 1915, 39). This meant prescribing what kind of movements are encouraged in lessons and festivals where Frances Elliott Clark, for example, insisted on the restriction of folk dance to Northern European styles.[85]

Radio changed the possibilities for regulation outside of school, making the public private and producing changes in patterns of music consumption.[86] The listener exercised freedom and privacy in musical selection—a happenstance that also made her a subject to be continually won over through the cultivation of critical attitudes. Focusing on musical choices on radio and phonograph, a conductor of the Minneapolis Symphony wrote, "Children know the merits of good music . . . It is a great credit to their intelligent discrimination that they not only know the difference between ragtime and real music but also that they demand that they be given . . . the music of the greatest composers" (Oberhoffer, as cited in F. Dunham 1929, 75). Acoustic technology enlarged the school's mission to keep watch over leisure. Assuming the radio's presence in nearly every home and its influence over music preferences as paramount, music educators were especially concerned about the effects of jazz.[87] By coupling the more beneficial sets of recordings and radio programs to the curriculum, the school embodied the values of turning those sources of distraction into a means to control unsupervised listening.

Reworking the Boundaries Between Civilization and Barbarism

The Victor Talking Machine materials for teaching music appreciation were highly organized, outlining nearly every step the classroom teacher should take in the interpretation of the musical samples. For some music supervisors, too much rationalism led away from the freedom encoded in both folk and masterworks.[88] Musical expressions, either folk or cultivated, should not be marred by an overanalysis that ruined its freshness.[89] For music educators, this meant that formerly rigid musical polarities (savage or civilized) of another age[90] were realigned to include samplings of primitive, spontaneous simplicity for pedagogical aims. They also saw benefit in aligning with the sophisticated tastes of a musical, art-consuming public that embraced primitive art as exemplars of the beautiful.[91]

Post–World War II educators witnessed a movement that objected to the isolation and holiness of masterworks, doubting their universal appeal and forming a consensus for support for what became known as a multicultural curriculum. "It is absurd to tell a person how sensitive he should be, ought to be or must be, or what music he should, ought or must like ... but that he should be provided with adequate exercise material through which his musical stature can attain" (Schoen 1955, 27–39). Reformers who gathered at what was known as the Tanglewood Symposium in 1967 to reevaluate the public school music curriculum had similar thoughts. "[Popular music does for people] just about they want it to do. Popular music celebrates love, conflict, yearning, and victory ... Do we secretly believe that 'good' music or 'serious' music or 'classic' music is better than popular music?" (Broudy 1968, 9). This was a stark change from several previous decades when most music educators agreed that "No one can claim social acceptability who does not know some of the works of all the great composers" (Clark 1929, 216). Clark was referring to the enormous array of recordings and radio programs that would convert listeners in homes and business apart from the school. Walter Damrosch's radio concerts for schoolchildren, for example, provided "an audience of untold numbers who listen" and the nation brought a cultivated taste to those who led "colorless" lives.[92]

The Tanglewood Symposium reflected a different view of aesthetics that were bound up with the changing constitution of the person in the postwar environment. The student who would make the most qualified adult was to be capable of not just screening out reprehensible sound, but also learning about *the world* of music offered to her (for example, through media, the urban street, and entertainment venues) and evaluating that world with guidance from a revised aesthetic position of music teaching.

Several attendees acknowledged that the modern listener had normalized a fragmented environment as a feature of her own attention.[93] In place of the early nineteenth-century notion of a unified mind, soul, or character, the modern listener maintained a diffused consciousness with the stipulation that she could focus on single aspects of the soundscape related to her own, and the community's, activity.[94]

The vision left by the Tanglewood and other reform conferences of a similar nature, even considering recommendations for openness to new aesthetic values, did not fundamentally change the primacy of maintaining concert etiquette a still body and a cultivated attitude toward introspection while listening to music. Several speakers acknowledged the beauty in rock and roll as well as other forms of pop music. What was left unmentioned were changing demographics and the desire among youth to experience music in closer relation to the body.[95]

Among the systems of thought and culture discussed at Tanglewood, none proposed that body entrainment was an issue. Embedded in the proceedings' papers are the values of reflection, analysis, and communication through teacher training that discussed whether or not to impose a hierarchy to various music. When the issue of "disadvantaged children" was raised, it reinserted the division between informed choices and those choices of children who have no background from which to choose wisely.[96] This, in itself, maintained the unconscious hierarchy of values that overlapped racial and ethnic divides in classrooms.

This outlook continued through the years that Pierre Bourdieu's 1984 critique about judgments of taste influenced researchers in music education to write critically of gender and versus the civilized listener or performer sustained itself in today's standards in the oppositions between "professional, authoritative knowledge" and popular practices with "unpurposeful" movement to music.[97] Looking back to Lowell Mason's "The Normal Singer" (1856), I want to examine how far the "normal" of whiteness penetrated into the twentieth-century literature for music teachers. This is the subject of the next chapter.

7
Reason, Ventriloquism, and National Music Memory Contests

Tameka stopped short of shouting out the words to a song she played on the violin when she noticed that she alone in that quartet was making her feeling known to the audience.

—*Ruth Gustafson, "Stories of Failure or Delight"*

The Grain is the body in the voice as it sings . . . I am determined to listen to my relation with the body of the man or woman singing.

—*Roland Barthes, "The Grain of the Voice"*

Roland Barthes's well-known essay, "The Grain of the Voice," establishes that singing is a relation between the voice and the body, or, more precisely, the voice discloses a bodily presence. This "grain" or imprint of the body, he goes on to say, are palpably evident in the singer's voice texture, her diction, and emotion.

In this chapter, I consider how the grain of the voice is relevant to thinking about early twentieth-century American singing and music appreciation. My argument is that the curriculum of that era spoke strongly for a particular set of social relations and values about the body divide the "lower" being from the more cultivated human type. As I discussed in previous chapters, the degenerate singer, or the grotesque, was banished form school songbooks and makes no pretense, in Bahktin's[1] words, of getting rid of the earthy body. Like many singers in William Billings's hymns, the lower voice represents the lower social body. "It is an image of impure corporeal bulk with its orifices . . . yawning wide and its lower regions (belly, legs, feet, buttocks and genitals) given priority over its upper regions (head, spirit, and reason)" (Stallybrass and White 1986, 9). The elite voice is citizenly, classic in its physical demeanor and modest temperament. And it continually takes the image of itself from the comparisons to what is plebeian and base.

My investigation attempts to demonstrate that vocal preferences are significantly entangled with ideas about the voice's gender and race. This entanglement reveals something quite germane to patterns of participation in school music than our historical research has so far considered.

The split emerges in a more explicit form than it had taken earlier, since music appreciation in the twentieth century was built on a cosmopolitan repertoire whose universal values brought reason, science, and art together in the worthy listener-citizen and in the music itself. Yet, insistence on this universal essence has also, paradoxically, disqualified individuals *as if their own capacities were not universal at all.* This occurs through something Thomas Popkewitz calls the "double" gestures of admiration and repugnance that we find in hierarchies of social life and social institutions, in the history of school subjects and in the operations of pedagogy: "The double gestures of pedagogy . . . are embodied in the recognition given to excluded groups for inclusion, yet that recognition radically differentiates and circumscribes something else that is both repulsive and fundamentally differentiated from the whole" (Popkewitz 2008, 6). It was what my cohort of students encountered in school music. To reconnoiter that peculiar and paradoxical mixture of admiration and repugnance, we have to envision the scene that praises black musical contribution, but, at the same time, sequesters it as "diverse"—second to the mainstream fare of light and heavy classical music.

The "double" lies in the overlap and near simultaneous operations of repulsion and embrace creating an ambiguous relation and limbo-like status for the questionable musical object, singer, and listener. But this very ambiguity is drawn from a rich, complex history of scientific experiments on the voice, ventriloquism, minstrelsy, and *bel canto* singing. The fabrication of the worthy versus the abject student in music instruction comes, to a great extent, from racialist thinking. I use racial, ethnic, or religious terminology here, not to heighten these rubrics' "realness" or essence but to focus on their use as semantic markers in music. Abandoning the historic role played by race in music would duplicate denial of its alliance with notions of whiteness.

In the first section of this chapter, I discuss some of the historical beliefs about reason, the nation, and singing that infused music education's view of the body in the early twentieth century. In the second section, I consider folk music and the poles of strangeness and familiarity they generated with regard to ranking the listener and performer. The third part of the chapter describes a psychoacoustic experiment that pursues the hunt for an essential musical difference between Negro song and the cultivated vocal tradition. And, finally, in the fourth section, I present some aspects of ventriloquism that are relevant to disembodiment of the black voice in the curriculum and on the concert stage.

Progress and Its Shadow

At the 1915 Panama-Pacific International Exposition, the following account is given of a music exhibit housed in the liberal arts classical Greek-inspired "temple" bearing the name of the Victor Talking Machine Company: "'The Victor in the Schools' is the miracle which has made it possible to bring real music of all the past ages and all lands to all the children ... and has caused the greatest advancement in public school music that has come to it since its beginning under Lowell Mason in Boston in 1836" (California Teachers Association 1915, 153). In addition to delivering "real" music to school audiences, the gramophone, recordings, and experimental science made it possible for music education to move toward defining ability and talent.[2] "[Psychoacoustic] research has been concentrated upon establishing norms for various measurements and interpreting the meaning of these norms so that when a pupil is measured we may compare his record with the various norms, which show what should be expected under various conditions" (Seashore 1913, 212). The new field of developmental psychology, notably the work of G. Stanley Hall, provided the framework for a calculable, age-based sequence for music listening in, for example, *Music Appreciation in the Schoolroom*.[3] Other educators were familiar enough with John Dewey's writing on the relation of action to inquiry to describe a "project method" in music appreciation. The method stressed the "doing" of music through classroom opera staging, appropriate to age,[4] while another engaged students in comparing types of music and their social benefits.[5] Translating opera scenes and other repertoire into classroom action, music appreciation realized the union of science (child development) and art that would plan for a democratic society. "The idea of foresight ... and intelligent action that Dewey spoke about were to bring the sensitivities of science into a mode of organizing daily life and constituting community" (Popkewitz, *Cosmopolitanism and the Age of School Reform*, 84).

Another social goal of music appreciation was to prevent delinquency. This aim brought the spirit of social reform into the classroom through attunement to the uplift in classical music. Crime prevention was consistent with the experimental findings of overexcitation and adolescent susceptibility to illicit behaviors. Teacher manuals and journals disseminated these ideas to music and general education teachers in a continuous stream of publications dating from 1901 to the present.[6] Part of the general music appreciation movement outside of schools included music organizers sensitive to the work of the Juvenile Protective Association in Chicago, for example, that initiated free public classical concerts to counteract the musical attractions of dance halls and cabarets.[7] Dr. P. P. Claxton, the U.S. Commissioner of Education, addressed music educators as standard bearers of a culture to spread democracy: "Our age is an industrial age and our Anglo-Saxon mind turns more to the appreciation of building and of industrial development that to music or the arts. Music should be made democratic through its presentation by groups of the people, and through groups who respond by listening. Choral work means more than solo work because it brings harmonizing influence to the masses" (Claxton 1915, 50). However, Claxton's recommendations also delimited how that harmony of democracy would be partitioned or marked off. In his view, Anglo-Saxons constitute the national need for listening to music as a performance of democracy provided by "other groups."

Early twentieth-century teacher manuals and journal literature give a more detailed picture of how music served national needs and how John Dewey's principles of learning through action provided the pragmatic framework to link music listening to national salvation. This psychological or philosophical outlook grew appreciably in the estimation of American music educators from 1910 to 1940 and was disseminated through a national network of conferences and journals. Calvin Cady, a well-known music pedagogue and founder of a music school in Oregon, taught at the Dewey School in Chicago. He had interpreted Dewey's "progressive" method, as it was called, as uniting the child's feelings with appropriate musical activity and de-emphasizing the intellectual tasks of the discipline until the body had experienced music's "feeling" mode.[8] From that point on, Cady, like Claxton, envisioned music education as a transcendence of the body and an attunement of consciousness to a secular vision of spirituality:[9] "The teacher, stated in the simplest terms, is to reveal to the child, and to lead it to exercise its capacity to discern, conceive and express in thought and action all pure idea, to lead the child of any and every seeming age to prove the scientific fact that 'the Kingdom of God'—the whole Kingdom—is within the individual consciousness" (Cady 1902, 5). However, the transformation of God's Kingdom also carried distinctions between listeners and speakers. A nation's reason was represented in speech and its musical voice was a sign of civilization:[10] "[God] gave his appointed ministers knowledge both to see the natures of men and also granted them the gift of utterance, that they might with ease win folk at their will, and frame them to reason by all good order" (Thomas Wilson, as cited in Nelson and Berens 1997, 52). Music, as Claxton had referenced it, was also a rhetorical device for performing consensus and harmony. Perceived control over the voice, sung and heard, distinguished the rational being from her opposite—one who could not be persuaded by reason. Demanding this regulation was an exercise in "soft compulsion" with a tradition in the democracy that reached back to Thomas Jefferson's use of music as an inducement to social harmony in his first inaugural address: every difference of opinion would not be a conflict over principle, he said.[11] Yet, built on the inequality of white and black, the democracy of Jefferson's reckoning perceived the grain of the reasonable voice as the opposite of slave song. In *Notes on the State of Virginia*, he drew stark contrasts between the social harmony of society and those who were not part of that society.[12]

Voices of (Un)Reason

The shadow side of democracy—exclusion—comes into view when we consider the historical ties between the voice, the body, and reason in the West. Speaking has been fundamental to hearing and hearing to understanding. What's more, the voice always occupies a space by virtue of its coming from the inside of a body. This has large implications for Western musical tradition in that the upper body is the "origin" of vocal sound. It is also the site of reason in contrast to speech and song associated with the "lower" body and nonreason. The voice, then, not only occupies a particular space in the body, it provides an ordering system for interpreting voices coming from other bodies.[13]

The preferences for particular kinds of vocal sound, bodily motion, and hearing were connected to theological debates over what it was possible to see, hear, and speak in relation to God. Supernatural phenomena, such as ecstatic speech, God speaking, or imitation of the Devil (glossolalia), made the person a fanatic, mentally ill, or "savage" with various religious, gender, and racial connotations. Reason was associated with secular reality while suspicion surrounded the individual who heard unauthorized voices. "I totally disbelieve that the Almighty ever did communicate anything to man, by any mode of speech, in any language, or by any kind of vision or appearance, or by any means which our senses are capable of receiving" (Thomas Paine, cited in Schmidt 2000, 586–67). Disavowal of the supernatural seemed necessary in twentieth-century efforts to expel divine or bedeviled voices from the radio and the phonograph.[14] In the 1930s, federal, state, and local institutions attempted to fully secure the telegraph, telephone, and radio airwaves as nonsupernatural, scientific entities.[15] By this time, auditory technology regulated and reorganized urban sound spaces into new categories of repulsion and meritorious speech.[16] "While local music radio helped to empower many community groups and to strengthen ethnic institutions, it excluded and marginalized ... most notably, African Americans ... [they] were typically barred from control rooms. Complicating the racialized situation further, minstrel dialect and musical entertainment featuring whites in blackface were a staple of 1920s broadcasting" (Vaillant 2003, 236).

Psychoacoustics became influential in music education in the 1920s and 1930s as it joined science in measuring ability, performance skills, and the effects of music on listeners. Through technology, thought completely objective, researchers would accurately, not subjectively, map the material qualities of the voice. This ability to map sound became central to manufacturing notions of race in singing. Disavowal of the "mystery'" of the power of the spiritual, it was thought, would come from a more scientific view of their acoustical qualities. The grain of the "Negro" voice would be clarified through scientific analysis. Reviewed in Carl Seashore's *The Psychology of Music* (1938), for example, Milton Metfessel's study on Negro song attempted to disenchant its aura and reveal its material essence. The investment in the truth-telling capacities of auditory technology in this study, an exemplar of many other studies of the voice and musical perception,[17] brought Negro song into the fold of artistic legitimacy. On the other hand, the psychological classifications of its voice, typed as Negro, continued the process of attributing essence to race itself and rating its sound, in comparison to the classically trained voice, as inferior. This dimension of psychoacoustic work made racial comparisons "scientific."

At the same time, the mapping of the essentially white voice was produced by the tradition of *bel canto* singing. *Bel canto's* ties to processes of exclusion in public school music teaching and postsecondary education lie in its authority to judge and rank comportment, diction, ornament, and register, among other qualities as musical values. Within *bel canto* style, the voice's grain or link to the singer is judged *as a quality of the soul* rather than as a message from the body.[18] *Bel canto* has no single origin or agent, but existed in a dispersal of vocal values connected to the churches and aristocratic traditions that reiterated the social hierarchy. Paul Robeson summarized the bifurcation of popular and artistic traditions when he wrote, "With the Renaissance, reason and intellect were placed above intuition and feeling" (Robeson, cited in Pencak 2002, 153).

For teachers in the nineteenth century who affiliated with *bel canto* style as it congealed, for example, in Thomas Hastings's *Dissertation on Musical Taste*, it embodied Teutonic vocalization, carriage, and gesture.[19] Inseparable from the racially inflected rhetoric about reason, the articulation of a Teutonic vocal style in the curriculum, in texts such as Hastings's *Dissertation* and in Lowell Mason's prefaces to songbooks and teacher manuals, the *bel canto* archetype formed a dramatic contrast to Jewish and Catholic rituals.[20] Chant involved whole body movement, whereas operatic singing stressed the subdued demeanor of the body's upper half. At the other end of the continuum, slave songs included pitch variance, extroverted body accentuation, and gestures viewed as foreign compared to the *bel canto* art that was embraced as native to American singing.

The most striking aspect of the *bel canto* aesthetic, further back in time, was that its preferences for the upper body are historically tied to the castrati practices of the sixteenth and seventeenth centuries. The practice of *castrato* voice required that a gifted male singer be castrated in his youth to preserve his ability to sing in the extended upper register. His voice connoted a unique, pure sound achieved without the apparent involvement of so-called lower instincts. It also served the practical purpose of representing the female in opera and oratorio without involving women in stage appearances that were customarily prohibited. The *castrato* comprised crucial aspects of the larger *bel canto* aesthetic as a whole, contributing to feminized and racialized notions of lower body involvement in so-called hot rhythms of black music.[21] Demand for *castrati* waned gradually, disappearing altogether in the twentieth century. However, the voice of the counter tenor, with similar high register and symbolic value, continued the artistic tradition minus the systemized practice of castration.[22] In relation to singing instruction in schools, *bel canto* brought "head tone" into a central place in its pedagogy.[23] In spite of many other tendencies and directions in vocal art in recent decades, this aesthetic system continues to inform reasoning about the voice *as the mark* of a particular human type. While the tradition is far from systematized across public school music programs, the emphasis on the head as the site of vocal control permeated postsecondary training of teachers. To this extent, concentration on the upper body went hand in hand with the discouragement of rhythmic gesturing or bodily accentuation of any kind.[24] "The work of teaching the proper use of the breathing muscles, tone-placing [pitch] pronunciation, enunciation, phrasing, etc. . . . For this and for class vocalizing, only simple exercises are used, that the pupil may concentrate the mind wholly upon the production of tone" (Bowen 1908, 177). The position taken in *The Music Hour*, a series of texts that circulated widely among music teachers in the 1920s and 1930s, embodies this ethos. The approach regulated pitch (singing in tune) and tone as central qualities. Its enforcement as in schools is variable since these techniques generally require individual study to perfect and are, in overuse, discouraging to many choral initiates.[25] Nonetheless, a dispositional bias toward classical performance pedagogy is found in recent teacher guides for general music teachers. "Children who cannot sing in tune share a common problem: They are unable to manipulate their voices so that they can match the musical sounds they hear. Before a child can sing comfortably, he or she must learn how it feels and sounds to use the *head voice*" (Anderson and Lawrence 2001, 85). Another significant partitioning of the voice with respect to bodily location is that the head tone appears in discussion of the technique of "unfinished singers" with respect to "belting," a characteristic of singers who have not been formally trained or those who cross over between vernacular and classical traditions. Popular performers often engage in pushing the voice beyond what is considered its "normal" ranges of volume, projecting the image of the whole body as the heart of the voice.[26] The definition of belting for some is vocal production that comes from below the neck. The "cultured" style of singing emphasizes restraint and polish that downplays the chest and vocal folds.

Vocal pedagogues are divided on effects of finished and "rough" technique. Some question the dictum that popular vocal techniques are dangerous and therefore require discouragement.[27] Others lament the discouragement and unfair evaluation that "belters" receive from teachers trained in the classical tradition.[28] Some of the discouragement for belting comes not from the perception of, strictly speaking, injury, but from visual judgments of the singer's motions that mark rhythm and emotion, overlapping patterns of race and class.²⁹ Decisions about the style of singing that qualifies as preprofessional in undergraduate schools of music are part of a historical knot of practices that have not only entwined singing styles with notions of racial "origins" of the singer but are increasingly competitive.

> The [recent] shift toward requisite early private training in voice indicates that more so than in the past, the possibility of becoming a public school vocal music teacher . . . hinges on early access to privilege and affluence, at least if a student's goal is to attend a relatively prestigious school of music. Affluence has long affected access to private colleges . . . but the access conundrum is increasingly evident at public institutions . . . Because the affluence gap has a racial pattern, the current admissions process becomes a racially discriminatory practice . . . One of the unchanged aspects of the auditions and the undergraduate program they precede is the emphasis on music from the European/American high art *bel canto* tradition. (Koza 2008, 147–48)

In public schools, there are sometimes different types of musical groups, such as gospel choirs, swing choirs, and so on. However, pedagogical texts, in general, prescribe the principles of voice production from the cultivated tradition with which music teachers are, very often, solely familiar with.³⁰ As Thomas Regelski notes in a recent text, the traditional training of the music teacher is the most familiar model. It produces a sort of half-conservatory atmosphere in which mastering singing techniques and repertoire sidestep the interests of students with different dispositional inclinations.³¹

So far, I have described the historical connection between the double gestures of inclusion and rejection that have been folded into the values and practices of the public school music curriculum. Those values are expressed through techniques to produce vocal sound that have been affiliated with whiteness and racialist premises in music in the United States. In order to expand this analysis to instruction in listening, I now turn to texts on music appreciation that allocated a specific role for indigenous music as an educative experience.

Folk Music in Public Schools

Music appreciation gestured toward the inclusion of what Dewey called "cultures of the world." For music educators, this was not an equal partitioning of culture but a rubric that classified cultures as "other." Surette and Mason's pedagogical manual, used widely by music educators between 1907 and 1930, took its main thesis from the idea that music differs by level of complexity corresponding to the racial nature of the societies that give rise to them.³²

Following early twentieth-century teacher manuals,³³ published by the main supplier of phonographs and records to public schools, chiefly the Columbia Grafonola and The Victor Talking Machine companies, we get an overview of how the double gestures of attraction and repugnance traveled as music appreciation.

The Victor Talking Machine Company had realized a huge opportunity in selling the schools material related to indigenous American music. Production expenses were minimal as performances were recorded, in most cases, without fees and without royalties. Sales and distribution were nationwide. For the Victor educational directors and the teachers who followed their curricular outlines, Negro and Indian songs symbolized a musical "past" of the human race that educated listeners should know about.[34] One Victor manual shows that Blackfoot Indian emotions were to be given a pedagogical role in contrast to the inscrutable stereotypes of Native American songs in an earlier era.[35] In the 1923 Victor manual, the oriental primitive is artfully showing thought and feeling: "Among primitive people and in the Orient, percussive effects are used skillfully to portray differences in thought and emotion: Grass Dance [from the Blackfoot Indian Tribe] record no. 17611" (Rhetts 1923, 6). But the space allotted was slight. Silver Burdett's *Music Hour* textbook series and C. C. Birchard's music appreciation texts for teachers assured that Northern European folk dance would model musical comportment at an early stage of development. Young children danced according to the beginnings of (white) racial experiences. The teacher was to regulate larger muscular movements and provide the "right" context and outlet for enlisting the child's control of mind and muscle.[36] Prescriptions for dance underwent similar assessments as to whether they communicated "low" practices and untutored comportment. Similar to the "grain" of the voice, the dancer approached sublimity or degradation.

Frances Elliott Clark, a major figure in music appreciation in the public schools, wrote that dance movements for young children recapitulated Western music's development from the "cries of savage warriors" to the "authentic and cultural" dances of the English, Danish, Swedish, and others. The primal, "first singing," she wrote, "was always accompanied by rhythmic bodily expression"(Clark 1913, 15). Of popular dance, she wrote, "These made up dances are set to music either wholly incongruous or some modern ragtime or other composition wholly unworthy and utterly devoid of educational value" (ibid.). Just as classical listening would save the child from "the dancing mad of the world," remarked a well-known educationist, it would also enable the music teacher to control dancers and to disallow ragtime, the tango, and "musical comedies" in school (Barnes 1915, 33–39). Some educators warned that the new dancing styles entrained aimlessness. "We distinguish between a purposive man and a drifter. We do not admire the latter who passively accepts what mere chance brings to him. We admire the man who purposes and plans to meet situations of moral responsibility. Such a person we say presents the ideals of democratic citizenship . . . [music listening] materials . . . acquaint children with the achievements of some of the greatest minds of the race" (Mohler 1924, 262). These two educators, Earl Barnes and Louis Mohler, were also involved in the progressive movement in music education that called for sensitivity to the child's experience. Educators recommended the sequenced logic of child development as a way to teach the history of primitive man's ascent to civilized adult. Anxieties about the savage persisted, but in the progressive view, notions of savage simplicity embedded the hope that "primitive" music furnished important knowledge necessary to the recognition of advanced civilization.

The philosophical grounding for teaching "culture of the ages" in music borrows from the broad summaries and interpretations of John Dewey's work.[37] What stood for Dewey's educational philosophy, as it was imported into the music curriculum, was alternately called the progressive method and the project method.[38] From a strategic point of view, music education's use of Dewey as an authority would put it on a par with other school subjects, connecting musical activity with being a member of a community in progress. As with the Blackfoot example, dramatization and dancing to drumming would make primitive rhythm not just a musical fact, but a part of the child's consciousness of the relation between culture and productive social action.

The "social" in music was linked to civic planning in some cities where parks, monuments, bridges, universities, hospitals, and concerts were financed by government and by philanthropy for the use of "everyone." Industry had a role in constructing the social realm as well. Steel manufacturers in Indiana, dependent on attracting African American migration to the North, treated the civic environment as a resource. Settlement houses, such as Stewart House in Gary, Indiana, spearheaded improvement in housing, schools, health, recreation, and employment, and also offered tuition-free music lessons for African Americans in the early decades of the twentieth century.[39] An all-black managed radio station in Chicago began, against all precedent, in the early twentieth century. The racial picture was not all in one direction and there was a rejection of social Darwinism among many cultural elites. Although segregation, racial violence, and political tensions persisted, many enterprises, too broad in scope to describe here, were committed to music education as a pragmatic means to regenerate and unify the social sphere that made the new citizen. Music would forge a commonality, like a shared language, to unify and democratize culture. "This [unified language of shared social concern] is the breaking down of barriers and rigid separation" in American society, wrote Dewey.[40] However, restrictions of African American participation remained the rule in most public concerts, lectures, and other venues, including most settlement houses.[41]

The paradox of social progressivism in music appreciation was that it remained tied to a view, however inclusive, that spoke to the premise of racial formation in music. This ambivalent acceptance was embedded in the Victor materials, for example, that recommended the dances of folk culture for renewal and recovery from an overly rational form of mental exertion in other academic activities.[42] However, if the body committed itself to too much exposure to "primitive" rhythmic and vocal expressions, it would produce primitive character. "Not only do Feelings constitute the inferior tracts of consciousness, but Feelings are in all cases the materials out of which, in the superior tracts of consciousness, Intellect evolved" (Spencer 1855/1899, 192).

This hierarchical scaling of intellect versus feeling coded the timeline for the listener's life span—the infant begins with "no higher rationality than a dog that recognizes its own name." The Anglo child embarks on a succession of experiences that eventually, with culture, "pass into human rationality" (ibid.). "There is an immense difference in abstractness between the reasonings of the aboriginal races who peopled Britain and the Bacons and Newtons ... There are records showing that the advance ... has taken place by slow steps ... the rational faculty of the cultivated European is essentially different from that of a savage or a child ... [such modes] of thought will become cumulative in generation after generation ... thus it happens that faculties, as of music, which scarcely exist in some inferior human races, become congenital in superior ones" (ibid., 461–71).

Spencer's work was, in many ways, consistent with other social or philosophical views on the physiological and psychological progress of the white race. The drift of their teaching was to caution against strenuous exercise. Bain reasoned that strenuous activity left the individual depleted and sickened, a consequence that Bain terms, "one of the most distressing that human nature is subject to." "The chase, the dance ... are prized for their stimulating character ... the rites of Bacchus and Demeter, yielding an enjoyment of the most intense and violent character ... this taste is often brought on among the Orientals ... by rapid dancing and music" (Bain 1874, 88–89). Comparisons to Bacchus, god of wine, and the fertility goddess, Demeter, were familiar points of reference for sexual licentiousness and the "uncivilized" impulses connected to ecstatic religion, supernatural spectacles, conjuring of spirits, and the physical exertion and physiognomy of irrational and primitive peoples.[43] Lack of discipline and impulsiveness were associated with popular and folk modes of dance in which the telltale exertion of muscle, mouth, and sexual organs bespoke images of excitation and exertion. One well-known figure in school music publishing used Alexander Bain as a reference to describe two kinds of musical listeners— the physically minded man versus the spiritual minded.[44] Taken together in the curriculum's "recapitulating" (Barnes 1915) pedagogy, they mirror the double gestures of admiration and repulsion.

Developmental schemas and psychoacoustic research exercised authority through various academic channels, methods texts, and the music supervisors' national conferences. Yet, as one reads across these documents, one sees a significant attempt by American music educators, especially Anne Faulkner Oberndorfer, to transcend stark dichotomies of the evolutionist language, inserting the hope that the primitive or civilized narrative was not the end point but the beginning of an era when one could see the beauty in both. As a feature of the musical or social evangelism of the times, particularly in the attempt to reach "all" children and families, the hope of transcending social boundaries was a response to the segmentation of urban life.[45] If the hallmark of nineteenth-century British natural philosophy was to chart the evolving progress of music from the ancient (primitive) to the modern and its perfect harmonies (Spadafora 1990, 179), American musical elites and educators adapted their schemas to form a contrasting, pragmatic outlook for music education that would create bonds between the cultivated community of listeners and those left "behind."[46]

Rhythm was a central feature of that message. Like the pedagogic exercises of Emile Jaques-Dalcroze, the rhythmic stresses taught in elementary school music classrooms would affect the listener's ability to bring the power of thought and action together. "Rhythm finds resonance in the whole organism. It is not a matter of the ear of the finger only; it is a matter of the two fundamental powers of life, namely, knowing and acting" (Seashore 1938, 143).

An article in the *Music Supervisor's National Journal* of 1915 characterizes music teachers as pastoral figures who are responsible for the progress of the race. Child centeredness in teacher manuals of the 1920s elaborated a full system of stepwise rhythmic exercises (action and a language) designed to bring the body's childlike impulses under control, but also to experience, and thus learn, the direction of development one was expected to follow.[47] Curriculum models such as Frances Elliott Clark's, in one of the first issues of *School Music Monthly*, provided an early guide to music appreciation for high school teachers. Her outline delineated "early" stages of music in which "Negro spirituals" appeared under the rubric of "foreign music" (Clark 1901, 5–7). Later, when Clark became educational director for the Victor Company, she wrote, "Music . . . must be presented . . . for the developing power of the child and . . . follow in some degree the stages of progress of the race using music as a means of expression" (Victor Talking Machine Company 1923, 67). Edith Rhetts's *Outlines of a Brief Study of Music Appreciation for High Schools*, published the same year, positioned Negro spirituals separately from "American" music. Anne Faulkner Oberndorfer's *What We Hear in Music*, written as a high school music appreciation curriculum series (eleven editions appearing between 1913 and 1943), includes spirituals under the rubric of "primitive races" (Oberndorfer 1943). However, the largest shift represented in her texts, in comparison to earlier listening guides, was the attention focused on Negro spirituals as art. The songs are not treated fragmentally, as similar artifacts in Surette and Mason's text, but given a place in the Oberndorfer text as a lesson in music appreciation. The aim was to make the spiritual a different, simpler aesthetic experience that would add to the appreciation of more complex classical works.

On the one hand, music education's rescripting of the "'savage" in *What We Hear in Music* was inclusive; on the other hand, it constituted a policing of how music should be heard and performed.[48] Many music educators, especially those familiar with body culture regimens and research on rhythm,[49] shared democratic visions that embodied a kind of Christian charity to "empower the United States to convert the world."[50] Yet, the paradox of this vision was that its transcendental values operated as gatekeepers.

The texts used by teachers to organize music appreciation lessons were comprised of comparative experiences with everything from folk music to opera. This sometimes involved listening to folk music as an expressive art, while paradoxically rejecting gestural qualities that are inherent to its meaning.

Psychoacoustics and the Double Gestures of Research and Teaching

In the early 1900s, the science that played a large role in the double gesturing came out of the overlapping disciplines of psychometrics and acoustics, forming "psychoacoustics." Equipped with more accurate phonographs and graphing techniques, researchers simultaneously recorded, measured, and graphed qualities in a particular sound event. These would provide tangible "proof" of physical and mental effects that were, only decades ago, intuitively grasped as impressions. While early psychometric research was marginal, at best, to music teaching, and its scope never approached that of testing in reading and mathematics, the possibilities of a science of music bolstered support for the construction of norms for musical ability. As noted earlier in this chapter, Carl Seashore, whose work was pivotal in attempting to align music instruction with science, attained a position of authority in music education through the cataloguing of abilities.[51] His work was used in recommending particular students for ensembles in the public schools and in some institutions; his tests were adjuncts for admission to postsecondary institutions. As a positivist method in music research, Seashore's studies became models for investigating musical perception.[52]

Seashore's chapter on Negro song in *Psychology of Music* provides an opportunity to observe the operation of the double gestures of acceptance and rejection in his review and comments on a study by Milton Metfessel.[53] In some sense, Seashore's commentary on that study is a music appreciation lesson with double inscriptions that compliment Negro singers, but reject it as substandard when compared to the finished quality of professional singers in the operatic tradition. Phonophotographic technology, as it was called, illuminated minute qualities of the performer *and* the listener. With this instrument, the listener was an ear and an eye together, focusing on the graphing of pitch and duration in simultaneous optical and audible dimensions. As Seashore explained, being able to "see" what is heard enables the listener to "gain a very rich basis for the comparison of singers and for a deeper realization of what any one singer actually accomplishes" (Seashore 1938, 284).[54] According to Seashore, Metfessel calibrated pitch and temporal dimension with phonetic translations of the African American dialect the singer used. The process added to the range of qualities that were understood as "racial" and further objectified what was heard as "primitive" as scientific fact.[55] "While this [Hampton Quartet] bass singer has appeared before learned audiences and thrilled musicians, he is still ignorant and sings by his primitive impulses with a most charming abandon. He was so lazy that it was difficult to keep him awake for the recording. The words of the song seemed appropriate to his character" (ibid., 352).

Seashore's ambivalent, to say the least, appreciation of the singer involved verification of the primitive as art through a technological measurement of its expressive qualities. The data establishes the primitive as the beginning of a developmental ladder of musical types, but it also produced categories to compare the "primitive" and the "complex" that revived older divisions between the two poles through scientific renderings of "Negro pattern" and "Negro license."[56] The ethnic differences referred to elements of wild singing, but also to the character of the singer: the lazy, the indulgent, and the licentious.

The research categories for comparison of sound were imported from conservatory practice: pitch precision, tone control, and diction, offered within a community of scholars and scientists for revelation of what made the songs artful when they were, formerly, either crude or incomprehensible.[57] The findings provide what appears as a scientifically verifiable set of characteristics for the spiritual and other songs that push against supernatural inspiration (conjuring, casting spells, and so on). Within this semigenetic framework, singing was wholly intuitive, without discipline, and without tradition in the sense that Western music had a forward-moving tradition. Yet, despite scientific rationales, the Metfessel or Seashore study funneled its science through familiar, stereotypical rubrics, activating the racial registers it had set out to exceed by hearing the songs as "art."

Seashore believed that the purpose of his work (and others in psychoacoustics) was to make to evident the fine distinctions between artists and musicians that were previously "mere" opinion.[58] In this respect, he was prescient in that the expansion of music education in postsecondary education would, as in other fields, align with experimental science in attempting to give solidity to teaching methods and precepts that had been backed solely by tradition and aesthetic preference.

With regard to democratic music education, Seashore believed it was wasteful. Currently under investigation is the extent to which he had aimed to use his research for eugenic purposes.[59] In terms of his interest in experimental methods, Seashore would have been at the center of the progressive program that would join science to art had it not been for the element of social Darwinism and eugenics in his work that others, such as Dewey, had determined ought to be left in the nineteenth century.[60]

"Hot Rhythm" and "Black" Music

The persistence of racial, religious, ethnic, and other social partitions dogged efforts for social unity and détente. As race and class were sometimes associated with ecstatic religious practices linked to gospel singing and communication with spirits, suspicion about the reason of communicants put them on the margin of acceptability.[61] Paradoxically, technology such as the phonograph and radio made it possible to enlarge the methods and demographic reach of supernatural communication. Some believers, for example, used telephone technology for spirit-talk through what was called the psychophone. As Leigh Schmidt remarks in his book, *Hearing Things: Religion, Illusion, and the American Enlightenment*, voices heard and projected on such "phones" took up "the double fate of the Enlightenment; they were not only signs of its failure, but also of its triumph" (Schmidt 2000, 244). With respect to vocal authority, another practice that eluded the universal reason of enlightenment was ventriloquism. I will take up the history of that practice in a later section of this chapter. I now turn to the way in which a notion of "hot" African rhythm was ventriloquized to speak for white difference.

For many, the mass immigration of African Americans to northern cities had regenerated anxieties about jazz and ragtime's effects on health and civic conduct.[62] The threat of social corruption through use of radio technology created a "jazz uproar" and became an obsession of some civic organizations and music teachers.[63] Carl Jung describes "African" rhythm on the American scene as insidious and ineffective with respect to its "primitive motility" (Jung 1964, 509). Similarly, recordings on the Black Victor label spread the idea of "hot rhythm" as a characteristic of African American sensuality. Historians of the African diaspora have argued that syncopated rhythmic patterns are not simply imports or retentions of "African" rhythms, but have arisen in the context of musical and social interaction that is neither black nor white, yet, for several centuries up to the present, music has been divided into black and white camps.[64] "As 'hot rhythm' American black music assumes the primary racial figure . . . That is not to say that dynamic percussive performance did not exist in eighteenth-century North American slavery . . . To these performances we should avoid casually assigning the designation 'rhythm' [that would be] inconsistent with the broad range of African and African-American performance concepts" (Radano 2003, 103).

Medical fictions about physiological differences are also related to the perception of Black rhythm. Social evangelism and medical science were combined to describe the affect and mood of jazz listeners as boredom and listlessness.[65] Educators decried ragtime dances that put the full body into motion, or what one educator called "gyration"; others were concerned for the mental health of young dancers.[66] Some hoped that scientific research would also lift music education's status to a level comparable to the science and mathematics curriculum.[67] While this has not happened, experimental psychology does continue to produce a large taxonomical grid for comparisons between one person's bodily response, vocal resonance, pitch accuracy, dialect, accentuation, personality, and so on.[68]

The hunt for musical or cultural differences underpins contemporary, experimental studies of race in relation to musical performances.[69] Some researchers argue that perceptions of race, either from visual cues or vocalization, are active factors in the judgments of teachers' criticisms as well. One study, in particular, took notice of the correlation of high mistrust of singing instruction among black female high school students in predominantly white institutions when their own singing dispositions were classified as gospel-style and African American.[70] What is noteworthy here is not the findings themselves, as they were inconclusive, but the role of research in the creation of racial categories. As one historian puts it, "music's place in the construction of difference traces a long history" (Radano 2003, 8). I now turn to ventriloquism and the treatment of (dis)embodied voices as another important link between the cultivated aesthetic tradition and the double gestures of attraction and repulsion, and the operation of racial logic in music teaching.

Ventriloquism

Ventriloquism is the practice of using the voice in such a way as to make it appear that someone other than the speaker himself is the one speaking. The ventriloquistic voice received its Latin name, "vent," or "stomach" speech, from the ancient belief that disembodied voices would emanate from the lower body of the prophet or seer. Eventually, the ventriloquist's ability to detach a voice from his body and "throw" it elsewhere or into an imaginary figure, incurred the negative connotation of a corrupt voice, one that is likely to be an imposter with evil intent. In short, the voice from below, in Christian times, was tied to the devil or heretic who represented false prophecies, lack of faith, and license to commit immoral acts. These pejorative terms overlap with the nineteenth-century, anecdotal references to Negro singing in which "the white observer identified a (feminized) object ... whose very existence as 'contagion' could inject the sanctity of the white body" (Radano 2003, 184).

In this sense, the history of Western voice and reason has an important connection to music education through the latter's pedagogical investment in the upper body. The valuation of "high" vocal art was linked to the racialist sorting of white versus black musical types. While ventriloquism itself has come to be associated with comedy and entertainment, music education absorbed the cultural history that enveloped it, organizing its pedagogy to assert control over the lower body and its "low" practices. Historically, public music instruction followed the Protestant churches' strictures, redoubling ancient suspicions about "tricks" of the voice, such as the simulation of God or the Devil.[71] However, what is of equal historical significance is the way in which, in the American setting, the voice from below acquired the specific racial polarities of black and white.

In addition, school music instruction and ventriloquism share the racialist dimensions of a colonialist past that are bundled together in negative representations of "black" music. The school subject promised to stabilize voice and create a national standard for its regulation. From the Classical Age to the European enlightenment, prophetic traditions that employed tricks of the voice continued to strengthen the circuit between speech and nonreason.

With respect to the template generating images of the worthy school singer and listener, the lower vocal type resides in so-called black music and serves as an alternately savage and mystically inspired expression, continually ventriloquized in performances of blackface that mourn white civilization's lost freshness.[72] Agnes Fryberger's book on music appreciation for schools, for example, recommends songs "having inspiration from the negro" in the person of "faithful, old mammy" (Fryberger 1925, 69), one popular stereotype, among others, that alternately represents an animal intelligence and capacity for extreme loyalty. Fryberger's "mammy" is also legible as an embodiment of a female sage within a domesticated, subservient role. In this sense, her performance as a social inferior reaches back to the split between instinctual seer and the figure of reason where the male represents "the divine essence, while woman is accretion, accident, or excrescence" (Connor 2000, 55). In addition to race, gendered divisions of high and low adhere to the commonplace black or white dichotomy in which primordial rhythms were "proof" that African American expressions were associatively "feminine," possessed by spirits and by promiscuous urges.[73] This gathering together of diction, emotion, and rhythms of one speaker in blackface minstrelsy was a kind of ventriloquism reminiscent of the witchcraft that threatened established churches and government. Other developments from the male or female dichotomy of voice and traditions of truth telling facilitated a division of mental functioning: female (often portrayed as hysterical) and male (represented as the reasoning individual) or primeval versus civilized[74] coded into versions of racial differences. About the time music came to play a larger role in defining nationalities and cultures, as, for example, in Herder's mythos of Teutonism[75] and Spencer's *The Origins of Music*. By the 1920s, extroverted rhythmic displays became representative of gendered and racial essences, engaging "rhetorics of the musically foreign and familiar" (Radano 2003, 90) as well as the idea of black "hot rhythm," which projected a prodigious sexual capacity. Added to negative images of so-called black practices was the reputation of the black church's extroverted practices. W. E. B. Du Bois described some religious singing as "stamping, shrieking and shouting, the rushing to and fro and wild waving arms, the weeping and laughing" (Du Bois 1903/1989, 135). Sermonizing traditions differed widely between black and white Protestant churches. Some black ministers expected their congregations to respond to sermons by raising spiritual energy while others favored a comportment that was more reserved. The history of black churches includes records of missionaries from the North attempting to change the more extroverted practices of ecstatic worship. On occasion, Negro music was heard as a vehicle for subversive meanings and a socially aggressive agenda. Songs such as "My Lord's Gonna Move This Wicked Race" were received as threatening the civil order of Jim Crow laws and northern segregation.[76]

With respect to the constitution of the music curriculum, the *bel canto* voice versus the lower type, continued to assert the division between black and white along with racialized theories of mental capacities. Musical impulses and sexual impulses organized an international exchange around the subject of what whiteness *was not*.[77] The bulk of the commentary directed to music teachers in the early twentieth century reflected music education's preoccupation with these themes and the consequent uproar over jazz that filtered into music education from a range of social enterprises. "The people of the world have gone dancing mad and old and young are whirling, hesitating,[78] bunny hugging and gyrating in private drawing rooms, public halls, restaurants and church parlors" (Barnes 1915, 34). "Circulating unpredictably, wildly, black sounds and their accompanying texts crisscrossed the public sphere . . . that solidified rhythm's appearance as a natural, essential form. When J. P. Wickersham, a University of Pennsylvania, Professor of Romance language and literature decried the scourge of 'jazz thinking,' he ascribed a sonic quality to a racially determined, irrational condition that was influencing the perception and behavior of urban whites" (Radano 2003, 236). This idea of a racially determined irrational condition that threatened whiteness fueled the mission of rehabilitation through music appreciation. Internal, logical contradictions abounded. Just what music was influencing what ethnic group was difficult to calculate since there were "essential" differences to be reckoned with. Yet, fear also contained that peculiarly American hope that education and science[79] could make out some future triumph for civilization. "America is possessed of music derived from two primitive races . . . Our great national music comes from these sources . . . In the study of this primitive expression can be definitely traced the influence of white man's civilization on the Indian" (Oberndorfer 1928, 91). And, notwithstanding the marginal place it held, increasing news of the success of black vocalists abroad enlarged the attention that the spiritual would receive in the music curriculum. Through association with the struggle for equality, the spiritual became a voice for the nation, while its singers remained second-class citizens unable to perform in the mostly segregated concert halls of its capitol and other major cities.

The Negro Spiritual and Popular Acclaim

International recognition dramatized what was at stake for the spiritual in terms of national politics, social justice, and reform as issues that would come to play a role in school music in the Civil Rights Movement of later decades. However, the songs themselves continued to bear the stigma of primitive voices and the lesser, suspect traditions of "black" sermonizing and music. In this respect, the career of twentieth-century contralto, Marian Anderson (and, in different ways, the actor and singer Paul Robeson), illustrate the double gestures of repulsion and acceptance that greeted African American voices in a variety of institutional venues and among many musical publics.

My focus on Marian Anderson is to bring out the qualities of her performance that came to the fore at the apex of her career in 1939, when she had been denied permission to sing in Constitution Hall in Washington, D.C., because of a complicated series of rulings on segregation of the audience. The outcome was that her performance took place on the steps of the Lincoln Memorial. The spirituals were seen as exceptionally eloquent and expressive and the atmosphere and rhetoric surrounding many of her famous concert appearances was tumultuous, driven by the undeniable stature and grace of the black performer who had to withstand the rebuffs of racist practices in her own country. However, the songs she sang and her use of the low, dark register prompted the double response of estrangement and acceptance for the cultivated audience, heightening the modern loss of innocence projected onto "others."[80]

The struggle for racial equality obtruded in a large proportion of the concerts she gave in the United States, whether it was in exclusion from ensemble work with white artists, concert hall segregation, or hotel accommodations. Civic and national turmoil over her engagements made her voice a sign of its special power to transcend the limitations of the "black" body and the social conditions that brought it nearer to inclusion in some circumstance but, at the same time, sequestered it as "other."

Anderson became, unwillingly according to her biographers, a symbol of America's unfinished deliverance. Interior Secretary Harold Ickes remarked in his address to the audience before Anderson's first appearance at the Lincoln Memorial in 1939, "Genius, like justice, is blind" (Keiler

2000, 212). Yet, the scope (an audience of 70,000) and its location underlined how much the attribution of genius and justice unfolded from the double tracing of Anderson's own talent and its projected "mysterious" origin. First, her own blackness was to perform national expiation for the denial of permission for her to perform in Constitution Hall. Racial internment was to provide a transformative moment of an unspeakable legacy of racial oppression in spite of the fact that her presence at the event itself was also a reminder of her being a "different" human kind.

The Lincoln Memorial concert was something she had only reluctantly assented to. Undertaken in her name, it culminated an epic struggle with the Washington establishment and the policies of segregation. She felt herself to be a vehicle of conflict, according to her biographer, Andrew Keiler, one whose body found it difficult to deliver what was promised by her own appearance there. Anderson wrote later that when she "stood up to sing our National Anthem, I felt for a moment as if I were choking. For a desperate second I felt that the words, well as I know them, would not come" (Keiler 2000, 212). Ickes spoke before she sang. "For genius has touched with the tip of her wing this woman [Marian Anderson], who, had it not been for the great mind of Jefferson, if it had not been for the great heart of Lincoln, would not be able to stand among us as a free individual" (Keiler 2000, 212). The dramatic staging of this Washington event to music education is that it provides an example, writ large, of the compelling power of black music and its inextricable, for the most part, unconscious rejection of the actual condition of black experience. Ickes, we see, represents Anderson as already liberated when her actual circumstances were precarious and uncertain.[81]

Attempting to overcome an apparently insurmountable difference between the spiritual and other art forms, Anderson's struggle to "find a place" on the concert stage expressed the troubled history of a white audience that sought a spiritual rapture that was felt, racially, to be out of its reach, except through euphemizing black experience. This dynamic reaches back to the historical attributions of the spiritual as a form that was beyond comprehension, its material and formative efforts forgotten, it motives simple and pure. It also represents the failure of American culture to acknowledge what appear as unspoiled qualities in the spirituals as artifacts of a continuing, historical nightmare. Ickes's speech embraces Anderson's genius but under the provision of white rescue.[82]

The split of body from voice was not a specifically American invention, but it is a dualism that gets called into action when beneficent politicians and reformers ventriloquize supposedly "liberated" subalterns to speak for the country. If Ickes's speech is an example of this habitual dislocation of Negro song as a transcendent object apart from the body, it is plausible that Anderson was heard as the slave in mid-flight from Egypt, the very literal meaning of many of the songs. Walter White, famous as the architect of the

Lincoln Memorial, afforded this interpretation when he noted Anderson in the crush of the crowd after the concert.

> A black, slender black girl . . . Hers was not the face of one who had been the beneficiary of much education or opportunity. Her hands were particularly noticeable as she thrust them forward and upward, trying desperately, though she was some distance away from Miss Anderson, to touch the singer. They were hands that despite their youth had known only the dreary work of manual labor. Tears streamed down the girl's dark face. Her hat was askew, but in her eyes flamed hope bordering on ecstasy. Life which had been none too easy for her held out greater hope because one who was also colored and who, like herself, had known poverty, privation and prejudice, had by her genius, gone a long way . . . If Marian Anderson could do it, the girl's eyes seemed to say, then I can, too. (Keiler 2000, 213–14)

The events surrounding Anderson's Washington concert illustrate, on a national scale, how the double gestures of a modern, cosmopolitan reason, whose crux is to unite humanity, produce exceptions to its embrace. Ickes's remarks on freedom recapture this irony of a universal, cosmopolitan ethos that would claim a transcendental world of music at the same time as they enact the enclosure of a pointedly "black" voice. On the one hand, Anderson is not free to sing in Constitution Hall; on the other, she represents the equality she does not have, a kind of transcendence of the conditions of her own body and its music. Disembodiment of the singer recalls the ventriloquistic maneuver in which "black" voice speaks for what "white" voice has lost in its capacity for feeling and emotion. This dynamic is crucial to the historical suspicion of vocalization that deviated from the traditions of cultivated drama, opera, art song, hymn singing, and genteel comportment.

Music appreciation in public schools did appear to provide the hope of progress by including the spiritual. Yet, it also enfolded the abjection of "black" voice as part of a historical process that made it possible for school music appreciation lessons to become a way of reasoning about race and citizenship.

8
The Listening Body and the Power of the Good Ear

> Yeah! We wanna dance to that. Jus' let the strings class do that beat and keep it goin'. And with that the three girls launched the music class into hip-hop.
>
> —Ruth Gustafson, "Practicum Notes"

> Only rarely was a black student accepted by a white teacher, and in every case there was the risk of unpleasantness. Agnes Reifsnyder, an experienced voice teacher whom many found utterly free of prejudice and sympathetic to the needs of black students, felt constrained to ask Agnes Pitts, a talented young black contralto, to go around to the back door on Saturdays when she came for her lesson. Only in this way could Reifsnyder hope to keep her white students.
>
> —Allan Keiler, Marian Anderson

Formal musical training in the Western classical tradition was difficult for African Americans to obtain. Even while Marian Anderson had established a reputation as an outstanding singer early in her career, many of her auditions failed for logistical and racial reasons. Permission to fill out an application at one Philadelphia music school was refused outright. Friends and advisors pressed her to continue to seek formal education because if she hoped to enter the classical arena, her voice needed the particular aspects of musical refinement offered by study of the operatic repertoire. Having started with a voice that had been largely shaped through gospel choir experience in an African American church in Philadelphia, Anderson's sense of comfort with operatic pedagogy took her through trials with several teachers well into her twenties.[1]

In the last chapter, I discussed how the historical instability over authoritative speech and hearing focused attention on the upper body, especially the head, as the source of reason, hearing, and voice. Today, many encounter the difficulty of changing vocal techniques that have been formed in a close social network. These socially acquired aesthetic values present substantial obstacles for many black singers seeking admittance to music schools and music teacher training institutions.[2] As with the alleged differences between Negro singing and the *bel canto* tradition discussed in Chapter 7, the preferences for operatic technique demands different constraints on pitch, diction, emotion, facial expression, movement, and gesture.

Controversy about which skills are teachable and which are inherited continued in a musical version of nurture versus nature within and around music education.[3] This chapter focuses on the role of teachers' training, the psychological categorization of talent, and traditions of body discipline, with concluding thoughts on the values we think of as perpetual and universal.

Ear Training

Formal study in the cultivated tradition, whether for music teaching or performance, usually involved some kind of ear training. The courses were offered by Normal Schools and elite colleges and various other institutions that were, in the main, segregated. The widespread use of "solfège," or pitch training, a system that uses a "do, re, mi" scale, made pitch discrimination a routine part of musical professionalism.[4] Frederic Louis Ritter, a nineteenth-century European musicologist and teacher, was well known in the United States for his numerous contributions to musical periodicals and pedagogical texts. Ritter collaborated with distinguished conductors and other musicians on several educational s projects of national significance, and his *Musical Dictation* contributed to making writing down music an important element of ear training. One feature of his text offers a series of exercises that address listening skills. "I am convinced of the fact that this course of musical dictation, when closely and rationally connected with the general course of music education will also help to form more intelligent, more appreciative listeners" (Ritter 1887, 6). *Musical Dictation*, published in the 1880s, had begun to circulate at about the same time as W. S. B. Mathews published his lectures on music at the Chicago Exhibition. Public school teachers exposed to Mathews's normal school curriculum for music appreciation, along with the extensive offerings of ear training designed in similar fashion to Ritter's, represented a small but influential musical public oriented toward developing an ability in students that was an important aspect of having a "good ear." In the set of vignettes from my years as researcher in music classrooms, I noted, "Mr. H. stressed that the songs sung in class, whether straightforward ballads or round type, were to be performed the way they were written down (in various versions) ... Katrel, an African American fourth grader persisted in altering the form even though this sometimes brought a correction from the teacher" (Gustafson 1991, 9). I further noted that Shirley Heath's 1983 study of black school children indicates that it is common in working class culture to treat songs of all kinds as occasions for spontaneous alterations: "Often a formulaic phrase expresses an essential idea, but this phrase, if for building form, and as such is continually subject to change as individuals perform and create simultaneously" (Heath 1983, 211).

The outlook of most music teachers trained in musical dictation and the authority of a musical score is that accurate reproduction is of higher value than spontaneous alteration. As a practice honored more by omission of spontaneous singing than by formal rule, the "good ear" became relevant to music listening at every level of instruction. In Roy Dickinson Welch's *The Appreciation of Music* (1927), he notes that appreciation of music is only possible if one works at training the ear. Mr. H.'s instruction, quoted from my journal, was in line with this view, but it resulted in the frequent exclusion of musical practices that diverged from the songs as written. Follow-up interviews with Mr. H. revealed how anxious he was to have his students sing "in tune," as he felt that was the mark of a diligent music teacher. Like others in his district, he felt home background was the key to attracting the concentrated attention of his students and the key to teaching what was in the curriculum. As he remarked at one point, "I mean to expose the kids equally, the kids from different backgrounds, but as they get older you see more difference in them" (Gustafson 1991, 17).

A typical lesson in ear training for future music teachers in colleges and universities involved writing down the rhythm, pitch, and some aspects of harmony of musical fragments as they sung or played. Similar courses form part of the certification processes for music specialists in public schools.[5] While ear training itself is not a significant aspect of the public school curriculum, except possibly for advanced music theory classes in select high schools, they are an integral part of the school music teacher's "tool bag" with which she plans her lessons and evaluates students. In the classrooms I have observed, attention to musical detail, such as the rise and fall of pitch in a melody or reference to a rhythmic pattern, underpins the teacher's evaluation of her students' performances in class. This skill set is hyper-developed in the professional, classically trained musician and provides an atmosphere in which "methods (e.g., research, teaching) and techniques (e.g., musical performance) are valued for their own sake and, thus, without reflection on whether or not practical results or needs are well served" (Regelski 2004, 36).

This direction intensified, along with the field of music psychology, in the early twentieth century, as shown in Chapter 7's description of Negro song. The larger significance of the trend was that it aligned with the strong current of racialism operating in studies on hereditary genius and interest in eugenic programs[6] as well as, more generally, the categories of the civilized versus the barbaric embodied in different musical audiences. There are few teachers of music in public schools who are willing or able to break from this tradition, resulting in the lack of recognition for the acculturated musical interests of many African American students. Musical inequity, being closely allied to embodied values and identity, is taken personally.

Appreciation, Structural Analysis, and Rhythm

Another aspect of training the musician and teacher involves the recognition of musical forms (symphony, concerto, etc.) and their internal structure (sonata form, rondo, etc.). Teaching public school students how to label and listen to these structural elements instilled principles so that the individual could compare the structure of various pieces of music. "All good appreciation teaching rests upon *comparisons and discrimination*" (Clark 1929, 217; emphasis in original). This was framed as an important step in developing the mind that would be able to go from the particular to the whole, the individual to the group. Quoting Dewey, Calvin Cady, wrote, "Dewey says that music and art are as 'valuable and important as any other work in the schools, not only in the development of the integral moral and aesthetic nature, but also to . . . better develop the power of attention, the habit of observation and of consecutiveness of seeing part in relation to the whole'" (Cady 1908, 151). Similar intellectual exertions should guide the curricular choices of teachers. "[Music] chosen for children [to listen to] should embody the same fundamental principles of design that distinguish all great works of musical art. These principles demand . . . recognizable contour, and recurrent rhythms. They demand further a certain tasteful economy of material . . . and the artistic blending of unity and variety achieved by repetition and contrast. For without unity, the listening mind is bewildered; without variety, it is benumbed" (McConathy et al. 1930, 13). The physical conditions under which one could best undertake these tasks are described in a section of a teacher's manual called "The Courteous Listener":

> One of the tasks of the teacher [is to direct listening] . . . Music, to be properly appreciated must be listened to without interruptions or disturbing noises. In school a quiet room is necessary to the success of any music lesson, and particularly of the lesson in music appreciations. Pupils should be taught to eliminate every sound except the music of the lesson . . . They are supposed to remain quiet and give their entire attention to the music . . . Those who fail to do this are discourteous to the performance and discourteous to those of the audience who desire to enjoy [the music in] absolute quiet. (Giddings et al. 1926, 31–33)

One teacher's guide describes the common bad situation as one in which students "do not pay attention" and "disturb the other children." "Some music is best appreciated in silence" (McConathy et al. 1929a, 14). These prescriptions also appeared in connection with the picture of the troubled child who became legible as the truant and delinquent with poor ability to listen to music.[7] The teacher must also be vigilant with regard to popular music. "[The use] of 'trashy and worthless' music for teaching has a harmful effect . . . undermining the splendid in building up tastes and ideals for good music in the ears and minds of children" (Clark 1913, 14). "The standard of musical taste in the majority of people is doubtless far below what it might be . . . the phonograph [provides the occasion for] the most sensuous, sentimental and even vulgar records" (Briggs 1925, 7). The failure of radio[8] to restrict the nation's listening was also addressed: "Our forward-looking country is making large plans for the future. In this future lies not only the promise of greater things to come, but also certain grave dangers to cope with. The radio with its untold possibilities for good is not yet controlled. Therefore, much that masquerades under the name of music is unworthy to be so designated" (C. Adams 1929, 85). The vagrant listener is "unable to take a reasonable point of view about music" (Surette 1906, 112). He was the too-interested listener, one who lacked the capacity for aesthetic *disinterest*. Drawing irrelevant ideas into music, his superficial understanding led to untoward body movements. These were signs that the listener did not comprehend the music. "[In] the feelings and passions aroused by music there always coexists a strong physical agitation . . . music produces in young people [whose natural inclination is not controlled by social restraints] a twitching of the whole body and especially the feet . . . whoever glances around in an opera house will notice the ladies involuntarily beating time with their heads to any lively or taking tune" (Hanslick 1854/1957, 84). Differences in response were linked to the dichotomy between the exercise of reason and the overexercise imagination: "Those . . . who ground the beautiful in music on the feelings it excites . . . only indulge in speculation and flights of fancy. An interpretation of music based on the feelings cannot be acceptable either to art or science" (ibid., 86). Teacher manuals reflected these ideas and glossed them according to developmental categories. "It is well for every teacher to have in mind the typical or average tendencies of children of every age . . . the average child . . . has as his most marked physical characteristic the love of free movement which involves the whole body" (Giddings et al. 1926, 27). Several handbooks encourage the very young child to experience the free play of physical response in order to engage feeling. *The Music Hour*, a handbook for kindergarten and first grade teachers recommends that children use percussion instruments while they listen. "This implies the ability to enter freely into rhythmic interpretation" (McConathy et al. 1929b, 4). Action enables "impressions [to] be followed by expressions, or in other words that reactions should be encouraged to the impressions brought to the children" (McConathy et al. 1929a, 14). However, as the child proceeds through the higher levels, free rhythmic response is discouraged: "[The student pays] attention to the flow of the phrases and to recognize the phrase . . . [as well as to] recognize also rhythmic differences as expressions of different moods and ideas" (McConathy et al. 1929b, 4). "[Teaching is directed to] a consciousness of musical structure . . . Consciousness is developed through rhythm play . . . by playing rhythm sticks, blocks, etc, to mark the fundamental rhythms. This phase of appreciation is further developed in the detailed observation of phrase, motive and figure relationships" (McConathy et al. 1929a,

14). Similarly, the *School Music Handbook* presents a typology of reactions to listening and ranks them according to level of sophistication: the most elementary level is the natural, undirected type, followed by more organized reactions: awareness of music history, discrimination, and imitative reproduction of musical phrases.[9] Finally, for the older student, "the pleasure resulting from beauty" comes from silent attention to detail and comparison of "good music to bad" (Cundiff and Dykema 1927, 172). Another teacher's guide lists similar goals. "If we are to become really appreciative listeners, exerting our intellectual and discriminative powers, we need to direct our attention to the factors which, touched with the spark of genius, contribute to making great music. If we wish to become less puzzled, more intelligent, more discriminating, we must become conscious of those factors which make good music and which by their absence cause the poverty and the unsatisfactory quality in music that is poor" (Giddings et al. 1926, 25). The early phase of bodily accentuation and percussive exercises is a temporary plateau leading to an understanding of great musical works. Motion of the body must be left behind as the child grows.

The prevalence of rhythm as a theme for national temperaments has led Michael Golston to call its influence "incubational," in that it linked up musicology, music education, anthropology, psychology, and aesthetics in a system of reasoning about across these fields that outline one's heritage and future prospects. At the center of this nexus was the idea that rhythmic practices shaped the modern psyche and character and that one coded and decoded character through its appearance on the body or in the body.[10] With a universal basis rhythm, the "normal" pattern of a strong beat proceeding a weak beat was underwritten by developmental psychology. Thomas Bolton, an associate of G. Stanley Hall at Clark University asserted, "The most common measures that occur in music are [duple and triple groupings]. In what might be termed the natural system of accents, the first note in each measure receives a strong accent . . . the strong accent occurs at regular intervals" (Bolton 1894, 166). The Dalcroze system of eurhythmics offered a similar ordering of strong and weak beats, favored as a natural, biological predisposition. Given the breadth of the involvement on rhythm and human types, rhythmic patterns propel racial signs.[11] Rhythms were read as character and therefore both social and antisocial impulses were subject to, and produced, particular rhythmic patterns.[12] Jarring, irregular musical rhythms, cries, café sounds, and street music were subject, increasingly, to local "noise" ordinances and zoning restrictions.[13] School music provided an antidote to the hearing fatigue of the city dweller, but was also an antidote to the abnormal and the abject rhythms of dubious quasi citizens.

As described in the classroom vignettes, and the Prelude to this book, the estrangement of syncopated idioms, motions, and sensations are symptomatic of racialist concepts of rhythms. The most common exercises in marking rhythm were those that normalized one set of musical possibilities as standards against which other patterns were aberrations. However, I was also able to observe teaching that got more equitable results. These were lessons that were less structured. For example, one lesson used a march from a classical piece of music for children to march to as they wished. Some marchers deftly stepped and clapped on the so-called weak or counter beats, marching in counter relation to the majority of the class. There was no "correction" for marking the counter beats.

Developmental Psychology

The principles that most frequently order rhythmic response also classify listening habits with age-appropriate behavior, enclosing listeners as types—childlike, immature, or more fully advanced. With reference to this system, the education of the listener appeared to be most promising in youth of "good" backgrounds.'[14] Even though there were wavering distinctions between the interior and the exterior of the child—that is, between what was alterable and what fixed[15]—developmental psychology oriented the curriculum. Developmental abnormality added a new dimension of unworthiness for the voice, rhythmic gesturing, and dance—the same aberration attached to syncopation, unreasonable spiritual practices, suspicious ventriloquy, minstrelsy, and jazz. Jazz was the chief anomaly of the 1890 through 1930 period (Blesh and Janis, 1966), along with ragtime in which stress on counter or off-beats is its chief characteristic. Linked to nightlife and black music, ragtime was off limits for public school music teaching.[16] A medical expert in the 1920s, employed by the Philadelphia High School for Girls, described jazz as causing disease in students and society as a whole.[17]

Familiarity with the products of genius, it was thought, would improve the common man. Disposition, culture, in short, all the habits derived from deficient musical experience, were to be "overcome" by a democratic outlook in music appreciation teaching. Yet, ironically and unconsciously, the effort to remove distinctions and build a national concert-going public inserted more distinctions between listeners. Having discipline as a listener meant that one paid attention to classical music as a national duty.

> The supposition that people generally can be brought to understand and like great music is part of a general theory of art that places the great man as the apex of a pyramid the base of which is the so-called common people ... we believe music belongs not to a select few—an inner cult or circle of special devotees—but to the many. We think that all kinds of people ... cultivated and uncultivated ... are susceptible to its Influence. The average person not musically educated, will get little out of one performance of Beethoven's *Eroica* Symphony, but when it is played in detail, bit by bit, until he becomes thoroughly familiar with it, he sees the connection of one part with another, the relationships that exist; it begins to make sense to him. (Surette 1906, 111–13)

Through repeated exposure, the *practiced* listener could develop the capacity to come to musical genius with the same sense as that of the cultivated person. "Repetition, repetition, again and again is an essential in teaching ... [the phonograph provides] a lasting tonal reference library of real music" (Clark 1929, 217). An increasing number of distinctions amongst listeners[18] produced childlike listeners and the deeper appreciators of the self-cultivated variety who had no need to gesture to the music. "Those adults who have never really known music will be stirred by the martial air of a street band and will respond by a tapping of the foot or a swaying of the body. The fact that many excellent dancers have no 'ear' for music is another evidence of the unconscious appeal made by the rhythm and the rhythm alone. Even the smallest child will respond unconsciously to the rhythmic pulsation of *Air De Ballet*" (Victor Taking Machine Company 1923, 21). Thomas Tapper, music supervisor in the public schools of Boston, stressed home life as the sign of potential and that "every pupil must be studied as a unique and individual center of future citizenship" (Tapper 1914, 33). "Our aim even with the very least who come to us, is to instill good habits of study, strict attention to the responsibility involved in becoming a student, love for music and reverence for the better things of life" (Tapper 1914, 129). Speaking of settlement houses giving lessons free of charge, Tapper hoped that the charitable provision of music lessons and concerts would benefit citizen-making: "Wherever schools like this spring up in the congested quarters of our large cities they immediately begin to mold character" (Tapper 1914, 48).

In an address to the Music Supervisors' National Conference in 1924, Louis Mohler stated that the individual "can rise no higher than his aspiration, and his aspiration can be no higher than his appreciation [of music]" (Mohler 1924, 262). G. Stanley Hall was often cited as an authority for teaching music appreciation during adolescence when music's intervention was purported to be most effective. His theories were adapted to music in an article on psychoacoustics by one of his colleagues at Clark University who wrote that a high-spirited affect induced by music was characteristic of primitives: "An assemblage of uncivilized people may be wrought by the mere rhythm of drums ... With some people these movements tend to increase in force until the whole body becomes involved and moves with the rhythm ... This known emotional influence stimulates ... the sexual passions ... The accents in the rhythm have the effect of summated stimuli and may even increase to a state of ecstasy and catalepsy" (Bolton 1894, 163). The twentieth-century narrative of musical fortune-telling hinged on the stages of growth each child passed through on her way to adulthood. An advanced appreciation of music signaled progress in the "natural" development of mind, just as it was the case for other school subjects. "[Music] serves to draw out the best powers of the developing child's mind ... relating mental images to reality and correlating and assimilating knowledge gained into the general plan of life" (Victor Talking Machine Company 1923, 19–23).

The good ear is recognizable as the listener who paid attention to rhythmic detail but made no indication of it. Similar to the recognition of merit in the early decades of singing instruction, the good listener gained his reputation largely through his comportment and physical appearance.[19] Normal listening in schoolrooms took the form of remaining quiet. The student who makes noise or falls asleep becomes the abnormal listener. Beliefs about mental attention and innate musical dispositions with respect to class, race, and nationality put many human types beyond the reach of music appreciation lessons.[20]

A student's ability to focus on the attributes of music and appear to concentrate functioned as a distinguishing quality that would propel him or sideline him in relation to his projected biography as a participant in the democratic society. Referring to rhythm's anticipation of continuity, Dewey recognized that listening to music was a manifestation of thought that made order out of randomness and chaos. This connection of music with self-discipline and modern self-governance reaches back, in the American Romantic tradition, to John S. Dwight's music journal[21] and Jefferson's rhythmic marking of the text of the Declaration of Independence.[22] Reading out loud for Jefferson's circle produced a rhythmic synchrony with his audience, just as for the music critic, Dwight, rhythm lent a harmonizing influence to the daily routines of life. Carrying on this tradition in the twentieth century, music appreciation was central to the synchronous harmony of communities across social classes. "Rhythm is rationality among qualities," wrote Dewey (1934, 169). However, the idea that "black" character was built on a different rhythmic idiom limited black participation in the vision of a harmonious democracy.

A Science of Rhythm and Attention

Psychoacoustics combines the physical aspects of music (acoustics) with the concept of an interior apparatus for perception of acoustics in the human being. In the early twentieth century, a growing body of literature on psychoacoustics gauged the effects of music, especially rhythm, on the body in order to observe subjects' mental makeup. The premise of psychoacoustics experiments was that observation of the mind was possible through reading "external" movement such as dance, inattentiveness, musical taste, excitation measure by pulse, and foot tapping.[23]

Studying musical effects endowed the field of music education a scientific aura. Earl Barnes, educational psychologist, sums up the then new attitude and union of music methods with psychology. "I am to speak to you today as an educational scientist and not as a musician. I am to deal with diagnosis and general education conditions. You are experts; and the making of prescriptions is in your hands. I am like the father bringing his children to the physician and lingering to explain their heredity and general family conditions under which they live" (Barnes 1915, 36). Continuing in this vein, the speaker outlined how the muscular system, over the life span, and with proper training, refines itself and imprints valuable musical subtleties on the mind. Similar to the idea of music's civilizing effects, Barnes's approach was to use music appreciation lessons to achieve a perfection of body, spirit, and attention. In his terms, the "prescription" given by the music teacher was palliative. The authors of the text, *Music Appreciation in the Classroom*, distinguished several types of attention: "Attention is of three kinds: passive attention . . . active attention . . . and automatic active attention" (Giddings et al. 1926, 31). The teacher is to engage the student's active attention through directed listening. "It is imperative that [attention] should be held for reasonably long periods if any real educational work is to be done" (Giddings et al. 1926, 27). As Jonathan Crary points out with respect to attention in the modern age,[24] fragmented attention was recognized as negative in the urban environment but, at the same time, it was also a biological and psychological phenomenon worthy of scientific exploration. Manuals such as *The Music Hour* for elementary school teachers reflect this preoccupation and recommend *silence* as a condition favorable to the listener. "We may move all we like provided no one hears us. Soft needles on the phonograph are also helpful in securing quiet listening. When the teacher plays a record it is important that she listen quietly as a model for the class" (Giddings et al. 1926, 32). Working against the teacher's efforts was a distracted listener, characterized alternately as a drifter, a name-caller, a gang-joiner, a juvenile offender, a joke-maker, a potential religious fanatic, one suffering from acute emotional stress and one intensely interested in sex.[25] The connection between the quality of attention and talent was assumed. Carl Seashore claimed that a full 10 percent of the children he tested for "hereditary" musical talent had none: "[They were] totally unfit for musical appreciation or production and . . . for that reason, should be excused from the school exercises which are not adapted to them" (Seashore 1916, 11). He described the musically beneficial home to be one in which the radio is not continuously playing; the mother is always present, soft-spoken, though animated, and always patient to answer her child's questions.[26] Sophie Gibling, an early twentieth-century musicologist, classified listeners as one of seven types; nearly half were simply passive "hearers" of music, and a few were "active listeners" (Gibling 1917, 385–88). "The joy of having a perfect listener, sensitively sympathetic and responsive . . . is rare . . . the power to listen well depends upon the quality of a man's personality; on his character, if you please, and on his mental make-up . . . a religious mood descends upon the ideal listener like an enfolding mantle. And the central quality of his listening is a great silence—the rich and wondrous silence" (Gibling 1917, 385–89). Experimental psychology used measures of bodily function and motion to gauge the quality of attention in the listener. "It is more than a figure of speech when one says we search among sounds. This hearkening search is a very observably a bodily activity . . . If obeying the drift of physiology we understand by attention nothing mystical, but a bodily disposition" (Ernst Mach, as cited by W. James 1890, 436). Similar studies attempted to show that an introspective demeanor and character, in sum, a kind of superior attention, *could* be acquired through the habit of listening to classical music:

> In listening to the unfamiliar classical music there is distinctly more tendency to lower the head, to avert the gaze, to assume a slightly puzzled, uncomprehending expression. There is also considerably less tendency towards smiling lines about the mouth. [After repetition of the classical selections] there is a greater erectness of posture, the greater directness of gaze and other subtler evidences of interest are definitely in favor of the classical records. So far as the photographic evidence goes, it tends to show that familiarization with classical music produces an attitude favourable to the best type of morale, whereas familiarization with jazz makes for a listless attitude . . . bored listlessness. (Gilliland and Moore 1927/1999, 219)

The effects of music and the changeable parameters of heritable attention were part of the atmosphere in which music appreciation took hold. These scientific ideas buttressed the figure of the noble listener against the images of a distracted, lethargic musical public associated with popular music. Psychoacoustics furnished evidence that attention was keenest when listening to classical music. Slight degrees of muscular tension were understood to be normal, but the excessive grip registered by electric nodes indicated too much tension and excitement, the source of which was overexciting rhythms.[27]

The search for the bodily disposition that would synchronize with the national body involved surveillance.[28] In schools, it required attracting the attention of children, disengaging them from either overexcitement or boredom, and directing them to ordered contemplation. In this respect, rhythmic movement stressing "fundamental beats" was encouraged as a means of holding students to task: "The natural appeal of strong rhythm . . . offers the skillful teacher [of young children] an invaluable resource for developing [further] training" (Giddings et al. 1926, 27). Stressing the "strong" beat's power to attract attention, an issue of the Victor Talking Machine Company's manual suggests alternating a three-beat pattern with a two-beat sequence to keep children alert.[29] For older children and adults, listening to the strong beat patterns was a byword. Waltzes, folk reels, marches, and polkas were used, but each phase of rhythmic difficulty also stressed selections from "great" composers: Grieg, Schubert, Beethoven, Brahms, Tchaikovsky, Elgar, Mozart, Gluck, and others, where these patterns could be found. Marks on the chalkboard represented strong beats in long lines, weaker beats in short lines, and, as noted earlier, ragtime and dances such as the tango were not represented.[30] Teachers held attention without descending to a level of "cheapening" the lesson through "exciting" actions.[31] "If then, through music we can secure that rapt attention . . . we have a real educational power . . . as music [stands] in the list of human needs—food, raiment, shelter," wrote Clark (1923, 19).

For many music educators, the recognition of this citizen in the making came from the observation of "correct" patterns of motion found in the movement of the body to folk dances or marches.[32] The ability to discern differences amongst listeners through their bodily rhythmic responses bought out the "internal" character of the individual with respect to the quality of attention and, therefore, his superior reasoning power or his need for rescue. Rhythm, whether regular and controlled or syncopated and uncontrolled, stood for a transcendental identity that was receptive to external discipline but, at the same time, showed resistance according to the life habits and racial or ethnic inheritance of individuals. Once again, there was ambiguity in pinning down whether or not rhythm was culturally acquired or inborn, who was amenable to instruction, and who it was who remained beyond redemption.

Paradoxically, as a nearly inexhaustible literature on the recognition of merit in the music listener was produced during the early twentieth century, music appreciation's effects as social discipline were marginal. Relying on the sense of a national emergency in which civilization was at stake, music educators wrote that popular idioms continued to threaten society in spite of the physical reach of a national curriculum such as the Victor Talking Machine music appreciation materials. The education of the meritorious listener on a mass scale in these decades failed, but it did succeed in identifying those *incapable* of adhering to the self-cultivation of musical discipline. Thus, school music never developed along the lines of a systematic study of music, but, instead, foregrounded a set of comparisons that outline deviance. These judgments overlapped demographic identities as those ill equipped for citizenship. This is a recurrent pattern that makes its way to present day music standards intimately bound up with and dependent on keeping that "lower" form of music within hearing distance.

Arrested Development, Dangerous Music

The developmental framework for school music curricula associated with G. Stanley Hall also mapped demographic differences. "For Hall the availability of thinking in terms of superman and the robust mother enabled such imaginings to inhere in the inscription of the child and the prescription of the ideal school . . . Development in this rendition was posited as an inherited unfolding . . . based on an appeal to morphological ideal-types that today would be recognized as nationalities, races, sexualities and genders" (Baker 2001, 478). Using Hall's stage theory of development, many music teacher manuals and other publications characterized the adolescent, that is, the high school student, as in danger of succumbing to "summated stimuli" and, therefore, more likely to join a gang, fall victim to religious proselytizing, or the seductions of gangs and violence.

> Morally this [adolescence] is a period of transition from the naïve acceptance of family ties to the powerful bonds of gang loyalty, which begin to appear prominently at twelve years. The growing impatience of family restraint is shown in the frequency of running away and in the decrease in the number of boys selecting the father's occupation ... Emotional development tends to outstrip the intellectual advance ... There is by far more uncontrolled violence of emotion than at any other time of life, as records of juvenile offence clearly show ... There is great increase in ... the frequency of [religious] conversion experience. (Giddings 1926, 29)

Dance music also posed the danger of regression by adolescents to earlier developmental stages. "Music is the speech of the half-buried racial soul" (Barnes 1915, 34). Inferior music offered only primitive rhythms. "Children differ in their attitude toward music. Theoretically there are three types of children: those who are auditory-minded ... those who are visual-minded and those who are motor-minded" (Giddings 1926, 30). The subtext was that rhythmic movement (the outward sign of the motor-minded child) signaled an inferior mentality—the child drawn to jazz or ragtime, or even to some degree, various kinds of folk music, was considered developmentally stunted. The link between embodied musical responses and social merit formed the core of one address to music teachers at the 1916 Music Supervisors' National Conference.

> Two kinds of value ... Of primary importance are intrinsic values ... To society however, as a unitary group of mutually stimulating and interdependent individuals, the behavior of its individual members, using the word behavior to mean all forms of *bodily expression* which can affect in any way the conscious experience of others ... are of *most importance*. One may have the most delightful experiences, may think the most sublime thoughts, but unless these somehow affect his behavior so as to influence ... the lives of others, such experiences ... are without social value. [The citizen's] tastes, appreciations, ideals, attitudes and mental perspectives [influencing behavior] are consequently a much better index of his true character than what he knows ... The aim, therefore, should not be to train great composers or great performers but to produce a community with refined musical tastes and discriminations and a disposition to appreciate and use the best music. (Withers 1916, 26–27; emphasis added)

Citing "sublime thoughts" in music as essential to social values means that musical taste indicates character. And this reading of taste, so clearly laid out as the equivalence of exterior and interior also allowed music appreciation to fortify social differences rather than diminish them. Instead of assuming equality, music appreciation transmitted levels of this specific cultural competence as a manifold of citizenly qualifications. That this is put in terms of the disadvantage of others does not appear as racism or classism per se, but the effect of the acquisition of cultural values that are *discounted as learned* in the family, community, and schooling in one form or another. When the source of aesthetic values is discounted, those values assume the status of the universal, overlooking the demographic and cultural factors that led to becoming the curriculum. For many, the judgment that is supposed to come after mastery and instruction comes before.[33] Selecting the templated comportment is a matter of verifying the body's claim over qualities of mind. "Stand in front of your pupils and watch their mouths if you really want to keep track of their brains. Can you by glancing over fifty faces at once see the ones who have that far-away dead-fish look in the eye?" (Giddings 1910, 15). Moreover, the mind of the child that is attuned or indifferent foretells mental acuity: "It is merely that conditions and requirements for quiet listening should [be made] . . . for children to listen to music with a quietness of body that leaves them freer for mental and imaginative activity" (Giddings 1910, 15). In contrast to a "bored listlessness" noted in Gilliland and Moore's psychological experiment or Giddings's observation of a "dead-fish" look in the eye, "quietness" of body becomes an important element in reasoning about the inner qualities of the future citizen and in distinguishing those at risk for antisocial behavior.

Ernest Hunt, author of a handbook for music teachers published in 1924, explains how the pupil's sensitivity and suitability for music study could be gleaned through physical characteristics. Sometimes, it is the shape of hands that gives away the personality in a way that is reminiscent of how build and carriage scripted medical and musical biographies in the early vocal curriculum. "Knowledge of what is shown by the types of fingers and hands is also of practical value. These types range from the conic, pointed-finger, highly emotional types through the artistic, philosophic square, rather blunt fingers, to the rudimentary animal-like . . . the emotional must be stabilized, the artistic welcomed, the philosophic stirred to action . . . and the rudimentary avoided . . . many manuals on the shape of the hands and fingers contain valuable information . . . In teaching John music, it is not only necessary to know music, but it is even more important to know John" (Hunt 1924, 119–20). There are contradictions involved in thinking about musical talent, physical makeup, and the teacher's psychological insight. On the one hand, the hypothetical teacher in Hunt's book asserts that only certain human types incline to a "high" musical culture. On the other, understanding the student depends on calibrating physique with mind and some are to be avoided entirely.

For other teachers, music searches for the soul. "Consider yourself called to the work [of planning music listening lessons] and as one of the chosen who would not for anything cease her effort to get music into the souls of children" (Fryberger 1925, 15). "This appreciative work . . . has been enhanced by a multiplicity of by-products. In investigating the nature and process of musical appreciation, certain incidental values appear ... from there to the collective social functions of the home, school and national life ... and into sympathetic consciousness of the solidarity of humanity" (Cady 1910, 53–57). Whether teaching one-on-one, as Ernest Hunt describes, or teaching public school music as Fryberger, Clark, Giddings, and Cady exemplified, the linkage between physical characteristics, reverent comportment, and mental acuity formed a template for musicians and cultivated listeners that bore the imprint of whiteness.

In Pursuit of the Charmed Appearance

The study of music and dance was historically entwined with what became known as "body culture."[34] Deliberate displays of aesthetic interests produced and made use of subtle comparisons of comportment and facial expression that could be acquired through bodybuilding and the imitation of classical Greek poses. Like Rodin's *The Thinker*, the posture of the "essential" reflective human being bears close relation to the ideal listener in music appreciation texts.[35]

The body culture movement had a special relevance to the perfection of the self the reading of character. For example, the Delsarte System of body training had an important following in the United States in the early twentieth century.[36] Like many European body culture programs, American Delsartism lessons stressed imitation of classical poses and provided graphics so that the individual could follow these principles on her own.

Delsartism, like other physical regimens, promoted health, strength, and grace through postural exercises. It thrived on the American appetite for elocution training in that it offered a parallel cultivation of expressive body movements that were perceived as enhancements to speech and one's social standing. Appealing chiefly to women, the main motions and poses were articulated by one of the principal leaders of American Delsartism, Genevieve Stebbins. In *The Delsarte System of Expression*, Stebbins stresses aesthetics more than bodybuilding or hygiene, being a method on how to comport oneself in order to leave a formidable impression on society.

> There must be simultaneous movement of all parts of body, from head to toe ... motion must be magnetic ... slow ... and as unaffected as the subtle evolution of a serpent ... every movement ... must unfold from within to without as naturally as the growth and expansion of a flower ... there must be no sudden seeking for opposition, no spasmodic attempt for sequence ... Every gradation of motion ... must be one beautiful flow of physical transformation ... To this I would add that every pose ... stands as the climax of harmonious poise ... it is not mere imitation of a Greek marble ... it is something infinitely greater ... a creative work of intellectual love ... a spiritual inspiration toward a superior and definite type of beauty. (Stebbins 1902, 459–61)

Such an "incorporating practice," contains "all the messages that a sender or senders impart by means of their own current bodily activity. The transmission occurring only during the time that their bodies are present" (Connerton 1989, 73). Delsartism premises percolated in milieus along with concert etiquette and extended to the popular readership of books such as Samuel Roberts Wells's *How to Read Character*.

Part of the social atmosphere accounting for the efforts to cultivate an aristocratic bearing after the Civil War had to do with the loss of dignity of the artisan class. With the increase in wageworkers from Europe and the South in industry, there was a coincident decline of independent artisans and more talk of a race "imperiled" through the expunging of white male independence. Analogous to the body culture programs of Delsartism, the bodybuilding world of gyms and exercise regimens offered middle-class men status through the fabled performances of escape artists such as Houdini, who demonstrated the agility of a male species reborn. Building personal strength attracted male schoolteachers and others interested in presenting a more robust, yet gentrified, image.[37] With the conspicuous absence of black laborers and the immigrant men and women whose strength was developed in the fields and coal mines, the membership of middle-class body culture adherents absorbed the myths of Teutonism (of which Tarzan might be considered an exemplar) that underwrote racial hierarchies in the context of military and national ambition.[38] Turn-of-the-century cultural preoccupations with manliness and national honor also included transformation of the boyish body into the godlike figure, a metamorphosis that combined physical feats with a classical body aesthetic modeled on Hercules and other mythical figures, emphasizing the sensual appeal of the perfected male body.[39] Like concert etiquette and other social formalities, body culture regimens promoted images of a heritage overlapping whiteness. Cultivated impressions of social rank carried enormous weight where disenfranchised groups had begun to challenge the racial or ethnic hierarchy and petition for wider participation in society.[40] Labor unions, race riots, and ethnically charged clashes of all kinds destabilized older social hierarchies and reorganized the stance of corporations toward labor. In *Social Work and Social Order, 1889–1930*, Ruth Crocker documents the interest steel companies took in the living conditions and educational opportunities of black workers in the North. Elites were anxious to impart genteel mores for the general good and many sectors of the population, including black families, took up the regimens of the middle class, such as child care classes, in order to be included in society. However, the common measure of gentility was, very often, comparisons made to migrating Southern workers, chiefly blacks, and new immigrants. Representations of blacks in *Amos and Andy* on radio and the still popular minstrel shows revolved around stereotypes of the clumsy, hapless body, uncontrolled elocution, comedic misuse of words, distended lips and tongues.[41] In contrast, the literature on classical comportment featured photographs of statuary, gestures, facial expressions, hand positions, and, most of all, the slight inclination of the head that signaled disinterest or disdain without activating large muscle groups.[42] Minstrelsy equated blackness with sexual anarchy in subtexts that titillated white audiences who took refuge behind the genteel façade. Delsartism and other body culture societies taught elaborate schemes with which to disguise "low" impulses. In concert with limbs and torso, the angle of the head could spell defiance or reflection, humility or eager interest, indecision, passivity, or attentiveness.[43]

Delsartism, in some respects, is reminiscent of the literature on consumptives in its interest in the outward manifestation of white nobility. Body culture regimens, like the portraits of tuberculosis sufferers, sound the register of soulfulness and stoic dignity, through the message of the "that certain something" that comes with aristocratic bearing. These visual cues coincide with the cultivated listener and the avoidance of rhythmic entrainment and emotion. With the Delsarte system, there were specific prescriptions for "concentric" facial expression and composure. A twist of the mouth would ruin the face's symmetry. Symmetry lent a certain impenetrability to emotion, the "Je ne sais quoi" that made it seem there was more depth to this immobility than its cultural value as a sign of coolness. As Bourdieu has noted, the effect notoriously masks not only the emotional investment of the individual but also the historical and social *means* through which the bearing was acquired. Twisting and other "rotary" movements of parts of the body, while they are part of many people's traditional observances of joy and grief, indicate a "childish impatience, spasmodic convulsion of the will" that was characteristic of inferior musical response and "a twitching of the whole body and especially the feet . . . [people] involuntarily beating time with their heads" (Hanslick 1854, 84).

Interest in body culture in the United States was associated with Northern European cultivation of the soul (*Bildung*). For American followers of Delsarte, bodybuilding, and music appreciation guides, these programs provided the distinction from the common laborer and the aimless vagrant who passively accepts what mere chance brings to him.[44] Discipline extended to music performances as the body and appearance of the players and conductors exhibited new forms of synchrony and discipline with stage protocols for tuning and movement on stage. Theodore Thomas, the conductor of popular series of concerts in Central Park and much acclaimed music director of the Chicago Symphony Orchestra, was notorious for his rigor, physical control, and emotional restraint. According to his biographer, Charles Russell, Thomas did not attend the funeral of his eldest son because it coincided with an orchestra performance.[45] Privileging duty over emotion, while extreme in Thomas's behavior, carried over to the middle-class audience's craving for rarefied discipline, such as what took place at the Wagner Bayreuth Festival. The lessons learned there eventually transferred to concert audiences in the United States. Mark Twain wrote from Europe.

> Yesterday the opera was *Tristan and Isolde*. I have seen all sorts of audiences ... but none which was twin to the Wagner audience of Bayreuth. For fixed, reverential attention ... You know they are stirred to their profoundest depths; that there are times when they want to rise and wave handkerchiefs and shout ... and times when tears are running down their faces, and that it would be a relief to free their pent emotions in sobs or screams; yet you hear not one utterance till the curtain swings together ... Then the dead rise with one impulse and shake the building with applause ... At the Metropolitan in New York they sit in a glare, and wear their showiest harness; they hum airs, they squeak fans, they titter, and they babble all the time ... Can that be an agreeable atmosphere to persons in whom this music produces a sort of divine ecstasy? (Mark Twain, as cited in Sessa 1984, 254)

European concert protocols left their imprint on music education as well. Educators urged music teachers to emulate this form of listening in the classroom by discarding sentimentality "for the silent, liquid and mysterious depths" (Briggs 1932, 39). This ethos congealed in music appreciation so that control over any kind of rhythmic entrainment was positioned against the chaos that would otherwise erupt. The values pertaining to comportment in public school music fall into one of two opposing views of the body: the body of order and the disordered, spasmodic body that resists the civilizing process.

> In order to appreciate music something must happen to us. We know that music affects our emotions through its outward manifestations of rhythm, melody, and harmony. No human being is immune to music. That has been proved by Dr. Willem Van de Wall in his experiments with people of unbalanced minds. I proved it once to my own great consternation in a class of ungraded children when giving an "Appreciation" lesson (and I use the quotes advisedly). I had a state of pandemonium on my hands after only sixteen measures of "The Hunt in the Black Forest" a wild orgy of descriptive music, was played for the class. I hope the composition is no longer in use anywhere ... We who are interested in music appreciation are mostly interested in the ear receiving it—or in other words, the person consciously listening ... How can we recreate in that consciousness which depends *wholly on the ear, what the composer had in his consciousness*? I think [studying music] history has to do with the development of music as a science. We can do as much as the mathematics teacher can do. We can prepare for listening. (F. Dunham 1929, 83–84; emphasis added)

"*Wholly on the ear, what the composer had in his consciousness*" is at the crux of the curriculum. If it instigates an exuberant show of feeling, it must also mean there is a lack of attention and loss of control and it must remain outside the classroom. It is the model most familiar to school experience and the symbolic power of Rodin's seated *Thinker*. We forget that it was not the only way to conduct oneself at a concert five decades earlier than the official beginning of formal listening lessons. In other words, we forget its historical relativity. "Neither the concert nor the opera was thought of as primarily an individual internal experience to be valued for its *Innerlichkeit* [inwardness], the feature that Romantics would later stress . . . people felt free to talk [at times]. Fisticuffs were not unknown . . . Those whose musical needs differed coexisted successfully enough . . . Some people listened and some socialized, but no one objected to their being together in one audience—as would not be the case today" (Weber 1984, 30–31).

From the point of view of a black audience at a blues program, we have a display of feeling and involvement that represents a sensibility outside the bounds of the standards of the school curriculum as well.

Because music educators are so familiar with present regulation of listening as the norm, its recognition of motionless and silent nobility seems to be a standard of universal value. A recently advertised poster for music classrooms asserts, "Silent and Listen are spelled with the same letters" (Anon., 2004). The directive forms the common sense of regulation that partitions the musical space of the school and produces the problematic condition of inequality described at the outset of this book.[46]

Traditional regulation has a history much longer than the modern democracy if we go back to Plato's *Republic*. There, although imagined, musical regulations were designed for the exclusive category of the citizen who was neither a slave nor a woman, more than half the population of Athens. These ancient restrictions, transmuted through time, continue to partition school music through present notions of the glorious and the undeserving. The music curriculum and the pedagogical expertise invested in its traditions generate a different, but nonetheless severely limited, model for musical merit. According to the music standards of the 1990s and the early twenty-first century, which I defer discussing in detail until Chapter 10, musical inadequacy disqualifies many from being the successful music learner. Call it the operation of the double gestures of educational policy or the nature of cultural capital, the word diversity in recent music reforms has the effect of making the "diverse" less than the music that is at the center.

My next chapter deals with biographical myth and autobiography. It looks at specific portraits and scenes by pitting the extremes of genius worship and total abjection against one another. Using German and African American literature as well as images of genius of Beethoven and Lincoln, I explore the uncanny similarity between the process of indenture undergone by students centuries ago and the enclosing of the "underprivileged" or "diverse" learner today.

9
Aural Icons and Social Outcasts
Beethoven, Lincoln, and "His Master's Voice"

> The snowstorm—students were very late on account of the buses and traffic that slowed to a crawl. Mr. Hoffman said he was tempted to reschedule the concert, but changed his mind when he spotted Carleton on his bicycle in the snowstorm, cello strapped to his back, making an end run around the right lane to get there on time.
>
> —R. Gustafson, *"Report on Minority Student Achievement"*

> The fifth symphony is a drama ... in which Beethoven gives us the entire struggle, tragedy and ultimate triumph of a human soul. If you will explain this work to an audience of older children ... you will be amazed at how much they can understand. Tell them to imagine Abraham Lincoln at the White House during the War of Secession, during the darkest age when it seemed impossible ... to preserve the Union ... You should tell them something of what a symphony is.
>
> —Walter Damrosch, *"A Lesson in Appreciation"*

As conductor of the New York Symphony Orchestra in the early decades of the twentieth century, Walter Damrosch gave a series of radio broadcast children's concerts in which he used an analogy between Lincoln and Beethoven as a pedagogical device for interpreting the symphony for an audience of young schoolchildren. The story of Beethoven's dedicating the *Eroica* Symphony to Napoleon, then changing his mind, and the famous "fate" theme of the Fifth Symphony appear throughout the literature on the music curriculum. Weaving the themes of moral suffering, creation, freedom, and sacrifice around visual images of Beethoven and Lincoln[1] marshaled the kind of attention that made classical a patriotic exercise.[2] Lincoln's oratory and Beethoven's deafness connected miraculous speech to hearing and both to grand sacrifice. Speech and music were linked their portraits to imaginaries of a united nation and cultivated public. In the post–Civil War era, oratory interlaced familiar notions of citizenly worthiness and whiteness that would command the airwaves of home and school a few decades later.[3]

In Chapter 7, I discussed questions about the sources of a voice preoccupied social institutions, acoustic science, and government. The long historical conflicts over authorizing the source of voice, hearing, and sound, and, in particular, racially inscribed sound such as Negro song, bolstered the idea that "head" tone, disciplined ears, and reasoned speech were the communicative organs of a modern democracy. Autocratic and primitive societies, with their irrational followers, formed the contrast.

Paradoxically, the spread of science's "miracles" (especially so in radio) increased its ability to command an audience standing for the nation. Radio shortened distance, collapsed time, and reenchanted the machine to trouble the mental acuity of citizen listeners and expunge their judgment.[4] In this sense, the large gestures of Beethoven's symphonies, as they were organized into public school lessons, were to transform the composer's lack of hearing into a symbolic triumph over deafness and irrationality that reputedly went with the deaf.[5] Likewise, Lincoln's ability to overcome and transform his moral suffering into triumph gave aural form to the Union in the Gettysburg Address and the Emancipation Proclamation. As Walter Damrosch wrote the *Journal of the Music Supervisor's National Conference*, it is through Beethoven's biography in particular that the student, as listener, will experience Lincoln as protector of democracy and identify overcoming social conflict with the unfinished work of self-governance. The nation, in American culture, following the remnants of the unfinished work of building the New Jerusalem, would always be a work in progress. Adapting religious vocabulary to the secular nation for which Damrosch designed his concerts for schoolchildren, the incomplete project was the conversion of the population, in a few lessons, to high culture. In this respect, Schubert's Unfinished Symphony was a symbol of the unfinished self and nation. Damrosch's concert series for youth also used memory devices such as, "This is- the Symph-o-nee that Schubert wrote, but never fin-ished," sung to its major theme.[6] The use of the musical mnemonic in the radio arena epitomizes the way in which appreciation lessons became mass events akin to the Puritan mass conversions of members of church communities. These religious events, and the radio programs, were peculiarly American in their pragmatic strategy that would astonish European observers who were accustomed to thinking of self-cultivation and salvation as an individual, lifelong journey.[7]

This chapter takes a closer look at how Lincoln's anguish became associated with the four-note theme of Beethoven's Fifth Symphony, focusing on the links between Lincoln and Beethoven that made the "good ear" a sign of the attentive citizen. Serving as both aural and oral ideals, Beethoven and Lincoln came to represent comparative models for those who struggled and overcame great difficulties such as deafness, disfigurement, and deep melancholia. Marginal individuals became the objects of radical social uplift that would prevent the national body from descending into chaos.

Music Appreciation, Beethoven, and Abraham Lincoln

The theme of Beethoven's Fifth Symphony was the epitome of what every music appreciator should know.[8] For music educators, Beethoven and the German "greats" represented a sublimity that allowed the listener to transcend the everyday through the contemplation of universal beauty. In one of his many transmutations, Beethoven was an American home deity. "In that legendary moon-lit room [it] came to [Beethoven] one of the most tender creations ... [giving] the listener the purest enjoyment of music ever experienced ... the ideal [is] set up before me as one eminently desirable and possible of attainment by many. But we may reach that ideal only by the road of simplicity and reverence ... of reverence, I doubt there is enough in all the United States to build one great temple" (Cady 1908, 148–49).

The imagery of the moonlight inspiration was extended to whole symphonies. Beethoven's standing as pastoral leader, comparable to Lincoln's image, culminated in several portraits. One shows him seated in a pose reminiscent of Michelangelo's famous *Moses*,[9] his hair resembling a lion's mane and his face set in an expression of ferocious determination. His deafness cast him as "amiable savage."[10] Early in his career, there were rumors and depictions of him as non-Teutonic, possibly "Negroid," gradually giving way to his portrayal as a "pure" German superhero after a series of sensational successes in the salon and concert world.[11] The most famous Romantic portrait of Beethoven, by Josef Stieler,[12] was often used as a point of entry into the music lesson: "The pupils have sung one or more songs by Beethoven. They have studied the picture of Beethoven in their book ... and have discussed the characteristics of the man and his music ... The teacher is standing near the phonograph which has been wound. [After playing a selection, there] follows a discussion of the composition in which the pupils are encouraged to express the idea which they have gained through listening to it" (McConathy et al. 1937, 109). Deafness was said to have sharpened Beethoven's powers of observation. His genius consisted of a telepathic mental ability in which the composer divined the sounds of nature, although he could not hear them.[13] "Before he was thirty years old he began to grow deaf ... Think of what that meant to one whose whole soul lived in the world of music, to one who understood that in his case the ears were the most precious organs in his body" (Victor Talking Machine Company 1923, 241). Deafness also intensified the image of Beethoven's body into classical proportions like Rodin's "The Thinker." At other points in his life, he is portrayed as Christ-like. Edith Rhetts and Mabel Glenn, frequent speakers at music supervisor conferences and writers for the educational department of the Victor Talking Machine Company, describe a legendary scene in their teacher's manual and lecture series for children's concerts: "Beethoven's . . . deafness was his greatest sorrow . . . After his deafness came upon him he turned more and more to music . . . He wrote music as he thought it should be . . . There were many rules in those days . . . Beethoven cared not for rules . . . but expressed himself through music in the way he felt it . . . History seldom paints a more pathetic picture than when he appeared . . . to conduct his great Ninth Symphony . . . [he] never knew that the loyal concert-master secretly directed . . . that it might sound as the great master would wish . . . He turned him around to see the Cheers" (Glenn and Rhetts 1923, 11). Beethoven's music was "feeling," as *Reading Lessons in Music Appreciation* described, yet it was more than that. Beethoven's inward genius delivered not *just* sound but *divinity* to educated ears. Legend had it that Beethoven thought of himself as the medium of divinity and that his music did not originate from him, but from God.[14] These stories occur frequently in the curriculum literature, demonstrating that Beethoven occupied the highest level of esteem in a secularized, yet devout, public. Insofar as this musical public connected the German greats and especially Beethoven to the mission of school music, the music teacher consecrated a mystical experience of the deaf leading the hearing and the deaf learning to speak. Overcoming deafness was rhetorically linked to the Protestant sermonizing about overcoming obstacles to salvation, which, in secular terms, served as shorthand for nation-building. The linkage of politics, religious reverence, and music was logical insofar as the Gettysburg and Second Inaugural Addresses marked the pinnacle of speech and Beethoven's music was also at the peak of music as a form of speech. The equivalences of language-music-speech were historically the channels through which interest in teaching the deaf and music marked the "soft" governance of nations.

"Across the 1600s to the 1800s, the service that the constitution of deafness provided to emergent national imaginaries . . . lay in relation to how governmental assemblages were being reorganized and given new boundaries and rationales . . . [it] is discernible in the pedagogic efforts of a charitable nature" (Baker 2003, 293). By reiterating the Beethoven and Lincoln link, music educators such as Walter Damrosch, C. C. Birchard, Mabel Glenn, Edith Rhetts, Frances Clark, Edward Birge, and others[15] constructed an imaginary of the aurally unified nation as the most important musical public in U.S. history. At the core of this national imaginary was public school music appreciation, regenerating a country ailing from cruder, suspect music making and makers.

The iconography and stories about Beethoven and Lincoln also revolved around making use of a melancholic tendency in their personalities. While portraiture and sculpture portrayed them as gloomy loners, this quality, like the images of the white martyr to consumption,[16] transformed disfigurement into positive attributes.[17] Both were long-suffering masters, especially Lincoln, who, in some illustrations, extended his hand to slaves who kneeled in front of him.[18]

With respect to the construction of genius that went on through the decades after their deaths, images made *during* the lives of both Lincoln and Beethoven record little sign of the kind of anguish that the posthumous iconography displayed.[19] Traits of derangement and suffering are only prominent in postmortem portraits and sculpture.[20] This suggests the extent to which popular mythologizing played a role in the martyr-like elements of their personas.

So, too, there are traces of fear in the awe many were to feel for Beethoven and Lincoln. Familiar postmortem images of Beethoven's mental and physical makeup make him an awe-inspiring figure midway between the "super-artist" of superior genius[21] and the grotesque, gothic or bizarre.[22] In this sense, Beethoven's reputation as melancholic[23] was part of an image industry of refinement of the Anglo-Saxon through spiritual and physical trials that depended on a partial view of the grotesque or bestial as limited rituals. These confirmed the idea that the struggle with "dark" forces can be overcome. As Stallybrass and White point out, what is left of the representations of the late eighteenth-century references to the lower body, such as those found in Billings's transgressive hymns, gradually disappears in the nineteenth century, cutting off the possibility of integrating instinct and thought. Equally the case for Lincoln, melancholy heightened the popular narrative of transcendence of war, death, personal loss, and pathological grief that Lincoln (and his wife) were reported to have suffered.[24] No doubt the posthumous depictions in theatre dramas and publications gathered momentum from an appetite for the same type of melodrama and myth that also buoyed Wagner's fame. These followed the parallel paths of "saving" schoolchildren and Teutonic culture from the low practices of the stage, nightlife, and airwaves.[25]

Admiration for the Teutonic virtues, despite the short-term effects of World War I on German artists, played a large role in fanciful stories about Beethoven's appearance. This is evident in differing accounts of his physical stature (he was actually not large), making his size as well as his music imposing.[26] The alternate use of "van" or "von" in his name assigned him alternately to the Flemish middle class or the German aristocracy. His renown grew with the identification of Beethoven as a "musician of the people, a socialist leader—indeed a hero of the French Revolution" (Newman 1983, 364).

Beethoven's loose connections with the fight for freedom of oppressed people joined with Lincoln's Emancipation fame. Portraits of the two figures arrived at schools all over the country from commercial venues such as school supply and music publishing houses. The phonograph would democratize classical listening and make it a right of freedom. "Too little has been said with respect to the right of the country child to cultural advantages more nearly comparable with those enjoyed by his city cousins ... Nearly every [classical] radio program, especially those presented for schools is based upon one central idea. It may be that the idea is the presentation of the works of some great composer ... The teacher plans to have the class listen to Beethoven's Country Dance from Victor record: 20451" (McConathy 1937, iii, 103, 109). The awe with which Beethoven was held was not unique to American music education. Adulation was the rule in the cosmopolitan circles of cultured audiences and in the international musical literature that sometimes referred to the composer as an "eagle" in infinite flight above the clouds of life. "The German soul ... a mystic prophet ... This deaf man heard the infinite ... If there is ever evidence that soul and body are not joined, it is Beethoven who proved it ... [with] crippled body and flying soul" (Victor Hugo, as cited in Newman 1983, 364). Feeding into this worship of the "flying soul" were the polarities of barbarism and the civilization or the opposition of mind and body. For music education, the identification of musical genius with infinite, divine insight reinforced the pedagogical "need" to *observe* students for their lapses in attention to these heights and the degree to which they interrupted concentration or were distracted while listening to "sacred music."[27] The reverent language of music appreciation insured that high taste transcended the body and manifested itself in the transport of the soul.

A metaphorical icon of the citizen with the good ear adorned every publication and phonograph; the label read, "His Master's Voice." Figure 9.1 shows one of the hundreds of advertisements appearing in the music teacher literature. Its sale to schools and teachers came with a detailed curriculum guide and supply of recordings. The interrelations of commercial interests and the aesthetic goals of the music program are difficult to entangle.

This emblem depicted Nipper, the loyal, spotted dog with his ear cocked to the Victrola horn on the center label of analog records and gramophones.

Figure 9.1. *Journal of the Music Supervisors National Conference*, March 1912, p. 21. Victor Talking Machine Company, advertisement.

The title, "His Master's Voice," implied a relation between master and slave. Some labels featured miniatures of exotic foreigners ringing the phonograph. Nipper's full frontal exposure to the phonograph horn was a metaphor for how the would-be citizen should position her still, attentive body to the Victrola.[28] Even if we consider some conscious or unconscious deprecation of the listener as dog, its eye-catching image had wide coverage in music periodicals and music teacher journals.[29]

So far, I have attempted to sketch a critical grasp of how the aural and visual icons in classrooms and in the music appreciation literature promoted the self-discipline of reverent listening. This ethos demands that the listener direct her attention not only to the classical and folk music "greats," but *away from* the attractions of popular, but lower, forms of entertainment. At this point, I want to consider the elements of the body culture movement that historically scaffold differential judgments in music classrooms. This turn in my story looks at older practices that shaped American musical publics and genteel values. It focuses on the aesthetic distinctions that emerged from widely dispersed geographic and temporal sites where literature produced a picture of the struggle for freedom and upward social mobility.[30] These appear in the form of musical methods, but also, most strikingly, in literary autobiography and in the novel of self-cultivation or "*Bildungsroman*,"[31] as it is called in German. Depicting unforgettable dramas of how bodily practices trigger acceptance or deny access to institutions and social advantage, the *Bildungsromans*, when read in conjunction with African American autobiography, reveal the fabrications of pariah status and their maintenance.[32] Given the European-American interchange of this literature, especially with regard to education and judging merit, they compel us to consider that the fabrication of whiteness as cultural nobility is relevant to our understanding of the present situation. There are great differences as well. Black slavery and oppression diverge in severity and scope from the indenture customs of German education in the eighteenth century.[33] However, the class codes in the German novels demonstrate the broad impact that comportment and language had on ranking the individual, a factor that powerfully intersects with the extremes of racialism tragically enacted in the twentieth century.[34] The point of connecting scenes from German and African American literature to musical practices are the uncanny resemblances between academic and musical exclusion centuries ago and the template for worthy participation in music instruction today. They also provide a vivid picture of the immense effort of cultural acquisition for the lower economic classes. In this sense, they are a warning not to take for granted the aesthetic values that pose as "natural" to those that have them.

Students who received charity in the German novels bear an important relation to students considered "at risk" in the United States, for example, those who participate in federally assisted school lunch programs.[35] The narratives are mutually informing in that each setting dramatizes how an array of postures and feelings, legible as signs of inferior personhood, disallow participation in education. And while the ethnological notions of rank discussed in previous chapters delineate a general framework for racial comparison, the novels show, with great precision and concreteness, how the stings of ingenious cruelties are a suppression of traits that "superiors" suspect are part of themselves. Psychological and social spitting of instincts and impulses in the German novels are analogous, in this sense, to the dysconscious injuries students with marginal status suffer today: judgments formed in relation to values about comportment, language and musical interests.

Historically, the indignities foisted on blacks and impoverished German students devolved from a shared, largely Protestant, reasoning about human types. Intra-German or Austrian state hierarchy was more about dialect, manners, religion, and dress than skin tone, since colonial racialism was limited in the daily functioning of German education. For African Americans, slaves and their descendants lacked a proper soul and so could not be saved through good works. Protestantism, as opposed to some Catholic tolerance in the Caribbean, accounted for fear of the "supernatural" in African American music and eventual repression of drumming and dance.[36] Yet, even with these differences, the resemblance between the German and the American cases speaks to an anatomy of dejection or a whole vocabulary of differences that isolate human types. Such discrepancies form a system of reasoning about cultural superiority that depends on taking possession of the other.[37]

Two novels of self-cultivation are discussed in this chapter, Karl Moritz's *Anton Reiser* and Frederick Douglass's *Narrative of the Life of Frederick Douglass*. These bring us face-to-face with the process of cultural or racial disqualification but also show us how withdrawal is an attempt to preserve integrity in the case of overwhelming rejection.[38] Because so much of the curriculum of public music instruction is borrowed from European themes of cultural nobility and its abject opposite,[39] these narratives provide a link connecting the larger traditions of comportment to school music. It is this historical entanglement between musical dispositions, a "universal" aesthetic, and judgments of bodily worth, as I discuss in relation to present reforms in the concluding chapter, that make more equal participation difficult to imagine.

Novels of Self-Cultivation

Anton Reiser is the story of the struggles of a student from the Prussian lower class in the late 1700s. The novel follows Reiser as he attempts to enter the ranks of a hereditary meritocracy (Lutheran ministry) through study at gymnasium and university. There he realizes that his material deprivation and social background will prevent him from incorporating the language and appearance necessary to success. Moreover, he has to indenture himself to a harsh master to meet the expenses of his education. Although he is adept at his studies, he cannot enter the circle of respect granted to more privileged scholars. In this sense, Anton's troubles are relevant to the problem of educational equity today in that physical stereotypes and social class often lead to negative impressions of ability.[40] "Terms such as *at-risk* become come words for students who are perceived by educators and the public to be 'problems' in the schools ... And it is no coincidence that status characteristics such as race, class and linguistic diversity become equated with '*at-risk-ness.*' So prevalent is the language ... that it is not unusual for urban teachers to define their entire class as at-risk" (Ladson-Billings 2001, 15).

The visual cues that signal the at-risk are generated from comparisons to the images similar to the classical poses in the Delsarte system (see Chapter 8) and in the traditions of aristocratic manners.[41] This code of behavior excludes "abject looks and countenances, consternations, prostrations, disfiguration ... disdainful postures ... beggarly tones, grimaces, cringings and unmanliness ... We [are] in our best state and [carry] our most becoming looks with us when we approach the gods" (Cooper 1923/1999, 391–92). In *Anton Reiser*, indentured labor in a hat factory stains Reiser's hands and face with dye, marking him as "unmanly." His general condition as a laborer implicates social inferiority, including what his music teacher perceived as an unbecoming use of his hands. "All he could master were a few arias and chorales ... Fingering was also very difficult for him ... [his music teacher] always found fault with the *shape* of his outspread fingers."[42] However, the most humiliating scenes for Reiser occurred during mealtime over resentment of the charity he received. This is reminiscent of tensions between some cafeteria workers and students who qualify for the federal lunch and breakfast programs.[43]

"Once he found himself ... [at the table] when the verger's wife started talking to him about hard times and ... finally burst into tears from worry about where their bread was going to come from and then Reiser, embarrassed by this tirade inadvertently dropped a hunk of bread on the floor ... she let fly and ... gave him to understand that people of his ilk ... were not welcome at her table" (Moritz 1785/1997, 116–17). "Now each of the people who gave Reiser a free meal, and on whom he was dependent, had a quite different way of thinking, and each of them threatened to withdraw his protecting hand from him if he did not follow his advice ... there were ... countless slights and humiliations to which Reiser was exposed and which are undoubtedly liable to befall a young person who while at school has the misfortune to seek support in the form of free meals" (Moritz 1790/1997, 95). Often, Reiser's suppers were accompanied by instructions in manners that spoiled the enjoyment of the meal. Lobenstein, Anton's master, always cut the sign of the cross into the bread and butter at the table, reminding the apprentices of the sacrifices made on their behalf. At one point, Anton gave up most of his meals. He preferred hunger to shaming glances. Occasionally, he had a burst of high spirits. When he was quiet, his master liked him. But when he lapsed from this somber demeanor, he was demoted from favorite to outcast. "Anton had been growing too clever for his liking, talked too much, argued back . . . he was growing too lively. For Lobenstein, this liveliness was the sure way to damnation . . . If Anton had been better at self advancement he could have put everything right again by assuming a downcast [look] . . . and pretending to feel fear and oppression in his soul. For Lobenstein would then have thought that God was drawing the lost soul back to Him" (Moritz 1790/1997, 51). "Growing too lively" had the effect of alienating the master. It was an indication of lack of respect. Once Anton was admonished at school for laughing and drinking wine at a party at the Rector's house the night before: "What! [He] lives on charity, even the Prince spends too much money on him, and when he is hospitably entertained in the house of his benefactor, who gives him a roof over his head, that is how he behaves—how vile, how ungrateful!" (Moritz 1790/1997, 146).

Narrative of the Life of Frederick Douglass, An American Slave presents a more drastic form of social shaming. Slave masters often staged scenes of humiliation around holidays as occasions to demand absolute subservience. One incident involved forcing slaves to swallow a huge amount of molasses. If the slave showed any objection, the consequences were grave.

In his own pursuit of learning, Douglass ran the danger of inviting cruel humiliation and physical punishment if he showed signs of ambition. "A mere look, word or motion—a mistake, accident or want of power—are all matters for which a slave could be whipped at any time. Does a slave look unsatisfied? It is said he has the devil in him and it must be whipped out" (Douglass 1845/2001, 58). These examples are extreme, but they touch upon dynamics that link the condition of indentured students and the social standing of students at risk today to a deep-seated, all-pervasive historical attitude toward social inferiors.[44] "Poor students . . . existed as a set of stereotypical images, encountered again and again in state regulations and in a wide variety of literature. The images reflect the social reality in some ways, they also . . . [function] as prescriptive devices, revealing a good deal about the problematic identity assigned the poor student . . . and the conditional terms for his acceptance" (La Vopa 1988, 46). These enactment of the social hierarchy offers a perspective on music instruction as well, illuminating the manner in which embodied dispositions were also part of the judgment of singers and listeners. As illustrated in previous chapters, some singers were overly effusive or resembled primitive, animalistic worshippers. These were the negative registers through which Negro song passed from nonmusic to spiritual.[45] These images accompanied calls for moderation in school songbook prefaces and listening guides that placed a high value on moderating affect[46] and monitoring the signs of strong feeling in singers and listeners.[47] This "moderation" existed in uneasy relation to the popularity of effusive operatic performances and melodramas[48] of all kinds in radio as well. "Degradation [through radio] is also filtered through other mechanical instruments . . . Raucous, noisy, blatant sounds conveying commonplace, even vulgar, sentiment in both words and music, tend to degrade the taste and more the morals of our youth" (C. Adams 1929, 85). Nothing in the teacher conference literature in the first three decades of the twentieth century indicates that a different attitude toward jazz and ragtime existed in schools, although it is more than likely that some teachers (not necessarily music teachers) might have brought recordings into schools or played one of the many sheet music versions of ragtime commonly available in those decades.[49]

Students "At-Risk," Language, and the Musical Home

Giving music lessons to the poor was sometimes characterized as raising ambitions for an undeserving class. Thomas Tapper describes this position in his manual for music teachers: "You take them out of their station and give them ideals difficult if not impossible to attain and make them dissatisfied with their condition" (Tapper 1914, 129). A more progressive attitude could be found in the press and in writing directed toward music teachers. This was the notion that the poor had a greater ability to appreciate fine music "by a kind of divine compensation" (Goepp 1910, 33) for their lives. Testimonies about the good that concerts did for the ordinary worker were presented at music teacher conferences. C. C. Birchard recalls that a woman who worked as a domestic had tickets to the Metropolitan Opera. He referred to her attendance there as "a pillar of support" (Birchard 1923, 73) and means of sustenance. But the tone of similar testimonials were patronizing and underlined the overall perception that the masses were uncultured.

Distinctions between the genteel and the laboring class became more urgent as Southern blacks migrated to the Northern cities. Stereotyped portrayals of black life and, in particular, African American dialect(s) conveyed negative images of blacks whose "misuse" of the language would lead to linguistic decay.[50]

With respect to well-known renditions of spirituals and poetry, there is a history of attitudes that reflect the attraction-repulsion dynamic, the oscillation between black and white (or mainstream) poles that W. E. B. Du Bois defined as the color line. Vocal dialects associated with abject bodies have important social and historical ties to cultural preferences for what is called Standard English,[51] a regulation that invites us to consider how its derived from the social machinery that regulates the attributes of citizens.[52] More recently, conflict over the teaching of Ebonics in California public schools brought national attention to debates over the status and use of

African American Vernacular English.[53] Media attention over the language in some hip-hop performances has also reawakened pejorative connotations of the black vernacular, leading to some recent attempts to use hip-hop to engage "at risk" youth.[54] This move to organize some aspects of the language curriculum around hip-hop pulls for standardizing the vernacular as well. "Since the speech community... is viewed as a monistic entity, a specific speech event is often presented as a generalized norm rather than characteristic of a particular style or genre."[55] The historical relationship between these social dynamics and a template that selects speech congruent with particular categories of acceptance allows us to consider the situation in music when dialect enters into judgment of the singer. Reading *Anton Reiser*, we stand in the shoes of someone who is an outsider himself but, nevertheless, one who identifies with the "higher" standard of the cultivated art. "Nothing sounded to him more sublime than when the choirmaster began to sing, 'vylo glorious sun'... The word 'vylo' was enough to transport him to higher regions—he thought it was some Oriental expression and to which for that very reason, he could attach as sublime a meaning as he wished; until one day he found... (the text read) 'Veil, O glorious Sun'... and now all the magic... vanished... the choirmaster... had a Thuringian accent."[56] From the perspective of schooling today, the events in Reiser's life are consistent with the predicament of at-risk students in that their performances in academics as well as music are often perceived and judged in the light of their speech.

For leading researchers in music education, standard diction and pronunciation were indicative of a mind that could appreciate fine music. Articulated as the mothers' responsibility, the quality of cultivation in the child came down to family background.[57] "Speech is an index to character, and the means for the development of character. Beautiful speech is musical speech... let the mother who worries about early piano or violin lessons first give thought to formal sympathetic cultivation of a beautiful speaking voice... The civilized world is just awakening to the possibilities and significance of beautiful and effective speech. Train the young child to the appreciation and development of power in beautiful and effective speech and you will have laid the best foundations for musical appreciation" (Seashore 1940, 17). Speech is an important player in the judgment of levels of intelligence and culture that occur in schools, as does an economic status that often travels with language as index of worthiness.[58] Remarks on the correlation of speech with music and music educators' references to the benefit of language through music listening occur throughout the literature on school music appreciation. Edward Bailey Birge, well-known for his history of music education published in 1928, writes for the *Journal of the Music Teachers' National Conference*, "If music is a language, the same laws govern reading it as govern any language-reading" (Birge 1913, 162). Sometimes this played out in relation to the study of musical phrases and periods whose perception was correlated to the formation of good grammar.

At the center of the correlation between music appreciation and language was the concern that dialect was a distortion of correct speech. Alice Keith, educational director for the Victor Talking Machine Company in the 1920s, wrote that radio programming for music appreciation instilled the standard speech of trained announcers, making it of service in language teaching: "Although radio . . . has appealed greatly to the imagination of school administrators, its possibilities have only faintly been sensed . . . [Music appreciation] programs will bring much correlative material for use in [the curriculum where] the enunciation, pronunciation and voice of trained orators and readers should serve as a means of overcoming the growing tendency toward dialect formation found today in all parts of America" (Keith 1929, 141–42). Disqualification is coded in the circumstances that position the student a priori and in the student-teacher interaction that is corrective and peremptory.[59] Lisa Delpit describes one such scenario in which the correction of a child's language precipitates a form of rejection. Quoting a study on linguistic difference, she writes,

> A teacher has been drilling her three and four-year-old charges on responding to the greeting, "Good morning, how are you?" Posting herself near the door one morning, she greeted a four-year-old black boy in an interchange that went something like this:
> Teacher: "Good morning, Tony, how are you?"
> Tony: "I be's fine."
> Teacher: "Tony, I said, How are you?"
> Tony: (with raised voice) "I be's *fine*."
> Teacher: "No, Tony, I said, *How are you*?"
> Tony: (angrily) "I done told you *I be's fine* and I ain't telling you no more!"
> (Krashen 1981, as cited in Delpit 1995, 51)

A similar dynamic takes place with regard to musical disposition. Accounts of "need" for great music make distinctions between the nurturing and the culturally impoverished home Instructing children in *consistently* "good" listening habits is similar in that it requires a conversion of the children's musical entertainment to the curriculum's values. In a section called "Listening to Music," published in the series *Our Singing World, The Kindergarten Book*, the author writes, "Expressive images and patterns of both speech and music can become familiar and useful only when continually heard . . . the capacity to get what is said in either music or speech arts . . . requires a mind well filled with meaningful musical imagery or else the listener has little to draw upon in reference to what is new" (Pitts, Glenn, and Waters 1949, 5). This assumes an empty mind of the child that may already be filled with imagery that he or she values highly. For Frances Clark, educational director at the Victor Company, the home environment was often an obstacle to the acquisition of knowledge about the Western art tradition where "one especially large lion in the way is the almost utter lack of music in the Kindergarten, and still worse in the home" (Clark 1929, 307). Calvin Cady described the model home for meaningful musical imagery: "[I entered] the humble home of a middle class German family, tempted from the fatherland by the lure of freedom and fortune. The home instrument was a small Hammerclavier [sic] with a tone of limpid and delicately luscious purity—just such an instrument as one can imagine Beethoven sat down to" (Cady 1908, 148).

The student without a beneficial musical imagery encounters continual obstacles. In *Anton Reiser*, ambition and accumulated debt conspire to create an image of subjugation put on uneasy terms with his aim of advancing in the educational hierarchy. In a circular dynamic, poverty propels a downward spiral of interchanges between patron and student: "Lobenstein's dislike of him often issued ... in [making him] do the most menial and humiliating tasks. Nothing was more wounding for Anton than when he had to carry a burden on his back ... This was one of the cruelest situations in his life [increasing] the master's distaste for his presence" (Moritz 1785/1997, 74). Corroborating Moritz's novelistic account in *Anton Reiser*, the eighteenth-century theologian and teacher at the famous Halle orphanage, Auguste Francke, describes his time extended to poor students as a gift to be bestowed only on those who displayed the signs of "*ingeniorum*" (La Vopa 1988, 140), or the show of inborn gifts. The mark of *ingeniorum* was a display of comportment intertwined with achievement but it is clear, at least for elites in the German university, that the student on charity would rarely pass muster.

> The censure of ambition fused personal, social, and moral attributes. In the habitual subordination to superiors, the student became "a slave of some people who make him some generous promises that they forget again a moment later." Baseness connoted servility ... Self-degradation—in accepting handouts and in tolerating the imperious treatment of benefactors was likely to involve a dishonest presentation of the self. Whereas proper defense was sincere, servility undermined the social order because it was, in this sense, calculating. The irony was that its very lack of balance also made it clumsily apparent ... An elite social consciousness ... exemplified a common tendency in the educated public ... to lump all uneducated families into a homogeneous and at least implicitly dishonorable mass. (La Vopa 1988, 56–57)

This German-specific distribution of merit, heavily favoring the appearance of a particular kind of knowledge, diligence, and nobility, took place under conditions that made it all but impossible to carry out the study essential for success. The practice has broad implications for the dominant cultural values in schools today. Those values are made plain in the judgments of the home background of the African American student (and others) that places her, for similar reasons in addition to the historical effects of oppression, beyond the promise of success or the old sign of *ingeniorum*, *which* favored those students who showed promise through stock, origin, or kind (in Latin, *genus*).

In contrast to *Anton Reiser*, the fate that befell the student fortunate enough to have an enlightened home is described in Johann Goethe's novel, *Wilhelm Meisters Wanderjahre*. There we have an intimate view of how the privileged were selected for an elite education that would imprint the signs of cultural nobility. In "The Province of Pedagogy" from that novel, Felix, a young boy, is learning to model reverence. The master tells Felix to take the position of arms across the chest and "'to gravely, but joyfully look upwards.'"[60] The schoolmaster explains to the boy's father: "These are not empty posturings ... Each is told to keep to himself and. must not chatter about it either with strangers or among themselves, and so our teaching can be adjusted to each individual. Furthermore, secrecy has great benefits" (Goethe 1825/1989, 200). The ultimate sign of "reverence" demands standing bolt upright. This was expressed in similar terms one hundred years later with reference to public school music: "Our general educational economy ... should be put forth in an effort to [unify] art music and science around the values of sincerity, individuality and reverence" (Cady 1908, 148). In Goethe's novel, a predisposition toward reverence is inborn but it is also a *fabricated* characteristic that results from long practice. [61]In Goethe's work, inner cultivation is taught through a program of exercises foreshadowing the body culture movements across Europe and the United States. "Well-born, healthy children ... bring a great deal with them. Nature has endowed each of them with whatever he would need for time and duration. Our duty is to develop these things, though they often develop on their own. But there is one thing that no one brings into the world ... Reverence!" (Goethe 1825/1999, 203). "Only with reluctance does a person commit himself to reverence ... It is a higher sense that must be given to him and that develops by itself only in certain specially favored natures, who from time immemorial have been called saints or gods on this account" (ibid., 204). But who has and who lacks this "given" quality from the outset? Goethe takes pains to differentiate this potential nobility from the "natural" inclination to show fear, especially in regard to "primitive" peoples who, he says, are enmeshed in fear and strive for freedom only to be driven back into fear, over and over. Fear indicates a lack of reason. It is synonymous with nonenlightenment, false religions, and uncivilized races. For Goethe, as for many intellectuals in his era, the "higher sense" is inborn; it encompasses religion in racial terms—the "well-born, healthy children" of Teutonic descent. For Frances Elliott Clark, director of the educational department at Victor, a similar stress on inheritance of prototypical sensitivity was the basis of a national curriculum for teaching music appreciation. "Good pedagogy ... should philogenetically and ontogenetically [be] the basis of all musical development" (Victor Talking Machine Company 1923, 21–22).

Marks of Subordination

In the role of producing refined citizens, the teacher, agent, or observer must appear to recognize students' merit with impartiality. Yet, the dynamics of disqualification, as we saw in the case of having an adequate supply of musical imagery and standard language, are already set in motion by their material circumstances and present values.[62] Anton's participation in a choir that took charity meals dramatizes how musical merit was gauged by the choir's poor circumstances and how it was a source of embarrassment for listeners: "When a large crowd of beggars approached the carriage singing hymns, the (clergymen) offered condolences (for this performance) to one of their colleagues who had remained aloof" (Moritz 1785/1997, 134). Many other scenes in the novel illustrate the instantaneous judgment of abjection and the process of *selectumingeniorum* used by the German academic clergy to advance the student with a particularly appealing carriage and speech. The *selection by kind* reverses this logic to disqualify those of "low" birth, such as the at-risk in schools in the United States today. Linking *Anton Reiser* to the present the reworking of the old system of charitable board in its present form of the federal school lunch program.[63] Here, many categories—racial, economic, and ethnic—are used for the distribution of federal assistance. Meal assistance is often read as an indication of eligibility for remediation and the condition of being "at risk" of school failure. Though logically in contradiction to the mission of the democratic, modern American school, music uses enrollment in the federal assistance programs to dissuade students from electing music instruction. The selection of the already-selected continues the old practice of insuring against moving students out of "their station" (Tapper 1914, 129) and beyond their place.   The sequestering of human types also occurs at the highest levels of achievement as well. Allan Keiler discusses this dynamic with regard to some of the touchy interactions between the contralto, Marian Anderson, and potential singing pedagogues who were prominent in operatic circles. Her experience was that "special conditions" were invoked to distinguish her difference from others who more easily met the usual selection criteria. In Anderson's case, in order for her talent to be fully recognized as equal to any operatic voice, a complicated web of racial politics would see to it that she *would* rise above her place, as it were. The effect of having a lower place was inherently problematic in that the very question of "place" for a Negro singer churned up doubt that she could securean equal place in the profession.[64] Such is the paradox of selection by human type and its paradoxical double gestures of acceptance and estrangement.

In my experience, the financial assistance given to promising African American students for private music study always came with the doubt that they would "rise" above their circumstances. More "special" advice always seemed to reinforce such doubts.[65] I remember one incident that took place when a Southeast Asian "at-risk" piano student was given a scholarship to study with a highly acclaimed teacher of Asian background at a professional musical institution. The teacher noted the student's talent but, while interviewing the student's mother, found fault with the home environment for her absence (at work) in the evenings. Obviously embarrassed, the mother, who held two part-time jobs, made no response and the student withdrew from instruction himself the following week. This mismatch made a large impression on my thinking about the dysconscious practices of selection for advancement that are not about race per se, but cluster around a chain of circumstances that links perceptions of worthiness to, for example, the routine of the student's household and the mother's presence.

The idea that a student is overstepping the (largely unconscious) boundaries of the selection of his human type was visible in the music teaching literature and autobiography long before the present day language of diversity and multiculturalism. As many scholars of race and education have pointed out, the language of diversity comes with the erasure of the "normal" or the nondiverse. In this respect, Frederick Douglass's *Narrative of a Slave* continues to be relevant to the socially unconscious operations of inequality. When masters taught slaves, it was often against the advice of others who feared slaves would seek something other than their "place." "Very soon after I went to live with Mr. And Mrs. Auld, she very kindly commenced to teach me the ABC ... As soon as Mr. Auld found out what was going on, and at once forbade Mrs. Auld to instruct me further, telling her, among other things, that it was unlawful as well as unsafe to teach a slave to read ... It would forever unfit him to be a slave. He would at once become unmanageable and of no value to his master" (Douglass 1845/2001, 31).

For enslaved African Americans, schooling had to be done in secret: "I held my Sabbath school at the house of a free colored man, whose name I deem it imprudent to mention; for should it be known, it might embarrass him greatly ... I had at one time over forty scholars ... all ardently desiring to learn" (ibid., 59). The sense of overstepping boundaries appropriate for class and race prevailed in free territory as well, where schooling was more readily available and permitted. Harriet E. Wilson provides another example in her novel, *Our Nig; or Sketches from the Life of a Free Black in a Two-Story White House, North*: "I have let Nig go out to evenings a few times and if you will believe it I found her reading the Bible today just as though she expected to turn pious nigger and preach to white folks. So now you see what good comes of sending her to school."[66] Vestiges of the master and slave or indentured servant tradition were never, until quite recently, far from the conditions and practices related to music study and teaching.[67] Nor were the lines of taste far from notions of genius and its followers such as the Victrola's obedient dog, Nipper. With respect to the early twentieth-century music appreciation curriculum, the pastoral role of music teachers made it plausible to ask, "How can school children be expected to appreciate good music, if they hear no good music? Of what value is it to tell them that ragtime [sic] is pernicious, if we give them nothing better?"⁶⁸

In this chapter, I have attempted to pinpoint how a reading of comportment and language assigns one to the pernicious class or to a class that has the grace of good musical taste. Its shadow side is a taste for the "pernicious" forms of music to which students without guidance fall prey. A key figure in the Delsartean body culture movement in the United States summarizes the calculable effect of social polish as "[the cultivation of body posture that is] at once the knowledge, the possession and the free direction of the agents by virtue of which are revealed the life, soul and mind. It is the *appropriation* of the sign to the thing" (Stebbins 1902, 91; emphasis added). Written in 1902, at the moment when music appreciation entered the curriculum, Stebbins speaks for a similar mastery of the signs of whiteness and cultural superiority without reference to prior acquisition.⁶⁹ However, she leaves out the core issue—that the virtues of "mind" are only recognizable on those who have already surmounted significant racial, linguistic, and dispositional hurdles. What *Anton Reiser's* indenture and Frederick Douglass's struggle with slavery show is the difficulty, if not the near impossibility, of whole classes of people being granted the recognition of intelligence to the image of "The Thinker."

10
Rethinking Participatory Limits
From Music Standards to Hip-Hop

Interviewer: Do you think the new curriculum guides will be followed by most music teachers in the district?
Mr. K., music teacher: You know, I think a lot of them will try to, but there will be a problem because some won't follow it at all and then the next teacher will have to pick up the pieces. To do the new curriculum well you have to work on it about 80% of class time—drilling all the elements and notation skills over and over. That leaves very little time for just doing music.
Interviewer: What's the most difficult aspect of teaching for you?
Mr. K. I think it's when the kids, especially the more outgoing ones, just want to sing along or take part of the lesson or interact. I am completely torn between reining them in and just letting them enjoy it. (Gustafson 1991, 16)

This final chapter focuses on more recent (post-1960) curriculum guidelines and standards. It is about the mood and tenor of those reforms and how they reflect contemporary *Nation at Risk* and *No Child Left Behind* movements for national educational reforms. In the last two decades, the music curriculum responds to national emergencies such as the child "left behind" with a profile of the child who prepares for the global economy. In this chapter, I analyze proposals made by the state of Wisconsin issued during the last decades of the twentieth century and curriculum guides from the period from 1997 through 2005. I also describe some of the controversy surrounding hip-hop that can be considered reenactments of preferences for particular dispositions. Toward the end of the chapter, I take up the process of transforming (alchemizing) hip-hop practices as music study and the double gestures involved in making hip-hoppers into school music exemplars. This is followed by some concluding remarks on proposals by other educators who are also working to break down the demographic boundaries so detrimental across subject areas as well as in music programs. I conclude with some thoughts on the difficult contradiction, even oxymoron, of planning democracy for others and the limits of progressive thought and "democratic" reform.

Science and Psychology, 1960–80

The Tanglewood Symposium of 1967 related reform in academic subject areas to music through the pursuit of the "sciences" of cognitive psychology and "scientific" musicology.[1] As an academic conference, the Tanglewood project had no legislative power, but it exerted a good deal of influence through position papers that called for diversification of the repertoire and teaching music in a more democratic and scientific way. The symposium proposed that the music teacher's role in contemporary society rested on the scientific analysis of musical objects. This meant that the work of the composer had to be given explicit definition through musical analysis. The general music student would be instructed in a graded system of skills on a par with many aspects of professional musical practices.

Following Tanglewood, the developmental sequence of choice for teaching young children borrowed from Jean Piaget and Jerome Bruner's theories of spirals of knowledge, replacing the grade level categories of G. Stanley Hall. Piaget's theory of conceptual learning was used in music education to rework Dewey's idea of knowledge derived from action. For music, this meant that concepts should be taught through exposure (termed '"discovery") in curriculum.

From the "object lessons" in the early nineteenth century, the twentieth century Progressive Era's project method, and from there to other curriculum movements, there is a noticeable shift to performance skills in the mid-twentieth century. The Kennedy administration's support for the performing arts was instrumental in bringing live, professional music into communities on programs funded through the National Foundation on the Arts and Humanities and the Contemporary Music Project. Assisted by funding in some local school districts, public music instruction attempted to foster skills connected to professional musicianship. For listening, The Tanglewood Symposium, coinciding with the Civil Rights Movement, is often cited as the turning point for bringing jazz and other genres into the curriculum.[2] The larger consequence was the formation of notions of diversity, the highlighting of formal analysis of music, and an emphasis on rather traditional notions of music's social significance. Most striking is the part to be played by popular music. A number of these shifts were outlined in a review of the Tanglewood Symposium: "The popular music of our youth embodies high art and a content, both cultural and aesthetic, that must inevitably receive serious attention . . . blues, jazz, folk music . . . and rock. Financially, and in the size of its audiences, it is the dominant music of our era . . . Anyone concerned with the arts [grasps] the social and aesthetic force of this music . . . An analysis of rhythm, cadence, fugue, of variation . . . ensures that it is an analysis they will be interested in and will remember" (McAllester, as cited in Keene 1982, 362–63). However, the Tanglewood reforms had even more conservative results. Professional terminology appeared in music text, but the popular genres occupied a marginal space. The widely used music teacher series, *Exploring Music*, in volumes published between 1966 and 1975, includes few out of approximately 130 selections that one could call the popular music of youth—for example, "La Bamba." The rest were folk tunes and international jazz classics. The omission points to the difficulty of making what is popular an object of mind when it carries the connotation of serving mass tastes and material interests. *Exploring Music* does fulfill the Tanglewood emphasis on professional knowledge with, for example, legato, polyphony, and rondo as named technical concepts the listener should take note of. Instruction is appropriate for the child who is not a just a child but a "young musician" (Boardman and Landis 1975, xi). The book's scheme for introducing musical concepts was adapted from Jerome Bruner's spiral accumulation of knowledge encompassing the ideas of a musical normality and abnormality.[3] The spiral schema was used as a curriculum model in the pilot program I had studied at the kindergarten through second-grade level. What is notable was that, like earlier frameworks, the spirals produced an overlapping pattern of the apt learner as growing out of the practice of bodily gesture to rely on music literacy.[4] Distinguishing the normal child from one behind, the comparative results are not deliberated but emerge as part of learning to forgo gesture and embrace mental operations. "A child may first communicate understanding of a concept through gesture. Later, he will demonstrate comprehension through increasingly precise visual and verbal images. Eventually he will use conventional notation and musical terminology" (Boardman and Landis 1975, xi). The diagnostic function of this approach creates a biography of sorts that is predictive of the child's academic performance. In the nineteenth century, experimental psychology and ethnology articulated various theories about "primitive" versus "complex" understandings of music and located these tendencies along a continuum of acceptable versus abject human types. Today, there is a different, though nonetheless comparative, way of reasoning about music in the intersection of music and cognitive science. With this in mind, the language in several curriculum publications—the Wisconsin State Music Standards (1997), a Wisconsin Department of Public Instruction curriculum guide, selected publications by music educators, and arts assessment guidelines commissioned by the U.S. Department of Education—deploy a variety of comparative categories. These are, for example, various forms of cognition, employability, self-assessment, and other terms that sort students as listeners and music makers. Although official texts in which these rubrics appear do not determine practice in a direct way, they provide a language to compare students that leaks from research, administrative, and policy documents into the classroom. This language attempts to rid itself of racialism through its objective concern for accurate and fair judgment, a process Pierre Bourdieu refers to as "symbolic violence."[5]

Current Curriculum Standards and Participatory Limits

An analysis of some of the embedded comparisons in teaching guidelines make up the gist of the dominant dialogue on this subject in cross-state collaboration and national discourse on arts education Web sites.[6] State standards for kindergarten through twelfth grade specify six modes of activity, graduated by complexity and difficulty for different grade levels. Here, I focus on standard descriptors numbers three through six:

3. Demonstrate perceptual skills by listening to, answering questions about, and describing music of various styles representing diverse cultures.
4. Use appropriate terminology in explaining music, music notation, music instruments and voices, and music performances.
5. Identify the sounds of a variety of instruments, including many orchestra and band instruments and instruments from various cultures, as well as male and female adult voices.
6. Respond through purposeful physical movement to selected prominent music characteristics or to specific music events while listening to music (Department of Public Instruction 1997, 12).

Recently issued curriculum guidelines[7] discuss how to implement standards. These explain a new strategy to incorporate professional judgment without testing. Though grades are given on report cards, there is a concerted effort to make evaluation consistent with self-perception. From informal observation in recent years (2003–5), selections used in protocols for general music classes, kindergarten through eighth grade, included classical pieces, traditional children's songs, rhythmic exercises with percussion instruments, and multicultural selections combined with dance. Diversity is handled in similar fashion as in *Exploring Music*. For example, some classes had West African drumming lessons and clapped to samba rhythms played on CD. From what I could tell, students joined in with more or less "free" movement. However, during the question and answer session that followed, the class divided along demographic lines. One group responded to questions about the characteristics of rhythms; the other group did not participate in the verbal session even when directly queried. Overall, the process resembled the events I described in the beginning of this book. I suspect that the traditional emphasis on separating thought from entrainment also operated in these middle school general music classes.

Purposeful Movement, Diversity, and Cognition

This seemed all the more important to me because by this time, state standards (see Standard 6) as well as curriculum guides specified "purposeful physical movement" as the catchall for limiting rhythmic marking to hands. The word "purposeful" is semantically tied to bodily restraint in instructions on listening and rhythm and uncannily reiterates the older warning that jazz would corrupt youth forcing them to lead purposeless lives. Like the early twentieth-century warnings against extroverted movements in ragtime and jazz, "purposeful movement" codifies what is proper and correct. Some contend, today, most notably with regard to rap and hip-hop, that its totality is negative. "Hip-hop exploded into popular consciousness ... rappers were soon all over MTV, reinforcing in images the ugly world portrayed in rap lyrics. Video after video features rap stars flashing jewelry, driving souped-up cars, sporting weapons, angrily gesticulating at the camera, and cavorting with interchangeable, mindlessly gyrating, scantily clad women" (McWhorter 2003). The similarity of this claim with older descriptions of ragtime are striking: "[The popularity of ragtime] heightened concerns that a dark terror, figured in terms of 'ragged rhythms' ... seducing an unsuspecting white populace ... [The city's] restless, bustling motion also communicated the worst of modernity's consequences ... the perceived threat of contamination of white women and men by syphilitic African Americans" (Radano 2003, 235). "Purposeful movement" also delimits what can be considered as "diverse" and, therefore, acceptable. Music standards credit the policy of representing "diverse cultures" with achieving equity[8] and the limited concept of multiculturalism foists an "around the world" strategy, sorting music according to high and low: or "our music" and theirs (Koza 2003, 101). This is complicated by forbidden desires projected on to low domains that create "ghettoes of difference" (Kowalczyk and Popkewitz 2005, 423).

Such competing definitions for music and music learners have been consistent features of critics' and educators' debates from the early 1800s. Recent reforms give us new cultural restrictions on the good ear and portray the student as "a life-long learner" (Department of Public Instruction [DPI] 2002–4) who plans her own musical development as a way to participate in the global society. Rhetoric is not explicitly about the "drifter" or the "dancing mad" as it was in the 1920s, but about obstacles to the achievement of success, needs, deficits, and the failure to prepare for constant change.[9] All of these inscribe the profile of a particular child who is worthy of participation in school music and in society in general.

Psychology and Music Standards

According to the most recent curriculum guidelines, the teacher is to create an atmosphere of collaboration and "supportive socio-emotional climate" for the cultivation of skills leading to employability (DPI, 2002–4). This "sociocentric" point of view, taken up by academics in music education, is, purportedly, to understand motivation: how [students] learn, with what eagerness or apathy they approach musical tasks.[10] Psychological competence in music now includes "metacognition": "For most of the students in our classes, metacognition [defined as the ability to think about thinking] is the hardest thing to grasp. Perhaps, for some of our students, it is impossible. The inability to monitor one's own thinking ... can be applied to disabilities ... Children dealing with the difficulties of their own diversity, due to either learning or culture differences, usually cannot monitor their own learning" (Welsbacher and Bernstorf 2002, 158). My focus on the above is on the correlation of deficits to "diverse populations." In this sense, the new psychological diagnoses are reminiscent of the psychoacoustic comparisons of listeners that divided the alert child from the insensible.[11]

Self-Assessment

The arts curriculum reforms in Wisconsin in the last decade have intended to level the ground between students from different backgrounds by avoiding bell curves and tests. Self-assessment and peer-assessment are new dimensions of Project Zero's *Arts Propel* that has been applied to music in Wisconsin for this purpose.[12] *Arts Propel* shifts assessment from teacher evaluation to student construction of knowledge in a "process folio" that contains the record of students' assessable work. Although intended as a way to represent learners' competencies without comparison to a norm, the "process folio" does not so much avoid norms as it inserts another normative procedure to replace the direct evaluation of the teacher. In this regard, the portfolio approach acts as a mental map for the selection of self-reflection, metacognition, and monitoring that overlap the student's demographic position and family background. This has the unfortunate effect of making the lack of self-reflection a deficiency expected of "at risk" youth in diverse urban or rural educational settings.[13] Distinctions between those whose thinking (metacognition) aligns with "purposeful movement," and those whose movements lack purpose activate the registers of disability. Without crediting differences of reasoning about music and the body's entrainment, the process folio reduces the standing of the child whose cultural community is not represented in this curricular language. While cognitive psychology appears to bring the promise of science to include how students reflect on their own performance, it produces another hierarchy with deficiencies that nonetheless reiterate what music is and how it should be listened to and performed.

The Rubric of "Authentic Work"

The ethos embodied in standards and curriculum guides such as the "A" word about so-called alternative curricula. It has had a very limited impact. While improvisational "jamming" experiences can offer some opposition to the traditional program, provisions and facilities suitable are, by and large, unavailable[14] and jamming is not part of the training of music teachers in the United States.[15] Aside from some isolated examples, it seems to be the case that students who display their enthusiasm *on the body* are considered lacking and are looked upon with suspicion.[16] In a racially antagonistic atmosphere, differences in musical tastes are used to reify boundaries between listeners within age groups, economic classes, and racial categories.[17]

Introduction of the process folio discussed previously, derives its authority from the assumption of whiteness as normative and blackness as aberrant.[18] The rubric of "authentic work" imparts stability and universality to a particular set of identifications such as exertions of the "good ear" discussed earlier in this book. Yet, whether affirmed as rules or suggestions, the use of the term "authentic" is historically situated in systems of reasoning that make it possible to endorse a particular set of values and cultural theses about how people should think, plan for, and act in relation to music.[19] This notion of "authentic" musical activity restricts ideas that might be brought to bear on the attrition rate of African Americans and others from school music programs. Meant to erase the top-down imprinting of official standards that educators felt were outdated, the terms "authentic work," "real-world concerns," and "adult practitioners" (DPI Wisconsin 2002–4) appear general enough to allow the autonomy of teachers in the classroom who are characterized as knowing "what works and what does not" (Pontious 2005) instead of asking, what is our work?[20] In this sense, the use of the word "authentic" is a self-referencing point of view that insures already existing borders. The restrictions to what is authentic flow from the concepts of civic-mindedness, family values, and "home and larger community"(DPI Wisconsin 2002–4) used elsewhere in the guidelines. The latter recalls the longstanding tradition of school songs, documented in earlier chapters of this book that functioned as performances of nation building and inoculation against degenerate forms of culture. My argument is not with the values of home and community, rather it is concerned with the way the music curriculum has fabricated ideals of community to compare the deficiency of populations rather than promote a general understanding of the complexity of culture in a democracy.[21]

Musicologists have written eloquently about the contests between aesthetic values and musical and social conventions that bear inscriptions of race, gender, and class. For the most part, however, the field of music education has not mined this rich literature to analyze to the demographics of school music programs nor intervened to provide a meaningful forum within mainstream national organizations where equity concerns may be heard.

The Case of Hip-Hop

The case of hip-hop presents an opportunity to observe the double gestures of inclusion and repulsion that remain unquestioned in the field. At this moment, a number of generalists and music specialists see hip-hop as a significant part of urban culture that has fallen victim to racial politics[22] at a point when it might impact attrition form music programs, high school dropout rates, and the enrollment levels in post-secondary education.[23] From this perspective, some argue that hip-hop has a unique capacity for the expression of social issues and problems. "One effective strategy to mobilize youth was the use of hip-hop youth culture ... For example, while youth organized to defeat Proposition 21 in California, youth organizations, community activists, and local hip-hop artists joined forces and organized hip-hop concert to conduct mass political education and distributed flyers with youthful graffiti art that encouraged disenfranchised youth to vote and participate in the political process" (Ginwright 2004, 127–28). "Culturally, a lot of young people do not read newspapers or even if you pass them a flyer, they might read it but it's not as real to them because it's an old way of organizing. So hip-hop can bring us new tools to organize" (Sydell, as cited in Ginwright 2004, 128). Above, hip-hop is presented as a tool to educate future citizens and takes precedence over its controversial value as entertainment. Objection to this idea comes from those who see the Afrocentric idea of curriculum[24] as a more positive, more dignified approach to school achievement. In contrast to the Afrocentric approach, which, in some cities, has had difficulty gaining a foothold, hip-hop centered instruction has met with positive, if limited, success.[25] "For Black youth, hip-hop culture is a vehicle for expressing pain, anger, and the frustration of oppression" (Ginwright 2004, 133). Combining hip-hop with Afrocentrism, in Ginwright's view, provides a powerful critique of the government and democracy as well as a point of entry for social action.

Like the treatment of jazz, rag, and the spiritual in the early twentieth century, hip-hop is often identified as black. This sequestering denies its percolation as a worldwide form of expression emerging from hundreds of different geographical and ethnic spaces. Many of these sites are virtual, occurring in cyberspace where they have no specific geographic location, ethnic, or racial affiliation—a cultural collage that makes it "the most multicultural past time ever" (University of Wisconsin–Madison, 2006). Yet, in spite of the electronic, global feedback loops of dispersal and production, hip-hop receives more severe treatment than, for example, white "death metal," which mimics violence and misogynist attitudes. In this discrepancy between the categories of white and black, the "hostile, deliberately ugly, violent, fantasy-drive music of white kids" (Jukebox, 2006) receives very little censure. Again, the comparison to early twentieth-century racial archetypes is relevant as a discourse that overshadowed the multiplicity of musical publics engaged with jazz.

In spite of widespread resistance to hip-hop in schools, some public music instruction sites have begun to represent hip-hop groups as "learning communities" that engage social relations in positive ways.[26] As I will describe, the use of this language establishes hip-hop's credentials at one level while articulating its opposite—rejection, a dynamic that is especially rife when any cultural genre is perceived as black.

Hip-Hop in Music Education

Ingroup critique by hip-hop artists has been described as both reflexive and competitive, entangled in conversations about past recordings and comparisons with other emcees (rappers), beat makers, deejays (synthesizer and turntable handlers), and visual graffiti artists. In place of the traditional analytics of melody and rhythm, there is another vocabulary at work: terms such as "swinging beats" as well as the concept of varying the speed of vocals when compared to the speed of beats. Lyrics are intertextual and they are full of allusions to recordings. Answering to and provoking new interchanges between dueling rappers is one of many complexities of hip-hop.

Along with that reflexivity and struggle to perfect one's performance come the aptitudes for "originality" and articulation of historical lineage. In this mode, hip-hop engagement begins to resemble the fulfillment of the imperative of self-cultivation and work on an "unfinished self" that remind us of the discipline of the classical tradition. Hip-hop even develops its own canon, and while entry into this canon requires knowledge of dance movements, dress, graphics, and hairstyling, it rises above the mundane quite easily in its craving for "that certain something" associated with cultural nobility. Not only that, but research on hip-hop validates its tradition of self-cultivation and refurbishes its bad reputation by underscoring narratives of biographical striving in which rappers, for example, overcome excesses and handicaps of various kinds. Spoken Word, the multifaceted organization, active in curriculum at secondary and postsecondary levels, promotes hip-hop as a form of personal and civic salvation to redress social injustice.[27]

I draw these parallels to the Northern European aesthetic codes, not to denigrate those efforts, but to point out some of the difficulty of conceiving of curriculum that *would not* reinscribe social hierarchy in the racial modes we are trying to avoid. Narratives of salvation are, so to speak, the bare bones of schooling and high culture. Just to pursue the analogy between hip-hop's rise in status as the fabrication of whiteness a little further, the analysis of classical iconography can be informative. As the Beethoven scholar, William Newman, points out, Beethoven became a hero of the French revolution, a socialist, and a symbol of solidarity with working people.[28] "This deaf man heard the infinite . . . If there is ever evidence that soul and body are not joined, it is Beethoven who proved it . . . [with] crippled body [and] flying soul" (Victor Hugo, as cited in Newman 1983, 364). This has its parallels in the mystique woven around hip-hop performers and social altruism. One could go so far as to say that the process of recognizing the composer or rapper as a popular icon, whether he happens to be in the cultivated tradition or in hip-hop, involves honing an image beyond the normal through extravagant claims, for example, of mythic nationalism (Wagner), altered consciousness induced by drugs, sickness, or, as in Beethoven's case, political symbolism, deafness, and rumors of syphilis. According to sociologists and philosophers of aesthetics,[29] the larger part of the recognition of artistic merit, including overcoming handicap, takes the form of a specific vocabulary, ideal, and pattern of reasoning in Western thought. Its logic is circular in that it proposes a diagnosis of a cultural symptom for which the symptom provides the cure. This symptom is *cultural need*.[30] For example, in an early twentieth-century teacher manual quoted in Chapter 9, you have the manufacture of the need to go against rules: "Beethoven's deafness was his greatest sorrow . . . After his deafness came upon him he turned more and more to music . . . he wrote music as he thought it should be . . . there were many rules in those days . . . Beethoven cared not for rules . . . but expressed himself through music in the way he felt it" (Glenn and Rhetts, 1923, 11). Although represented very simply in this manual, the authors establish a necessary link between rule breaking and the *cultural need to transcend rules*. Value is formed in the linkage between perception of handicap, language, and the performance of need that it is not, strictly speaking, in the music per se but in what is said *about* the music. Sensational or sensory aspects of events or persons are not, in themselves, important aspects of the awe with which we hold them. (Soul and body remain disjoined, paraphrasing Hugo.) This is not to deny the experience of awe one might feel when listening, but to locate that feeling within the "social" so as to understand what makes our judgment possible. Beethoven's *Eroica* Symphony, for example, is a material fact in itself but it is not the gist of the manner in which we honor it. If we apply Pierre Bourdieu's theory, the recognition of the sublime in the *Eroica* and other artworks relies on seeing its invention as fulfillment of the deprivation inherent in the ordinary.[31] This is why the handicapped Beethoven writes music as "it should be." Similar phrases come to light in connection with hip-hop in educational contexts.[32] "Need" is untranslatable into any other medium than the music.[33] Even though the text is geared to teaching children, it contains that essential "need" that we also find in contemporary philosophies of music education, distinguishing the person (listener, performer) as having a superior form of taste compared to the listener who gratifies feeling itself (by crying or any other bodily sign). Deciphering of high value is in contrast to coarser forms of appreciation such as crying or laughing. High value emerged from a cultivated need for mindful activity with the simultaneous means to satisfy it. In music pedagogy, it creates the very need it satisfies and, while it presents this quality as natural or universal, it is wholly dependent on the cultural, historical conditions from which it comes. Art forms are deciphered by those who are a priori disposed to value them.[34]

Sociological studies on taste locate the source of this mental exertion in a general background or a specific education, whether formal or not. Education bestows a language of cultural need and knowledge (and variations of this lexicon). It provides a mastery of the language for a hierarchical system of distinctions, locating the work of the mind as primary and the techniques of artistry, its material aspects, as secondary. In the hierarchical arrangement of mind and body, the "cultural need" is simultaneously created and found as the mind grasps the depth, not just the materiality of the artwork. A "true" appraisal is "linked with the cerebral cortex" (Ruyer, as cited in Bourdieu 1993, 219). As Dewey wrote, the immediate expression of emotion is not aesthetically rewarding. "What is sometimes called an act of expression might better be termed one of self-exposure; it discloses character—or lack of character—to others. In itself, it is only a spewing forth" (Dewey in *Art as Experience*, as cited in Reimer 1989, 44). What we have here is a blueprint of an aesthetic dichotomy, one of high character, the other without, the same one that splits blackness and whiteness. This process made Beethoven extraordinary and the spiritual a mere marginal undertaking. Historically, located in affect or the bodily production of emotional signs, "black" expression was, and continues to be, burdened with a need or deficiency that is filled by the values associated with whiteness.

Much of the literature on whiteness centers around this historical split with many writers agreeing that the dichotomy was never complete but has always required renewed effort to sustain "whiteness" as a separate state of being. Ralph Ellison describes this phenomenon in ways that resemble the reception given Marian Anderson as the savior of freedom discussed in Chapter 7 of this book. Ellison thought that stereotypes were much more than "simple racial clichés introduced into the society by a ruling class to control political and economic realities" (Ellison, in Lensmire 2008, 309). In his view, stereotypes require the continual labor of projecting parts of envy and blame for that envy onto blackness. As I noted earlier, concern for broader participation in music education frames hip-hop as a rescue of youth through music. Soderman and Folkestad (2004) note that hip-hop furthers group cooperation so that many versions of a piece are scrutinized before it passes on any given product. Similarly, as hip-hop becomes an object of curricular debate (and analysis) by critics and educators, it is also undergoing a transformation that will make it more than just hip-hop, bringing it closer to the curriculum via the language of the disciplines connected to education.[35] However, it is the ingredient of cultural need that will transform hip-hop from an embodied, abject practice to a "cerebral" one. As discussed earlier with its capacity for inserting itself into a lineage of rappers and originality, music education itself supplies the language through disciplinary organs that feed formal and informal policy. Music journalism also makes an impression. A recent piece of music criticism from *The New Yorker* magazine provides model through which popular music becomes concert fare. "Composers are genius parasites; they feed voraciously on the song matter of their time in order to engender something new ... I don't identify with the listener who responds to the symphony by saying, 'Ah, civilization.' That wasn't what Beethoven wanted: his intention was to shake the European mind ... It [the *Eroica Symphony*] knows which way you think the music is going and veers triumphantly in the wrong direction" (Ross 2004, 148–49). Ross (above) casually describes shaking the European *mind* by taking "song matter of the time" into high culture. He argues that these blur in distinction as one fuses with the reputation of the other. "Mind" signifies the high aesthetic value of a cultural need to veer off in an original direction. Like the description of Beethoven in Glenn and Rhetts's teachers' manual, material composition can only work its magic if its significance goes beyond the predictable in fulfilling a need for the "original" artwork. Its formal and technical aspects are important only insofar as they can be related to the "cultural symptom"[36] of the unexpected turn the symphony takes. Naïve impressions do not disclose intimate knowledge of the style or notice of the work's cultural symptoms. If all that is said of the *Eroica* over time is "We love the melody," its reputation would lack an adequate symbolic significance. Now to return to hip-hop to examine how the mixture of aesthetics and social concerns transform its material overtones into matters of the mind.

The educational debates about hip-hop's inclusion in the curriculum revolve, primarily, around its social significance and its possibilities for teaching the language arts. In the music education literature—hip-hop's translation from street (electronic street as well) to school—the attribution of cultural meaning falls within the curriculum's historical concerns for the proper use of leisure, relief from the negatives of urban life, and the engagement in musical processes. The debate follows similar controversies in the selection of school songs (Chapters 1 through 4) and in the development of informal policy concerning music appreciation. The pedagogical alchemy that brought spirituals and Negro folk songs into general music (and other subjects in school music programs) reconciled association with the black body in motion with aspects of "artistic" merit. However, as I have argued, their placement in the space of "diverse" musical culture enacts the double gestures of approval and repulsion in curriculum conference proceedings and in the literature on music teaching.[37]

The alchemical production of hip-hop, according to notions of cognitive and mental processes such as metacognition, retention, decodification, and problem solving, is occurring across the country in literacy workshops for teachers. In one course on hip-hop for practicing and future teachers, presenters noted the verbal complexity and vast index of images that a hip-hop emcee typically masters.[38] In a recent issue of *Psychology of Music*, the cognitive aspects of hip-hop were explored among 250 subjects.[39] The collective weight of this conversation across professional venues contributes to making hip-hop an object of *mind* that elides the coarseness of its reputation in other spheres. "The national hip-hop political convention is a gathering of the hip-hop generation. This is a political event . . . We are using this historical/cultural framework of hip-hop as a rally [sic] cry for organizing and mobilizing young people to develop a political agenda and get them involved in electoral politics on the local, state and federal level" (The Freechild Project 2008). Establishing a historical pedigree of hip-hop has molded its embodied forms of action—dance, graffiti, fashion—into a classical tradition of its own. In an article published in the American journal *Music Education Research*, classical Greek aesthetics translates hip-hop in Scandinavia into more than a streetwise species of pop music. "With hip-hop, the ancient Greek word *mousiké* has re-emerged. In 'mousiké' the four art forms, *dance, visual arts, music and literature* are integrated. Being *mousikeal* means to be able to act in an aesthetic way where the four art forms are intertwined. Hip-hop is not only music: it is something much more, an integrated artistic expression" (Soderman and Folkestad 2004, 325). The language makes it possible to organize hip-hop as a bona fide curricular entity. To use Bourdieu's terms, the cultural need for hip-hop comes to fruition in combination with the discovery of its "universal" qualities, both social and aesthetic, that make it relevant to urban life today. A counterdiscourse interprets the debates over hip-hop as a form social control. In any case, its true essence, beyond debate, gives it uplift and nobility. "[Regarding debate on hip-hop] politicians and media folks love to play with our heads . . . They know that by controlling the language used to discuss an issue they can shape the way that issue is delivered to the public . . . shape the boundaries of debate on that issue." (jsmooth995, 2003). My reading of the reshaping of hip-hop as an object of study attributes is candidacy to mutually reinforced and widely distributed cultural needs and pastoral beliefs. This organizes power in a circular network rather than top-down and it allows for both reception and rejection since it evokes the level of cultivated exertion necessary to ensconce hip-hop in the '"diverse" category and *not* completely within art music per se. It maintains a distinction between those who "hear" hip-hop differently, let us say as *unrelated to* politics, art history, and social institutions, and those who use the terminology of social concern and aesthetics to contain its value. Above all, the association with blackness encloses it within forms of artistry that are similar to figures of the past—Marian Anderson, Duke Ellington, and Scott Joplin—who represent a form of black cultural nobility apart from universal greatness, but can be referenced in a knowing way. "What we call ease is the privilege of those who having imperceptibly acquired their culture through a gradual familiarization . . . can maintain a familiar rapport [with academic culture] that implies the unconsciousness of its acquisition."[40]

The Partitioning of Culture and Aesthetics

A familiar rapport with aesthetic codes, as Bourdieu states, facilitates self-selection. Lisa Delpit, in *Other People's Children: Cultural Conflict in the Classroom*, urges educators to communicate these aesthetic codes in more explicit ways to students who are unfamiliar with their demands and need more explanation. "There are codes or rules for participating . . . [these] relate to linguistic forms, communicative strategies and presentation of self; that is ways of talking, ways of writing, ways of dressing, and ways of interacting. This means that success in institutions—is predicated upon *acquisition* of the culture of those who are in power . . . If you are not already a participant in this culture, being told explicitly the rules of that culture makes acquiring power easier" (Delpit 1995, 25; emphasis added). While I do not doubt that the ground rules can be made more visible, their efficacy depends on whether the issue at hand is relatively superficial or whether it requires that the child *be someone else*. As I have argued, he or she finds it very onerous to become intimate with that which is antithetical to his or her core.[41] Delpit's "explicit presentation" assumes there is a state of receptivity that is there to reprogram. This thesis underestimates the extent to which a systematized *inequality* in the predominant aesthetics associated with whiteness denies that very receptivity.[42] In this sense, the template deselects those who will remain incapable or unwilling to enact aesthetic codes by way of imitating what is not theirs.[43] In the context of the integrated school in which the child is palpably and newly aware of being on strange ground, the defense of culture by withdrawal is a reasonable response in the face of perceived threat. This, I believe, was the situation in the music classes that delivered a rejection of the child's disposition. Judging from many years of observation, there was not only much explicit instruction (too much) but also great redundancy that rehearsed injury to the embodied musical experience most meaningful to many children.[44] In Figure 10.1, one of the students in the cohort mentioned in the prelude to this book has retained his memories of Mr. Hoffmann's class as one of the best of his school years (he is now twenty-eight years old). He was discouraged from strings at the middle school level for reasons that I explain in this book.

Figure 10.1. Michael Jendrisak, violinist. Duwayne Hoffman's fifth-grade strings, Randall Elementary School, Madison, Wisconsin, 1992. Photograph by Jane Armstrong.

In Figure 10.2, Brian Kenndy is shown in the rehearsal room before a middle school concert. Although he played his bass through most of high school, lack of support for African American students and their struggle with the issues I have outlined in this book led to his dropping out of the music program before twelfth grade.

Grappling with the Problems of Change

The rethinking of the history of music education in this book is dedicated to questioning the classical image of listening and participating but it also poses the difficult problem of changing the aesthetic rules. This is not about diversity but about equality. Replacing the "around the world" approach to diversity and multiculturalism with equality is,[45] as Gloria Ladson-Billings puts it, "not merely to 'color' the scholarship" (Ladson-Billings 2000, 271) but to find the linchpin that keeps inequality in place. My argument has been that unequal apportionment of cultural nobility, almost certainly germane to other school subjects, is at the core of black attrition in music. I have seen attempts at more equal treatment of musical dispositions in several fourth and fifth grade music classes led by a veteran teacher, Duwayne Hoffman, in Madison, Wisconsin. Mr. Hoffman's regular teaching practice provided music from popular venues and the classical repertoire that encouraged children to use instruments, voice, and dance in different combinations so that everyone would participate, at some point,

Figure 10.2. Brian Kennedy, bassist. Velma Hamilton Middle School Orchestra, Madison, Wisconsin, 1995. Photograph by Ruth Gustafson.

in leading a musical activity in each class. In some lessons, Hoffman pointed out structural characteristics of music that made the pieces easier to play, sing, and remember. He would include such well-known themes as those from, "Jaws," "The Pink Panther," hip-hop, and Beethoven's symphonies. In this way, he infused the curriculum with possibilities for different modes of listening and encouraged students to embody and express the music that excited and interested them. The result was that even students who had not had private lessons on a string instrument, for example, were encouraged to move beyond rudimentary levels to sixth grade ensembles at the middle school level. He also offered private lessons to teach specific techniques, showing that he recognized the role they play as stepping-stones to further opportunities in schools and careers and that there are many musical publics that represent similar, vital possibilities for musical practices.

The most salient difference between Hoffman and other music teachers I observed was that he provided equal time and space for various genres with performances of beginners as well as accomplished student musicians in his classes. At no time did he discourage extroverted movement or sing-along. The themes of famous symphonies, facts about the lives of composers, and the melodies of popular songs were cross-cultural events in which it was just as valuable to gesture energetically with head arms and legs along with the opening theme of Beethoven's *Symphony no. 5* as to perform a movement from a violin concerto by Vivaldi. In not following the traditional curriculum, he also refused to "weed out" students in their sign-up for middle school ensembles. In this regard, he backed the participation of some of the African American children I had seen withdraw from their general music classes several years earlier, but who wanted class to continue music in middle school after experience in Hoffman's classes. At the middle school level, some teachers tried to make appropriate alterations in their priorities, picking up, instinctually, on what Hoffman had been able to accomplish.[46] However, pressures in the district to shape ensembles in the traditional mold for the national organization of public school music educators contests and local events made it very difficult for teachers to prioritize a nontraditional approach.

The curriculum guides that Hoffman and others worked with and, at times, against, is best known under the umbrella terms of Comprehensive Musicianship through Performance (CMP), an approach to curriculum that, in the 1980s, captured the ground of teacher education through the promotion of conservative outlooks on aesthetics music history, theory, and performance traditions.[47] Presenting CMP as an approach to music that would not be overly analytical, these texts were seen as a remedy for the fragmented musical knowledge of the past. The CMP curriculum initiative divided music into "two aspects—concepts and skills—all related music courses were grouped into three categories—composition, analysis, and history" (Mark 1986, 184) to integrate knowledge across what were seen as distinct disciplinary divides.

Through continual appearances in an array of twentieth-century teachers' manuals, journal articles, and conference proceedings, especially at the Music Educators National Conference (now called the National Association for Music Education, Education), CMP (an initiative that had its start in 1965) reinforced traditional approaches to rhythm and meter that privileged accentuation of main or strong beats. Teaching rhythm and other musical concepts at the higher grade levels called for less physical involvement. The continuity between the pedagogical concepts in CMP and the earlier music appreciation manuals is striking, making it difficult to assess what change occurred in its wake. Handbooks for teachers of young children that followed the precepts of earlier decades, for example: "By listening ... carefully the ear will be sharpened by nuance; taste will develop as to sound, rhythm and form" (Rosentrauch 1970, 4) are similar to Agnes Fryberger (1925) and Anne Faulkner Oberndorfer's (1943) recommendation for listening methods. In Thomas Regelski's *Teaching General Music*, a teacher's manual published in the 1980s, "listening [is] an intention to aurally discriminate ... the key elements in learning to listen are awareness and interest. [Interest] provides the intention of seeking to become aware of refinements" (Regelski 1981, 9).

Like the music appreciation texts of the pre–World War II era, general music in the CMP mode continues to stress listening skills that were formulated in older music appreciation texts.[48] Change that could possibly be attributed to CMP came in the form of increased detail of instruction in general music. Where the music appreciation texts downplayed dependence on music literacy and other kinds of technical knowledge (see, for example, Oberndorfer, 1921), by the late 1980s, educational reform under the umbrella and political atmosphere of Ronald Reagan's speech "A Nation at Risk" (Reagan 1982) encouraged close evaluation of students in subject content. Smaller, testable items became the focus of curriculum designs that garnered state and national attention. An elaborate grading system for competence in dozens of musical items furnished the grounding for teachers' lessons and the outcomes I observed in kindergarten and early elementary general music.[49] At higher levels of proficiency, performance took precedence over the whole-discipline CMP design. Through the next two decades, the underlying premise of general music classes was that the student with the good ear shows a disposition for silent, analytic listening, similar to the slogan on a poster in a popular catalog that reads, "*Silent and listen are spelled with the same letters*" (Anon., Music in Motion Catalog, 2004). In the 2004 edition of Regelski's *Teaching General Music in Grades 4–8*, we find rhythm taught through the traditional emphasis on main (strong) beats and the recommended method of clapping on those beats. "Clapping, if used, should be done properly. Hold on-hand stationary like a drum and 'clap' it with the other hand using only a little controlled wrist and forearm movement. Avoid large movements from the shoulders and moving both hands" (Regelski 2004, 96). This clearly demonstrates the minimal role that the body is to play in musical entrainment in the classroom and its connection to the listening guide of the first half of the twentieth century in terms of the traditions of comportment associated with music literacy and the cultivated art. Multiculturalism and diversity, represented by the ensemble, "Stomp," and improvised drumming music make the gesture to include the "importance of rhythm as its own musical interest" (Regelski 2004b, 98) In music curricula, there has been no change in African American patterns of participation. The prominent gap is now evident in general music classes at the middle school level where chorus is the default for students who decline band or orchestra. There is also a division between those who look toward rap and hip-hop as a cultural affiliation and those who group themselves in relation to school music programs in more favorable ways. Validation for both groups appears to come through peers, but while one is represented as the group with high academic status, the other expresses the racial and social sorting that remains opaque in the academic culture.[50] "The use of hip-hop culture . . . establishes the boundaries of who is and who isn't popular, who gets the most support and encouragement and who is and isn't Black in [particular]" (Clay 2003, 1356). In the above assessment of behavior strategies of black youth, the construction of blackness from comportment, that is, from clothing style and other visual and aural affiliations to hip-hop, differentiates itself from the "white" culture pervasive in schools. In hip-hop culture, "race" is not a simple matter of skin color. In some cases, it ignores skin color altogether to allocate to blackness an assemblage of various cultural signs and embodied tastes.[51]

Conclusion

This book began with the high attrition rate of African Americans from school music classrooms. Watching students drop out of bands, choruses, and orchestras over a decade led me to ask what it was about the curriculum that contributed to their leaving. Many of the young children I observed withdrew from active enjoyment just at the point when the ear became the focus of the lesson. Others were increasingly isolated from their peers as singletons in school music programs. The students who sustained participation resembled *The Thinker*, a motionless body symbolizing reason. Throughout the book, I discussed the cultural traditions (some academic, some popular) that made musical differences into social and racial qualities. Rejecting the idea that schools and teachers were the root of the problem, I focused on comparisons of motion, speech, and singing that made one child's entrainment superior to another's. From this perspective, I was able to trace a web of aesthetic and social values in the curriculum that activate a biography of the child and an index of her (dis)qualifications as citizen.

Several points deserve additional emphasis. The first is that the history of racialist practices that underpin music illustrate the shifting nature of categories of revulsion and acceptance along with their political and social contexts. A prime example is metacognition that focuses on the child who does not "think about thinking." It is an example of current preoccupations of music with educational psychology but it also fits into the educational theme of the "child left behind." My method of approaching rubrics such as "metacognition," the "reflective" listener, and the "musically mature" has been to tie them to the larger historical project of the unfinished self who is to save society from a fall into a barbaric state.[52] Second, while all music practices are cultivated in some way—the verbal pluckiness of hip-hop and the attitudes of reverence for classical music are learned over time— "coarseness" takes its historical associations from fabricated notions of white sounds versus black that are evident in many facets of society and are particularly relevant to gateway educational situations such as admissions to teacher training programs. Current auditions continue an unwavering emphasis on music from the European or American high art *bel canto* tradition,[53] reminding us of the past in which "her voice, with its keen, searching fire, its penetrating vibrant quality, its 'timbre' . . . cut its way like a Damascus blade to the heart. It was the more touching for occasional rusticities and artistic defect, which showed that she had received no culture from art" (Stowe, as cited in Lott 1993, 235). Third, the distinction between deliberate interpersonal racism and the operation of what some educators call a "dysconscious"[54] performance of racial regulation is crucial. Starting from what has become a truism—i.e., race matters—I have attempted to focus on racial incitements that underpin how one is supposed to think about music, learn it, listen to it, and perform it. Although these are presented as pedagogically *necessary* to participation, they are merely prescriptive of what has been heard and seen as racially specific to fabricated qualities of whiteness.[55] This means that the interleafing of different musical styles, or multicultural curricula, appear to provide an equal footing for difference while its marginal status persists.[56] Fulfilling the school's promise of education "for all" is the kind of phrase that inserts an anonymous consensus and comparative rules that make some forms of music superior to others. The fear of failure in this mission creates "the narratives of danger and dangerous populations" (Popkewitz, *Cosmopolitanism and the Age of School Reform*, 130). I have argued that public music instruction follows this pattern by assigning admirable characteristics to "all" while simultaneously rendering them insignificant.[57] And finally, one could ask, "Isn't teaching supposed to be about 'high' values, especially those experiences that one is not likely to acquire outside of school?" The answer is both yes and no. Yes, because music is part of a socialization process and its study includes the classical genres. But the answer is also no, since the exclusive recognition of one form of civic virtue has to question those "monuments of culture" that hinder democratic participation (Popkewitz 1998, 105).

At present, significant challenges to inequality are afoot. Among many who struggle with this issue, several music educators raise questions related to the universal listener and performer. There is a vibrant movement to repartition the opportunities for music making in schools.[58] Some call for change "according to unique variables between learners, teachers, and all the many locally situated factors typically involved." (Regelski 2004, 34), Regelski suggest this option "Rather than following a rationale based on a controversial speculative and metaphysical doctrine of musical value, a doctrine with class-based repercussions that ironically excludes more people in its claims for musical redemption than it includes" (ibid.). The norms of embodied practices related to the selection and regulation of teachers are also under discussion.[59] Julia Koza argues that music teacher education absorbs and reproduces images of "attractive" femininity with ties to racialist themes of hairiness, licentiousness, and gender roles.[60] Many equity researchers suggest that the recognition of what constitutes bona fide knowledge in one culture versus another can do much to clarify the imbalance between what are largely taken-for-granted values embedded in school knowledge and less credited ways of thinking about and reacting to the world.[61]

However, the reform process itself has to face the paradox of democracy. Change must come from those who experience the walls of exclusion as well as their allies. In this sense, my work in the history of the music curriculum is an attempt to recognize a different kind of ethical responsibility as opposed to redividing the curriculum between the dominant culture and subalterns. It is a start at leaving off my own vision of social justice for an uncertain future that I cannot plan or speak for. The difference is not rhetorical. It is a matter of who decides what and for whom. Although this view is stark, the problem of reform *as it usually* happens lies in the fabricated consensus cobbled together by academics, psychological rubrics, social anxieties about substandard participation, and political slogans. The flight to consensus in public education is in part a reaction to the myth of the "monstrous reign of adolescence" (Rancière 2007, 26). In this scenario, the educator can unwittingly become the "last witness of civilization" who looks continually backward to see what has been lost. She refers to a cultural perfection that was once certain about what counts as music and the necessity for reasserting its timeless values.[62] Looking for that consensus, whether from history or a future event, passes over the crucial negotiation of interested parties and the debate among equals that would decide what is not decidable at present. Consensus cannot replace the fact of a clash of interests and loyalties.

But I am not without responsibility. My descriptions of Teshawn and David, like the black working-class girl who strained, over the crowd, to touch Marian Anderson's hand at the Lincoln Memorial, relive the enchantment of finding one's own sovereign being, one's full body and gestures, mirrored in music. Simply put, this book finds the limitations to their participation as sovereign beings a violation of democracy. In *The Ignorant School Master* and *Hatred of Democracy,* Jacques Rancière argues that democracy involves a process of political conflict over how resources in the sensible world, as Rancière puts it, are divided. Speaking for others, no matter how well intentioned, he writes, reestablishes the shepherd's guidance over the sheep. It reaffirms inequality by policing what is possible.[63] What if we could assume equality? *Might we be able to* open an area of "undecidability" about these crucial matters?[64] Would we then see (and hear) musical space divided more equally and the aesthetic of mind and body up for examination by parties who do not share its value? While *I* cannot change things *for* my students, I have put up some resistance to what encumbers their inclusion.

Finally, Thomas Popkewitz, Jinting Wu, and my husband, Jim Gustafson, have reminded me that what I say in this book about music has broader implications for education, writing history, and thinking about the limits of democratic discourse. If so, it may be because each school subject, like music, is built from a whole world of relations obscured by a veil of "common sense" about what schools ought to do and the notion that we ought to tell others what is good for them. Seeing that common sense as paradoxically odd and finding interconnections to still odder justifications makes history, for me, an exciting journey. It shows us that *what is* did not have to be and was not inevitable. It means our schools form and change under, perhaps, different circumstances than we wish. And it means we can think about schools in ways we thought unthinkable. If I can indulge the language of music for its resonance with this journey, it is that "we seem to be . . . so caught up in the rhythm of machinery that we seem to have lost our own core rhythm . . . This is the 'secret of education . . . It has to return on its trail, and recover, if it can, its line of force'" (J. Gustafson, 2008, 134).[65]

Notes

Prelude

1. King, "Dysconscious Racism."
2. Radano, *Lying Up a Nation*.
3. See ibid., 18–20.
4. Thomas Popkewitz has done extensive work on social exclusion and inclusion. See, for example, *Struggling for the Soul*. For a different way of analyzing this problem, see Lee, "The Centrality of Culture."
5. The scope and complexity of representing blackness as an identity in cultural and postcolonial studies have been deeply plumbed by bell hooks, Sylvia Wynter, W. E. B. Du Bois, Franz Fanon, and numerous others. See also King, "Dysconscious Racism."
6. See works by Michel Foucault, *The Oder of Things, An Archaeology of Knowledge, Technologies of the Self*, "The Subject and Power," "Different Spaces," "Docile Bodies," and "Nietzsche, Genealogy and History."

Chapter 1

1. Stellings, "Music Cognition Theory," 285.
2. Janson, *History of Art*, 503.
3. Bourdieu, *The Field of Cultural Production*.
4. King, "Dysconscious Racism."
5. Fitz, *A Child's Songbook*, 6.
6. I am using this song as a forerunner of the method that was called the "Pestalozzi object lesson" by American music educators, teaching moral and academic principles through emersion in song.
7. Mason, *Manual*, 132.
8. Mann, *Report*, 151.
9. The literature on the discipline of school/church behavior in both England and New England loosened strictures against music and dance in the late eighteenth century. Even though prohibitions became increasingly rare, extroverted dancing remained a social evil. See Southey, *The Doctor and Etc*. See also Preston Cummings, *A Dictionary of Congregational Usages* (Boston: S. K. Whipple, 1853).
10. What it meant to be a slave had not fundamentally altered from the terrifying brutality of the early colonial days. While there were widely differing conditions slaves had to contend with, there was always the possibility that one's circumstances, even if relatively favorable, could, through sale of one's body, be as murderous, brutal, and dehumanizing as many accounts have documented. Ronald Radano discusses several histories of the African Holocaust with regard to the pervasiveness of an oppression and cruelty that expunged family ties and cultural forms in the period up to the American Revolution.

Slave narratives such as that by Frederick Douglass were eyewitness accounts of these practices. See Radano, *Lying Up a Nation*, 59, 309n25.
11. See Schultz, *The Culture Factory*.
12. Mason and Ives, "Preface," in *The Juvenile Lyre*, 2–3.
13. Prussian public schools were to enlighten a new generation for service in the state bureaucracy. See La Vopa, *Prussian School Teachers*.
14. See Michael Broyles's description of Lowell Mason's musical aims in his *"Music of the Highest Class,"* 88.
15. Attempts to delimit this concept, for example, in the early twentieth-century work of Carl Seashore ("The Measurement of Musical Talent," 1913, and "Talent in the Public Schools," 1916), have been odious in their connection to the reasoning implicit in the eugenics movement.
16. Ment, "Racial Segregation in the Public Schools."
17. Popkewitz and Gustafson, "Standards of Music Education."
18. See Michael Broyles's description of Lowell Mason's musical aims in his *"Music of the Highest Class,"* 88.
19. See Popkewitz, *Struggling for the Soul*, 71.
20. Hastings, *Dissertation*.
21. Broyles, *"Music of the Highest Class."*
22. Ment, "Racial Segregation in the Public Schools."
23. See Bercovitch, *The Rites of Assent*.
24. J. Adams, "A Dissertation," 464.
25. Hammer, "Puritanism." See also J. Q. Adams, *An Oration*.
26. Wood, *Radicalism*, 181.
27. Shaftesbury's *Characteristics of Men, Manners, Opinions, Times* elaborates the British tradition of aristocratic manners. There is also a large literature on the profile of cultural nobility in the *Bildungromans* texts called novels of self-cultivation. This genre was brought to America via German philosophic thought circulating in music and literary circles.
28. Eric Lott describes the racial mixing and sexual cross-dressing that made blackface both a scandal and alluring theatre success in the pre–Civil War period. See Lott, *Love and Theft*.
29. Hutton, *Curiosities*, 34.
30. Botstein, "Listening through Reading."
31. Antoine, "The Rhetoric of Jeremy Taylor's Prose," 75.
32. Mann, quoted in Brooks, *The Flowering of New England*, 181.
33. For an expanded discussion of the dynamics of political power in secularism, see Talal Asad, *Formations of the Secular: Christianity, Islam and Modernity* (Palo Alto, CA: Stanford University Press, 2003), 134–40.
34. Aspects of performance that convey memory and reinforce ideas and sentiments are covered in Radano and Bohlman, ed., *Music and the Racial Imagination*. For a more in depth analysis of performativity, see Judith Butler's *Theories of Subjection*.
35. Tia DeNora, in her book *Music in Everyday Life*, discusses this process of entrainment in relation to social role in chapter, "Music as a Device of Social Ordering," 109–50. This subject is explored, in its many ramifications, for social status by McClary in *Feminine Endings*.
36. For a theoretical discussion of "difference" between subjects (individuals), see Homi Bhabha, *The Location of Culture*. This issue is one of the main differences between neo-Marxist interpretations of music, such as Adorno's work

and scholars such as Tia DeNora, who attempt to understand the social significance of music from both the production side *and* the interpreter's or listener's point of view.
37. Kay K. Shelemay, in *Let Jasmine Rain Down: Song and Remembrance Among Syrian Jews*, explores the capacity of song to solidify geographical and ethnic belonging.
38. In his analysis of public school textbooks, Walter Jones writes that from 1845 to 1865, patriotic songs were markedly increased. After 1869, nature songs again became the most common type. Jones, "An Analysis of Public School Textbooks."
39. See DeNora, *Music in Everyday Life*, 62–78.
40. See, for example, Shelemay's *Let Jasmine Rain Down*.
41. See John Sullivan Dwight, "Introductory," where he wrote that the elevating style was a model for the tempo of democratic life. Dwight was referring to the concert music of elites, contrasting the social value of cultivated art with popular forms of music. While no mention is made of the social hierarchy, the leadership of the elite class is taken for granted.
42. In the early 1800s, the term genteel was not pejorative; rather, it was a badge of belonging that indicated a high degree of social standing and social manners. See Broyles, "*Music of the Highest Class*," 296.
43. See, for example, Giffe, *The New Favorite*; Leslie, *The Cyclone of Song*.
44. Keene, *A History of Music Education*; Mark, *A History of American Music Education*. See also Reese, *The Origins of the American High School*, and Jones, "An Analysis of Public School Textbooks."
45. Songs were, for the most part, a special type of ballad; sometimes, they were not harmonized, but most often they were written out for three voice parts or with piano accompaniment in accordance with the harmonic practices common to music published in the Western art tradition; tunes from whatever source were encased and transmuted to fit the moral and musical specifications of the school setting. See Jones, "An Analysis of Public School Music Textbooks," 68.
46. Esterhammer, *The Romantic Performative*.
47. Butler, *Theories of Subjection*, 99. "Within speech act theory, a performative is that discursive practice that enacts or produces that which it names. According to the biblical rendition of the performative i.e. Let there be light! it appears that it is by *virtue of the power of a subject or its will* that phenomenon is named into being." The notion that language summons a fictive world that can displace or augment the "real" has been explored extensively in literary criticism as well. See Esterhammer, *The Romantic Performative,* and Wolfgang Iser, *The Fictive and the Imaginary: Charting Literary Anthropology* (Baltimore: Johns Hopkins University Press, 1993).
48. Mason and Ives, *The Juvenile Lyre*.
49. S. Schultz, *The Culture Factory*, 32.
50. See Mann, *Sixth Annual Report*.
51. Emerson, *The Golden Wreath*, 184.
52. Crawford, "Musical Learning."
53. For a description of "others," see Fitzgerald, "The Origins of New York's Child Care Systems," in "Irish Catholic Nuns," 394–477.
54. See Radano, *Lying Up a Nation*, 81–82. See also Lott, *Love and Theft*, 77–78, and Gould, *The Mismeasure of Man*.

55. Ment, "Racial Segregation in the Public Schools."
56. "Do They Think of Me at Home?" and "When the Swallows Homeward Fly" and "We're Tenting on the Old Camp Ground" are examples from Butler, *The Silver Bell*.
57. Perkins and Perkins, *The Nightingale*.
58. Parry, *The Evolution of the Art of Music*.

Chapter 2

1. See Morris's discussion of the formation of social administration in *Cholera, 1832*.
2. There is a large body of literature documenting an international hygienic discourse that impacted the common or public school movement and in the formation of singing and exercise societies. See Lempa, "German Body Culture." See also Hultqvist, "The Future Is Already Here." Hultqvist discussed the history of education in Sweden and the widespread use of metaphors of blood circulation in the early 1800s.
3. The task of making physical distinctions between the intellectually inclined and others was not new to the theorization of the educated individual in the 1830s and 1840s. John Locke's interest in evaluating the child, previous to instruction, was a physiological estimation—a reading of his body as one with a particular temperament or "humor." See Locke's "Some Thoughts Concerning Education." See also Baker, *In Perpetual Motion*. See pages 197–99 for a discussion of Locke's interest in the four humors; also see pages 132–33 for a history of the body's interior in relation to mechanical theories of anatomy that provided a stable map for the operations of pedagogy.
4. L. Mason, "Manual," 127–33.
5. See McMurry, "'And I? I Am in a Consumption.'"
6. Mann, "Report for 1844," 148–49.
7. See Broman, "The Transformation of Academic Medicine." See pages 148–60, where Broman discusses differences between the principles of *Wissenschaft*, or scientific research, and the art of medical practice. For a discussion of the status of *Wissenschaft*, see Emch-Deriaz, "The Non-Medicals Made Easy."
8. Mann was a follower of George Combe, a prominent Scottish phrenologist and author of *The Constitution of Man*, which was a whole system of thought about man's physiology in relation to his general health and social ideas.
9. The British study in Mann's *Sixth Annual Report* claimed that the gentry lived to an average age of forty. Persons in trade and their families lived to thirty-three years and lower-class laborers had a substantially shorter life span. Mann wrote that a physiological education would improve this statistical situation.
10. One indication of the general influence and status of the new medical knowledge of this period can be gleaned from the fact that it was allocated a large space in the French *Encyclopedie* of the late eighteenth century as a strategy for broadening readership and sales. See Emch-Deriaz, "The Non-Medicals Made Easy," 134–59.
11. Mann's recommendations followed George Combe, whose "Preface" to his long essay, *The Constitution of Man*, attempts to make one coherent system of moral and organic functioning.
12. Mann, *Sixth Annual Report*, 65. See also Tharp, *Until Victory*.

13. See Lempa, "German Body Culture," for discussion of intersection of health regimens associated with high status culture. As a phrenologist, Combe's premises were likely to be misunderstood as mere "readings" of cranial bumps on the skull, but phrenology actually worked against the summary notion of a single ranked intelligence per human being in distinction to the practices of skull comparison. Clarifying phrenology's differences with craniometry, Stephen Jay Gould writes that "phrenologists celebrated the theory of richly multiple and independent intelligences ... the major challenge to Jensen in the last generation and to Hernsstein and Murray today ... By reading each bump on the skull as a measure of domesticity or performativeness or sublimity ... the phrenologists divided mental functioning into rich congeries of largely independent attributes. With such a view, no single number could possibly express human worth." See Gould, *The Mismeasure of Man*.
14. See Broyles, *"Music of the Highest Class."*
15. See Stephen J. Gould, *The Mismeasure of Man*, on the LaMarckian precepts that underpinned a notion of inheritance of acquired traits and often a very blurry distinction between the two in this period.
16. See Schultz, *The Culture Factory*.
17. Ibid., 297.
18. See Rosenberg, *The Cholera Years*.
19. See Morris's discussion of government in New York City in *Cholera, 1832*.
20. See Anderson's *The American Census*, 22. In 1839, a National Bureau for Statistics was established in Washington for the development of statistical methods to compare characteristics of different populations and locales. The statistical procedures classified whites, colored people, immigrants, and Indians for purposes of legislative apportionment, calibrated on a fractional basis; the Irish were counted ambiguously due to their frequent official status as "indentured" or as "servant." Another result of fractional counting, Anderson writes, was that the conflation of "colored" persons with the category of insanity, solicited on the census of 1840, lowered the count of African Americans who had representation in legislatures.
21. Alain Desrosières, *The Politics of Large Numbers*, discusses how the importance of census data grew exponentially as a tool for change, dating from the "birth" of modern statistical reasoning with Quetelet's work in the 1830s. With the founding of the American Statistical Association in 1839, described by Margo Anderson in *The American Census*, statistics produced population profiles, creating the standard for health of the "normal" public.
22. Giffe, "Wine is a Mocker," 137–38.
23. Rosenberg, *The Cholera Years*.
24. Schultz, *The Culture Factory*, 32.
25. Of the many hygienic practices popular in this era, exercise was central. More organized in Europe, and especially in the German territories, Prussian gymnastic societies offered quasi-military training. General interest in cultivating the body's strength and appearance permeated the discourse about cultivating the mind resulting in the compression of mental endowment and fitness with Teutonic heritage. See Lempa, "German Body Culture"; see also McMillan, "Germany Incarnate," 136.
26. As late as the beginning of the twentieth century, popular remedies and superstitions persisted, making it difficult to convince the population at large that

germs were the prime causes of disease, not sin or poverty per se, but bacteria. See Morris's discussion of government in New York City in *Cholera*.
27. Horace Mann's notion of phrenology was attributed to his reading and meeting with George Combe, author of *The Constitution of Man*.
28. Tharp, *Until Victory*.
29. Mann, "Report for 1844," 149; see also Anderson's *American Census* for a discussion of the use of statistics in this period.
30. Mann, *Sixth Annual Report*, 36.
31. See discussion of body culture movement in Lempa, "German Body Culture," 179.
32. Mann, "Report for 1844," 149.
33. Mann's imagery is a reminder that mechanistic rather than, strictly speaking, biological theories underpinned notions of the body's vital energy. Quantities of that energy were seen as both belonging and lacking, in various degrees to different populations and human types. See Combe, *The Constitution of Man*.
34. Mann, "Report for 1844," 149.
35. See Kemper Davis's *Report to the Boston School Committee* on the merits of vocal instruction, and John Locke's comments on music as mental relief in "Some Thoughts Concerning Education," both in Mark, *Source Readings in Music Education*.
36. Gould, *The Mismeasure of Man*, chapter titled "American Polygeny and Craniometry Before Darwin," 62–105.
37. McMurry, "'And I? I Am in a Consumption.'"
38. See Toni Morrison, *Playing in the Dark: Whiteness and the Literary Imagination*, for a theorization of whiteness in relation to an Africanist absence.
39. See Alan Kraut, *Silent Travelers: Germs, Genes and the Immigrant Menace*, for description of the prevalent mythologizing that surrounded disease in this period.
40. See Rosenberg's discussion of contagion in *The Cholera Years*, 16–36. See also Morris, *Cholera, 1832*.
41. See Rosenberg, *The Cholera Years*, 24–30; see also Kraut, *Silent Travelers*, 13.
42. Mann, "Report for 1844," 144–54.
43. McMurry, "'And I? I Am in a Consumption.'"
44. C. E. Leslie, "Give the Boy a Chance," in *The Cyclone of Song*, 52–53.
45. See Popkewitz, "Pastoral Power, Redemption and Rescuing the Soul," in *Struggling for the Soul*.
46. See Lott, *Love and Theft*.
47. See McMurry, "'And I? I Am in a Consumption.'"
48. Ibid.
49. The images that pervade medical texts, often written in high literary style with allusions to images from the classical age, inscribed a regal and elevated level of civilization on the consumptive persona. They took a tone of awe or reverence in describing the ennobling effects of consumption on the victim and the power of the disease to release the highest quality of aesthetic and emotional feeling.
50. McMurry, "'And I? I Am in a Consumption,'" 105.
51. See Morris's discussion of government in New York City in *Cholera, 1832*.
52. See John Greenleaf Whittier, "Snow-Bound," in *The American Tradition in Literature* (New York: W. W. Norton, 1961), 724–41.

53. See McMurry, "'And I? I Am in a Consumption.'"
54. For a discussion of medical approval of minuet over waltz, see Lempa, "German Body Culture," 251–68. Lempa gives an account of the history of dance and its relation to body comportment in the late eighteenth and early nineteenth centuries. She argues that German social life featured all kinds of line dances, reels, and folk dancing; hornpipes, gavottes, and reels, with the minuet appearing in some contexts to consecrate gentility and insure the status of the company. By the 1820s, the waltz was entering a new phase of acceptance. Its former association with sexuality challenged the notion that dance was an act of self-cultivation at all. Where the minuet resembled the stylized movement of the ballet, the waltz had dropped the methodological deliberateness of body movement, especially of the upper body, to increase the activity of the lower body. Moreover, the waltz was a free-moving dance, where the male of the couple determined the direction and path of the twosome's progress across the dance floor. As the century progressed, the waltz disentangled itself from the projections of a negative, explicitly sexual comportment to become more a more fashionably romantic coupling.
55. See McMurry, "'And I? I Am in a Consumption.'"
56. Invitations and manners at balls and social events recognized a particular register of social comportment through limited access. Taking off one's hat was a minuet-like gesture of cultivation between members of the upper social ranks, while lower-class people did not merit any special gesture; the degenerate were to be ignored. Where the minuet was a moral education in itself, the waltz enacted an association with the signs of aristocratic refinement, in greetings and the formality of social introduction, now lodged in a broader segment of a rising middle class. See Lempa, "German Body Culture," 253, for a discussion of European manners that were also emulated by the upper and middle classes of the United States.
57. See Amy DeRogatis, *Moral Geography: Maps, Missionaries and the American Frontier*, for a discussion of the planning of Oberlin, Ohio, as a replica of the American Puritan community. Also relevant to American exceptionalism is David Nye's *The American Technological Sublime*.
58. Thomas Jefferson, as quoted in Bercovitch, *The Rites of Assent*, 84. See also discussion of Representative Robert Winthrop in Bercovitch, *The Rites of Assent*, 85.
59. See Wood, *The Radicalism of the American Revolution*, 135–37, for a discussion of emulative consumption and luxury as corruption in need of oversight through moral education. See Elias, "The Civilizing Process," on the uptake of aristocratic manners. Pierre Bourdieu discusses the importance of manners in a democratic society as an individual's distinguishing traits when other forms of cultural capital are equal, for instance, academic standing or lineage. See Bourdieu, *The State Nobility*, 214–17.

Chapter 3

1. Eric Lott gives an overall view of the pre–Civil War audiences for popular entertainment in his *Love and Theft: Blackface Minstrelsy and the American Working Class*.

2. In the early 1800s, the term genteel was not a pejorative; rather, it was a badge of belonging that indicated a high degree of social standing and social manners. See Broyles, "Music of the Highest Class," 296.
3. Orsini, *Coleridge and German Idealism*, 162.
4. See also Bradbury and Sanders, *The Young Choir*, and "Preface" in that songbook by S. W. Seton.
5. In Bernadette M. Baker's *In Perpetual Motion: Theories of Power, Educational History and The Child*, 193–95, there is a discussion of schooling of the citizen underwritten by Locke's views. Baker writes that education was not so much to rescue as to provide a system of appraisal that would make it possible for the child himself to choose between proper and improper impulses, wants and pleasures.
6. Friedrich Froebel, founder of the kindergarten in 1837, viewed instruction and experience of the child in singing, as bringing out the divine aspect of his being. See Froebel, "The Education of Man," in Mark, *Source Readings in Music Education*, 95–97.
7. Lott, *Love and Theft*, 86.
8. See Douglass, *Narrative of the Life*, 20–21.
9. See Radano, *Lying Up a Nation*.
10. For a discussion of some of genteel views on social dancing in the early nineteenth century, see Broyles on the Pierian Society of Harvard College in "*Music of the Highest Class*."
11. See chapter titled "A Motley Crew" in Ernst Krohn, *Music Publishing in St. Louis* (Warren, MI: Harmonie Park, 1988), 37–40.
12. The subject of overstimulation is mentioned by John Ruskin in his remarks on music to the Senate House in Cambridge, England, in 1867. See Ruskin, "On the Relation of National Ethics," 98–103. This was a shadow theme in the documents from the 1830s and 1840s on public music education in Boston, corresponding to the public interest in phrenology that sought to avoid both under-exercise and overstimulation. See Mann, *Sixth Annual Report*.
13. Among the various social causes to which Eve lent a hand was the argument for equality under the assumption that the white race, as a whole, degenerates under the deprivation of women's ability to rise, conjointly with men, in the progress of civilization. See Bederman, "Not to Sex—But to Race!" and the "Return of the Primitive Rapist," in *Manliness and Civilization*, 121–69.
14. The Primary School Board of Boston made claims for the physical, moral, and mental inferiority of "colored" children. See Ment, "Racial Segregation in the Public Schools," 13.
15. See page 99 in Eric Lott's *Love and Theft*, 257n21.
16. See Chamberlain and Gilman, *Degeneration*; see also Jefferson, *Notes on the State of Virginia*, 132–38, in which Jefferson discusses what he called the inferior traits of the Negro race.
17. See Gould, *The Mismeasure of Man*, chapter titled "American Polygeny and Craniometry before Darwin."
18. The crossing of what we now consider separate disciplines was part of an exchange of ideas between musical elites, school songbook composers, educationists, and European scientific and intellectual thought to a degree that is, by comparison, unusual today. This is illustrated most strikingly in Goethe's *Wilhelm Meister* novels with respect to the Mignon story and the author's interest in plant morphology. Also, see Goethe, *Wilhelm Meister's Journeyman Years*,

203, and Carolyn Steedman's commentary on Goethe in *Strange Dislocations*, chap. 3, "Figures and Physiology," 44–62.
19. Goethe, *Wilhelm Meister's Journeyman Years*, 203.
20. For a complete analysis of educational theories of the child in this period, see Baker, *In Perpetual Motion*.
21. See Pemberton, *Lowell Mason*, 63–66.
22. Herder is close to Johann Gottlieb Fichte's writing with regard to what Fichte refers to as the national destiny of Germany.
23. Boston School Committee, "Report of Special Committee, 1837," 135.
24. See Scholes, *The Puritans and Music in England and New England*, 69–79.
25. See discussion of the idea of nation in language and music in Schulze, *The Course of German Nationalism*, 65–66.
26. The proliferation of music societies that were devoted to Wagner was an important aspect of the spread of the notion of the "German" greats (Leslie Blasius, personal communication, February 23, 2005).
27. Charles White, from his *Account of the Regular Gradations of Man*, as cited in Gould, *The Mismeasure of Man*, 73–74 .
28. See Charles White, quoted in Gossett, *Race: The History of an Idea in America* (New York: Oxford University Press, 1997), 49.
29. See Jefferson, *Notes on the State of Virginia*, 135.
30. Ibid., 137.
31. Litwack, *North of Slavery*.
32. See Daniel Pick's introduction to *Faces of Degeneration*.
33. Lott, *Love and Theft*.
34. Carlyle, "The Nigger Question," 12.
35. Herder, *Kalligone*, 44.
36. Shelemay, *Let Jasmine Rain Down*; see also Bohlman, "The Remembrance of Things Past."
37. See discussion of fraternal Teutonism in McMillan, "Germany Incarnate."
38. There was an ambivalent relation to Europe that made school songs, on the one hand, representative of an honored educational tradition in music, as mentioned in music text prefaces. On the other hand, a cultural rivalry that had so long put Europe in the lead in this regard made the antebellum publication of numerous American school songbooks—numbering over thirty, according to one study—a significant factor in breaking the shackles of Europe's lead in the domain of public school music education. In Bradbury and Sanders's *The Young Choir*, several Swiss and German songs were published along with songs titled after American holidays such as "Independence Day" and "Columbia's Natal Day"; these fostered a convergence of American and Northern European tastes by putting European borrowings in service of a vision of forming a republic, rather than an enlightened monarchy. See Jones, "List of Textbooks," in "An Analysis of Public School Textbooks Before 1900," 150–54.
39. The dactyl is embedded in a 4/4 scheme of meter: a quarter note followed by dotted eighth note and sixteenth, then a quarter followed by two eighths. There are several "architectonic levels" of poetic musical rhythm that capture the manifold nature of the song's rhythmic trajectory. See discussion of contradictory and overlapping patterns of meter and rhythm in Cooper and Meyer, *The Rhythmic Structure of Music*, 1–9.
40. Emerson, *The Golden Wreath*, 119.
41. Butler, *The Silver Bell*, 213.

42. See Perry Miller, *The Errand into the Wilderness*, for a scholarly history of the use of this image in Puritan New England.
43. Weekley, *An Etymological Dictionary of Modern English*, 543, 594.
44. Gove, ed., *Webster's Seventh New Collegiate Dictionary*, 815.
45. Popkewitz, *Struggling for the Soul*.
46. This offers one take on Althusser's allegory involving a policeman hailing a subject and that subject turning around in obeisant recognition of his authority. See Butler, *Theories of Subjection*, chapter titled "Althusser's Subjection," 106–31.
47. Butler's use of psychic resistance allows for a diversity of hailing's effects. In this sense, her theory is consistent with strategies for racial identity described by Frederick Douglass, Franz Fanon, Sylvia Wynter, W. E. B. Du Bois, and others. What these strategies boil down to is the construction of a "double" that performs a subservient role while the individual holds another idea of herself.
48. Gordon S. Wood describes this dynamic throughout his *Radicalism of the American Revolution*.
49. This is what Norbert Elias makes clear in his *History of Manners* insofar as formalized interactions were part of the internal pacification that occurred in nation building in Europe. Elias, "The Civilizing Process," in *The Norbert Elias Reader*, 48.
50. See Popkewitz, *Struggling for the Soul: The Politics of Schooling and the Construction of the Teacher*, chapter titled "Pastoral Power, Redemption and Rescuing the Soul."
51. See Nye, *The American Technological Sublime*.
52. Jones, "An Analysis of Public School Textbooks."
53. See Painter, *Standing at Armageddon*.
54. Jay Fliegelman gives detailed attention to Thomas Jefferson's concern about the music of the "Declaration of Independence" as he was preparing it for oral delivery in the Continental Congress in *Declaring Independence: Jefferson, Natural Language, and the Culture of Performance*.
55. See Robert Ferguson, *The American Enlightenment, 1750–1820*, for a discussion of the uses of literary and political strategies as expressed in the Declaration of Independence and the Constitution.
56. Schelling was a follower of Kant's *Critique of Pure Reason*, in which the products of the artist are self-organizing; its teleology is pleasure. See discussion of Kant and Schelling in Orsini, *Coleridge and German Idealism*, 161–62.
57. For a discussion the late nineteenth-century coordination of language with music, or what von Humboldt termed *Naturlänge*, is close to Riemann's use of sonority. Rehding, "Nature and Nationhood."
58. See Ferguson, *The American Enlightenment*, 26–27.
59. See "Geographical Song," in Bascom, *The School Harp*, 63.
60. See Hughes, *Winter Pollen*, 334–35.
61. Quoted from Samuel Taylor Coleridge's *Biographia Literaria* in Orsini, *Coleridge and German Idealism*, 38; also see discussion on Burke and political performatives in Orsini, 162–63.
62. Coleridge, "Eolian Harp," 1381–82.
63. Mann, "Report for 1844: Vocal Music in the Schools," 149. For a discussion of the elite class's interest in musical matters, see Broyles, "*Music of the Highest Class*"; also see Lempa, *German Body Culture*. Lempa discusses the

intersection of health regimens with high status culture. See also Mann, *Sixth Annual Report*, 65; and Tharp, *Until Victory*.
64. This was part of the organic relation between language and thought that had traveled in the work of the English poets, chiefly through Coleridge, whose signature rhythms were heard in the early nineteenth century as a revolutionary break with eighteenth-century rationalism in Pope's work, for example. See discussion of the intersection of performative language at the end of the eighteenth century that brings together politics, a philosophy of language, and literature, in Esterhammer, *The Romantic Performative*, 61–67.
65. This early "literary" pragmatism is discussed in Henry Adams's essay, "The Dynamo and the Virgin," in *The American Tradition in Literature*, ed. Bradley Sculley (New York: W. W. Norton), 1136–45.

Chapter 4

1. See Lott, *Love and Theft*.
2. Coleridge and Wordsworth's joint groundbreaking edition of *Lyrical Ballads* was "to share the archaic strengths of the people" chiefly through the trope of childhood innocence and domestic tranquility; in their view, the poet would speak more spontaneously, conveying a less formal tone than the sonnet or long elegiac poems.
3. See "Emergence," in P. Miller, *The Transcendentalists*.
4. See Bederman, *Manliness and Civilization*.
5. Simon Schama, a prominent twenty-first-century art historian who has written extensively on the topic of culture as understood through painting, observes that the characteristic traits of a given landscape are imagined as well as mapped onto "real" space. See Schama, *Landscape and Memory*.
6. See Eliot, "Third Annual Report."
7. See Birge, *History of Public School Music*. One of the latest versions of the trope is found in a comprehensive history of music education published, Michael Mark's *A History of American Music Education*.
8. According to Edward Bailey Birge and Michael Mark, the first singing school was established in Boston in 1717 as the initial step toward reversing the degeneration of the hymn. Birge, *History of Public School Music*; Mark, *A History of American Music Education*.
9. See disapproval of degenerate music in Mason and Ives, "Preface," 2–3. *The Normal Singer's* introduction points to the negative features excessive "levity, frolic or idle mirth" in popular songs of the time. See Mason, *The Normal Singer*, iii.
10. Hastings, *Dissertation*, 20.
11. The line is set with the metrical stress on the word "save." Hastings, *Dissertation*, 158.
12. In *Lying Up a Nation*, 92–93, Ronald Radano writes, "In the form of the sonic projection named 'African' and then 'black' music, noise became monstrous, reinforcing European assumptions about music's profound cultural significance."
13. Hymn attributed to Rev. Christopher, *Poets of Methodism* (n.p., n.d.) in Perkins and Dwight, *History of the Handel and Haydn Society*, 21.
14. The widespread popularity of Billings's fuguing tunes and transgressive versifications indicated that, for nearly a quarter century, many churches and

musicians were captivated by his music. Michael Broyles comments that there was a deep divide between those who were both religious and musical conservatives and those looking for a way to renounce what they saw as oppressive social and musical practices: "Within the service psalms seemed at the very least to have been an outlet, an opportunity to vent feelings in a generally repressive and emotionally constricting environment. As such . . . psalmody was subversive." See Broyles, *"Music of the Highest Class,"* 42.

15. For a fuller political description of the era, see Wood, *Radicalism of the American Revolution*, 124–45, in which he discusses the contestation between the "gentlemanly" oral contract and the spread of Jacksonian, legalistic forms of contract.

16. See, for example, the preface in Emerson, *The Golden Wreath*; also see Leslie, *The Cyclone of Song*.

17. See Lawrence Levine's "The Sacralization of Culture," in *Highbrow and Lowbrow*, 83–169. See also Ronald Radano, *Lying Up a Nation*, 90: "Theories of [musical inferiority] could not exist in isolation but were posited in their relation to (and difference from) the civilized expressions of Europe, and this conversation radically reshaped the understanding of European music's history and character. The engagement of rhetoric of the musically foreign and familiar, the 'low' and the 'high' reveals, in turn, the importance of elite critical concepts."

18. Loosely led and ill-performed hymn singing appeared to be a very old concern. Both John Cotton and Cotton Mather wrote treatises on the proper singing of hymns noting the state of degeneracy circa 1700. These and similar writings mentioned the jarring of the ear that was a common church experience when parishioners engaged in purposeful disharmony. It seemed that when one hymn was ordered, some in the congregation would choose to sing another.

19. For a discussion of the westward spread of education, see the chapter titled "The Moral Garden of the Western World: Bodies, Towns and Families," in Amy DeRogatis, *Moral Geography*, 90–127.

20. It is apparent that the hymns themselves were not the sole targets of censure; one target was William Billings, for example, a late eighteenth-century merchant who composed verses similar to "Ye monsters . . ." Billings had the reputation of being the chief creator of musical mischief and relished the reactions to his irreverence and flaunted his "untutored" style to make fun of what he saw as the erudite pretensions of church hymnody. See Perkins and Dwight, *History of the Handel and Haydn Society*.

21. See James, "A Survey of Teacher-Training Programs."

22. Such divisions overrode the continual intermingling of musical genres in the sonic space of the nation insofar as public music instruction was concerned. Even while a variety of music occupied the public space, the choices for the music curriculum reflected the elite habitus. After long neglect of the importance of "folk" music to elite tastes, it was not until the early twentieth century that scholars began to document this interconnectedness. See discussion of American music historiography in Crawford, "Cosmopolitan and Provincial: American Musical Historiography," in *The American Musical Landscape*, 3–40.

23. See Tapper, *The Education of the Music Teacher*; also see Giddings, *School Music Teaching*.

24. Tia DeNora describes a turning point in the musical salons of Europe when performing artists would wait for silence before beginning their performance. See DeNora, *Beethoven and the Construction of Genius*, 71.
25. Dwight's disgust at this sort of entertainment earned him the name John Sebastian Dwight; a similar disposition led the chief patron of the Boston Symphony Orchestra, Henry Lee Higginson, to oppose the mixing of the masterworks with mere virtuosic displays by instrumentalists and vocalists. See Levine's "The Sacralization of Culture," in *Highbrow and Lowbrow*, 104–5, 120.
26. See Burke, *Enquiry into the Sublime*, xxxix. The anvils also suggest the awesome spectacle of mass industrialization. See Nye, *The American Technological Sublime*.
27. Hurley, *The Gothic Body*. See also Stallybrass and White, *The Politics and Poetics of Transgression*.
28. See Ronald Radano, "First Truth, Second Hearing," in *Lying Up a Nation* for a vivid account of revival camps that draws on a variety of primary sources as witnesses to the breech of societal conventions in revival meetings.
29. See Broyles, *"Music of the Highest Class,"* 88. Irish Catholic devotionalism was captioned as "the Beast," and the rituals of mass and catechism were heard as the undermining of civil authority, especially since many Catholics resisted public schooling for their children. See Fitzgerald, "Irish Catholic Nuns," 171–79.
30. Revivalism sometimes took the form of protests against the repression of the human spirit and the inhumanity of slavery. Evangelistic Christians saw the enslavement of the Negro as a national sin and slave owners as anti-Christs. See Radano, "First Truth, Second Hearing," in *Lying Up a Nation*.
31. See Table 3 in Jones, "An Analysis of Public School Music Textbooks.
32. See Painter, *Standing At Armageddon*, xxxii.
33. See Anderson, *The American Census*, for a discussion of this issue; see also Litwack, *North of Slavery*.
34. The power of the Northern revivalist movement was reflected in its view that the South's political leaders were anti-Christs, aiding Lincoln's defeat of Douglas and the ascension of a young Republican party in the 1850s. See Bercovitch, *The American Jeremiad*, 170–75. See also Ronald Radano's extensive discussion of the connection between the abolitionist cause and revival singing in his *Lying Up a Nation*
35. See Priscilla Wald's discussion of demographic changes in her chapter "Neither Citizen Nor Alien," in *Constituting American*.
36. By the 1990s, W. S. B. Mathews and Mary Regal were teaching music appreciation in normal schools and high schools, respectively, each using a pedagogical method that required close attention to music's formal characteristics.
37. See Cooper, *Characteristics of Men*.
38. Kant's influential *Critique of Judgment* articulated some of the salient aspects of the relation between art and universal value that came to embody a cosmopolitan outlook. Beauty's universal appeal rested on the same dimensions of mind that enabled an exercise of moral reasoning and self-governance. See J. H. Bernard's "Introduction" in Kant, *Critique*, xx.
39. See Rose Subotnik's discussion of this historical attitude in her "Introduction" to *Deconstructive Variations*.

40. See Hanslick, *The Beautiful in Music*, 72. This point of view had important racial dimensions in that it followed the contours of an aesthetic philosophy linking music to the notion of national expression. Hanslick's idea of musical disinterest required that the self-interest of the listener be converted to a kind of aesthetic and social altruism. See Fendler, "The Educated Subject," for a discussion of social altruism in this period.
41. See Leppert and McClary's "Introduction" to *Music and Society*, xiii.
42. Henderson, *What Is Good Music?* 186; see also reference to Henderson in Fryberger, *Listening Lessons in Music*, 218.
43. Anon., "Notes from a Professor's Lecture," 19.
44. See "Preliminaries" in Parry, *The Evolution of the Art of Music*, 6.
45. Goepp, "Musical Appreciation," 33.
46. Wagner, "*Artwork of the Future*," 217.
47. This subject was also discussed in Chapter 4 of the dissertation as popular entertainment. "Jim Crow" and "Zip Coon" were published in the antebellum period in *The Ethiopian Glee Book* and several versions of these, some by Thomas Rice and Stephen Foster, circulated among a wide public. See Marrocco and Gleason, *Music in America*, 263–64.
48. Part of that message was represented in the opening bars of each "sorrow song" that Du Bois used to introduce each of the fourteen chapters of these essays. In *The Souls of Black Folk*, the guiding metaphor is a journey up a mountain that performs the substance of a jeremiadic sermon. See Du Bois, *The Souls of Black Folk*.
49. Du Bois, *The Souls of Black Folk*, 29. Recent multicultural approaches represent attempts to amend this situation, but this is problematic given the absorption of a history that is limited to discussions of music per se.
50. Walter Jones, "An Analysis of Public School Music Textbooks." The "Preface" to the school music book *The Juvenile Lyre* stated that "to distort the child's proper growth by exposing him to the music of degenerates"; *The Normal Singer's* introduction states that the collection avoids songs that feature excessive "levity, frolic or idle mirth," a veiled reference to the minstrel songs and ballads popular at the time. See Mason and Ives, "Preface," in *The Juvenile Lyre*, 2–3. But the line between what was a "school songbook" and what was, for example, a collection for the general public, was not always firm. Songs such as "Nellie Gray," concerning a slave woman, may have appeared in some collections.
51. See, for example, Aiken, *Aiken's Music Course*, and Knowlton, *Nature Songs for Children*.
52. See Ronald Radano's discussion on historical collections of Negro music in *Lying Up a Nation*, 206. Radano also discusses the historical inflection of sublimity and transcendence attached to "slave music" or "Negro music" in other decades and circumstances.
53. For example, see Eric Lott's descriptions of church singing in Lott, *Love and Theft*, 16.
54. Clark, "Outline of Music History."
55. Indian death songs and minstrel ballads were harmonized and published as exemplars of Native American and black stereotypes. See Marrocco and Gleason, *Music in America*, 213.
56. Victor Talking Machine Company, 22.
57. Said, *Culture and Imperialism*, xi–xiii.

58. With regard to an epistemological frame for the Negro's "natural abilities, it falls within the trope of using Nature as a source of the capacity for language where, to extend Bernadette Baker's linkages of music, mother, language, and to music, there is a construction of Negro musical abilities as child-like; this locates its 'naturalness' in a permanent caesura along the stages of development, but also ties it to family and discourses of 'learning,' rather than putting it on an equal footing with the art-song." See Baker, "Hear Ye! Hear Ye!" 287–312.
59. This point of view is at its most virulent in Richard Wagner's writing. See Wagner, "Jews in Music," 51–59.
60. See T. Morrison, *Playing in the Dark*, 26–27. See classification of "baby and childhood" songs in Jones, "An Analysis of Public School Music Textbooks."
61. Rickford, *African American Vernacular English*, 4–5.
62. See John Rickford's summary of Wolfram Labov and others' views of linguistic traits that indicate an African American speaker in the chapter titled "Social Contact and Linguistic Diffusion: Hiberno English and New World Black English," in *African American Vernacular English*, 174–220.
63. T. Morrison, *Playing in the Dark*.
64. The large canon of songs appropriate for school use represented the speech of the nation as genteel rhetoric, diction, and sentiment. The mixture of African American vernacular usages and Standard English in "Yeep" is similar to the colloquial speech of a broad working class and some Southern dialects as well.
65. Quoted from *The Liberator*, October 21, 1846, in Ment, "Racial Segregation," 45. Similar accusations were flung at Irish Catholics; moreover, public school was seen as a way to rescue Catholic children from parochial indoctrination. See Maureen Fitzgerald, "The Origins of New York's Child Care Systems," 394–477, cited in Fitzgerald, *Irish Catholic Nuns and the Development of the New York City Welfare System*, 176.
66. See Bosco, "Introduction," in *The Puritan Sermon in America*. The paradigmatic degeneration upon which other forms of the degenerative state were imposed was the falling away of the native born American Puritans from church laws and moral code. This type of degeneration had inspired the First and Second Great Awakenings of evangelicals, the former taking place in the 1740s and latter taking place in the second decade of the nineteenth century. But this disappointment with the forms of community life was carried over into the post-Revolutionary society after the initial euphoria of independence made clear that establishing a constitutional union would be difficult.
67. For an overview of these cases, see Litwacks, *North of Slavery*.
68. This is the thesis of Ronald Radano's *Lying Up a Nation*.
69. As Toni Morrison comments in relation to the novel, the denial of black humanity denied whites full expression as well through a restriction of what could be considered as music. See also Koza, "Rap Music," 171–96.
70. See Scholes, *The Puritans and Music*.

Chapter 5

1. See Vaillant, *Sounds of Reform*, 40–90.
2. Ibid.

3. For several decades, black announcers were banned from radio stations in Chicago. See Vaillant, "Sounds of Whiteness."
4. See discussion of Carl Jung in Golston, "Im Anfang War Der Rhythmus."
5. Erskine, "Adult Education in Music."
6. For analysis of the child study movement, see B. M. Baker, *In Perpetual Motion*.
7. K. Miller, "Americanism Musically."
8. See fuller discussion of the performance in Mantle, "Mantle Hopes," 14.
9. See Lott, *Love and Theft*, 112.
10. Quotation from Joaquin Miller's "Song of the Centennial" in Rydell, *All the World's a Fair*.
11. See K. Miller, "Americanism Musically," appendix.
12. See Vaillant, *Sounds of Reform*, 79.
13. See Sovetov, *Aunt Jemima*.
14. The class was taught by Mary Regal in Springfield, Massachusetts. See Keene, *A History of Music Education*.
15. Burney, *A General History of Music*, iii.
16. W. Mathews, *How to Understand Music*.
17. See Botstein, "Listening through Reading." See also DeNora, *Beethoven and the Construction of Genius*.
18. See Stallybrass and White, *The Politics and Poetics of Transgression*, 2–3.
19. See Immanuel Kant, *Critique of Pure Judgment*, 165–220, for discussion of standards for beauty's universal value that came to embody the reasoning individual on pages.
20. See Fliegelman, *Declaring Independence*.
21. See Radano, *Lying Up a Nation*, for a discussion of racial formation and characteristics of black music making.
22. See Koch, "The History and Promotional Activities of the National Bureau for the Advancement of Music."
23. See Derek Vaillant's description of the Chicago race riot of 1919 in *Sounds of Reform: Progressivism and Music in Chicago, 1873–1935*. I have also referenced Carl Seashore's music ability tests, put to use in many public school districts as part of a larger historical project on classifying musical characteristics of races and nationalities that can be traced back to Frances Galton's research on eugenics and hereditary traits.
24. See Levine, *Black Culture and Black Consciousness*.
25. See Blesh and Janis, *They All Played Ragtime*, with respect to the upper- and middle-class makeup of audience and dancers to rag music.
26. See Vaillant's discussion about musical control aimed at juvenile delinquency in *Sounds of Reform*, 199–213.
27. Genteel taste and definitions of the sublime were assumed to be universal, but distinctions between types of listeners intersected, at times, with contradictory views of the sublime. Classical musical enthusiasts were divided into two camps in the late nineteenth and early twentieth centuries—Wagnerites and Hanslick supporters. For examples of the enmity between these two factions, see Surette, "Musical Appreciation for the General Public" and Francis York, "Report of Appreciation Conference"; see also Henderson, *What Is Good Music?*
28. Louis Dumont, *German Ideology*, 85.
29. Ibid., 264.

30. Tröhler, "History of Language of Education." In a paper delivered at the University of Wisconsin–Madison on March 1, 2004, Daniel Tröhler drew important distinctions between the language of German educational theory and philosophy and an American democratic outlook.
31. For a description of the social anxieties over immorality and entertainment in this period, see Vaillant, "Sounds of Whiteness."
32. See Dalcroze, *Rhythm, Music and Education*, 95, where the author quotes Lionel Dauriac as the source for the phrase "faculty of the soul."
33. See Gehrkens, "Rhythm Training."
34. The *Bildung* tradition had a Lutheran genealogy, but it also had aspects in common with American Calvinism's predestination and notions about what man could do on earth to please God. As Wingren, in *The Christian's Calling*, puts it, paraphrasing Luther, "God himself will mild the cows through him whose vocation that is." The sense of duty, *Amt*, and calling, *Beruf*, constrained the individual to his role in the world. *Bildung* entitled one to the highest respect in the eyes of God and one's fellows in the process of following a high calling. See Wingren, *The Christian's Calling*, 9. The secularization of American Calvinism provided avenues to improvise a covenant with God in which the individual's chosen path would show its rightness in one's outward success.
35. Satires by Twain and insights from Henry James's novels provide some sense of this issue. For further explication of *Bildung* in various settings and eras, see William Pinar, "Bildung and the Internationalization of Curriculum Studies," *Transnational Curriculum Inquiry* 3, no. 2 (2006), http://nitinat.library.ubc.ca/ojs/index.php/tci.
36. See Koselleck, "On the Anthropological and Semantic Structure of *Bildung*," 199. See also Von Humboldt, "A General Introduction to Language," 257. For discussion of this concept as it pertains to the American scene, see Tuveson, *Redeemer Nation*, 97–109. Von Humboldt's vision concerned the public school as a vehicle for achieving an enlightened society in Prussia. See also Anthony La Vopa, *Prussian School Teachers*, with regard to how the concept applied to educating schoolteachers.
37. Crocker, *Social Work and Social Order*, 30–33, 69–70.
38. See Du Bois, "Of Mr. Booker T. Washington and Others," in *The Souls of Black Folk*, 30–42.
39. These statistics appear in an article titled "Somebody is Getting this School Business," *The Voice of the Victor* (October 1924), as cited in R. Dunham, "Music Appreciation in the Public Schools," 140.
40. Kingman, "The Place and Importance of Music," 30.
41. See Chapter 2, "Jim Crow's Triumph," in Spear, *Black Chicago*.
42. Perusal of the Internet and library catalogues online reveals a lack of literature on black students and music teachers at all levels of public school music education.
43. See R. James, "A Survey of Teacher-Training Programs," Appendix B-5.
44. See Reuben, *The Making of the Modern University*.
45. See Tröhler, "The 'Kingdom of God on Earth.'"
46. Friedrich Schiller, "Letters on the Aesthetic Education of Man" as quoted in Swales, *The German Bildungsroman from Wieland to Hesse*, 15.
47. Lippman, ed., *Musical Aesthetics*.
48. Weber. "Wagner, Wagnerism, and Musical Idealism."

49. See Surette and Mason's extensive treatment of Beethoven's Fifth in *The Appreciation of Music*, 181–221. See also Spalding's discussion of the human soul in Beethoven's work in *Music: An Art and a Language*, 128.
50. Thompson, *College Music*. See also Fleming, "Music in the High School."
51. Rydell, *All the World's a Fair*.
52. R. Thomas, *Memoirs of Theodore Thomas*.
53. See S. Green, "Art for Life's Sake"; Vaillant, *Sounds of Reform;* Lasch-Quinn, *Black Neighbors*.
54. K. Miller, "Americanism Musically," 148–49.
55. Spear, *Black Chicago*, 29.
56. See B. Baker, *In Perpetual Motion*, 465–67, for an overview of the problem of identifying the "progressive" in the aggregation of practices and philosophies traveling under that rubric.
57. See Popkewitz, *Cosmopolitanism and the Age of School Reform*, 96.
58. See Radano, "Hot Fantasies."
59. Emery, *Black Dance*.
60. See Michael Golston, "Im Anfang War Der Rhythmus." Available from http://www.stanford.edu/group/SHR/5-supp/text/golston.html.
61. See Crary, *Suspensions of Perception*.
62. Thaddeus Bolton, "Rhythm."
63. Numerous studies in Max Schoen, ed., *The Effects of Music*, are concerned with the effects of music on the muscles, emotions, and pulse.
64. Earl Barnes, "The Relation of Rhythmic Exercises to Music." See also Mohler, "The Project Method in Teaching Music Appreciation."
65. See Birchard, "Music for Individual and Social Life."
66. The reason framing music appreciation is a rendering of the alchemy of dispositions of an educated elite. See Popkewitz, "The Alchemy of the Mathematics Curriculum."
67. Thomas Bolton, "*Rhythm*."

Chapter 6

1. See DeNora, *Beethoven and the Construction of Genius*, 24–25.
2. See Botstein, "Listening through Reading."
3. Goodman, *Radio's Public*.
4. See Habermas, *The Social Transformation of the Public Sphere*. See also DeNora, *Beethoven and the Construction of Genius*.
5. Edward Hanslick, eminent music critic of the nineteenth century, wrote, "It will always, however, be a matter of course that the . . . different parts of a sonata are bound up in a harmonious whole and that each should set off and heighten the effect of others according to the aesthetic laws of music . . . It is the frame of mind bent on musical unity which gives the character of an organically related whole." Hanslick, *The Beautiful in Music*, 60.
6. See Fliegelman, *Declaring Independence*, 220.
7. Vaillant, *Sounds of Reform*, 75–76.
8. Jay Fliegelman quotes Jefferson in *Declaring Independence*, 190–91.
9. From Jefferson, *Notes on the State of Virginia*, 288.
10. See Fliegelman, 194–95.

11. See R. Dunham, "Music Appreciation in the Public Schools," 184, as well as a more general theory of fabricated communities in B. Anderson, *Imagined Communities*.
12. The National Bureau for the Advancement of Music, "Music Memory Contests," cited in R. Dunham, "Music Appreciation in the Public Schools," 184.
13. Nye, *The American Technological Sublime*.
14. See Vaillant, "Introduction," in *Sounds of Reform*, 1–9.
15. See Damrosch, "Music and the Radio," and Goepp, "Musical Appreciation in America."
16. See Crawford, *The American Musical Landscape* and "Musical Learning in Nineteenth-Century America."
17. See Kasson, *Rudeness and Civility*, and Eric Lott, *Love and Theft*.
18. Leon Botstein, "Listening through Reading." See also reference to Tremaine's efforts to increase declining piano sales in Robert Koch's "The History and Promotional Activities of the National Bureau for the Advancement of Music."
19. See Koch, "The History and Promotional Activities."
20. Musical tastes were all over the map with the advent of local stations in Chicago, according to Valliant, "Sounds of Whiteness."
21. Kasson, *Rudeness and Civility*.
22. One of the teacher training programs for music appreciation was at Iowa State University. The first text of major circulation on music appreciation was Surette and Mason's *The Appreciation of Music*.
23. These ideas are discussed eloquently in Derek Vaillant's work. See "Peddling Noise," *Sounds of Reform*, and "Sounds of Whiteness." Also see Rydell, *All the World's a Fair*.
24. Robert Rydell's *All the World's a Fair* follows the implication of comparative ethnology in the large exhibitions occurring from 1876 through 1916.
25. See Rydell, *All the World's a Fair*.
26. Robert Koch describes Tremaine's effort to transfer the National Bureau for the Advancement of Music to the sponsorship of the Music Teachers National Conference. See "The History and Promotional Activities of the National Bureau for the Advancement of Music."
27. See Claxton, "The Place of Music in National Education."
28. See James, "A Survey of Teacher-Training Programs." Also, with regard to ear training and music appreciation, see Alchin, *Ear Training and Teacher and Pupil*.
29. Fryberger, *Listening Lessons in Music*.
30. On P. T. Barnum, see Lott, *Love and Theft*, 113.
31. The black contralto Marian Anderson pursued a search for training that is a case in point. See Keiler, *Marian Anderson: A Singer's Journey*.
32. Julia Koza, professor of music education, personal communication, February 4, 2008.
33. See Spear, *Black Chicago*.
34. See Kasson, *Rudeness and Civility*.
35. Briggs, "Music Memory Contests." Susan Cook, professor of music, mentions so-called crossover artists who got their start on the Victor black label—Caruso, for example. Personal communication, October 11, 2004.
36. See James Keene, *A History of Music Education in the United States*, on the subject of Ada Fleming and Will Earheart, both teachers in the Midwest who wrote on music appreciation for public schools.

37. France Elliott Clark's work, published in journals of the Music Teachers National Association and Music Supervisors National Conference, made mention of these practices with no specifics.
38. See Keene, *A History of Music Education*, 271.
39. Notable exceptions were John Sullivan Dwight, a champion of the Negro spiritual, and others who saw in folk music the spiritual essence of the nation. See Radano, "Denoting Difference."
40. Evidence of Surette's acclaim appear in the *New York Times*, for example, in an article on his Wagner lectures; see Anon., "*School Lecture Recitals*."
41. This would include the musicologist Sir Hubert Parry and philosopher Herbert Spencer.
42. This expression is from one of the papers published by the conference. See Coffin, "Report of the Sub-Committee on Music Appreciation."
43. See Koch, "The History and Promotional Activities of the National Bureau for the Advancement of Music."
44. For example, Glenn and Rhetts, *Reading Lessons in Music Appreciation* as well as Rhetts, *Outlines of a Brief Study of Music Appreciation*.
45. For example, one series of music teacher manuals by Giddings et al., *Music Appreciation in the Schoolroom*, references the psychology of child study associated with G. Stanley Hall.
46. See Victor Talking Machine Company, *Music Appreciation with the Victrola for Children*.
47. See Mussulman, *Music in the Cultured Generation*, 109.
48. Popkewitz and Gustafson, "Standards of Music Education."
49. See Vaillant, "Sounds of Whiteness."
50. Ibid.
51. For an overview of segregation controversies in Washington, D.C., New York, and Chicago in particular, see biographical texts Keiler, *Marian Anderson: A Singer's Journey*, and Dorinson, "Paul Robeson and Jackie Robinson." See also Savage, *Broadcasting Freedom*.
52. One speaker at music teacher conferences headed the National Bureau for the Advancement of Music. See Tremaine, "The Music Memory Contest."
53. R. Dunham's "Music Appreciation in the Public Schools, 1887–1930," 66, provides an account of the overlapping of the concert world, music teachers, and wholesale commercial gramophone manufacturers.
54. Clark, "Outline of Music History."
55. Rhetts, *Outline of a Brief Study of Music Appreciation*.
56. For a thorough treatment of transcription and interpretation of Negro song, see Radano, "Denoting Difference."
57. Anne Shaw Faulkner, *What We Hear in Music*, ed. Victor Talking Machine Company (Camden: RCA Victor, 1928).
58. Morrison, *Playing in the Dark*.
59. See H. Thomas, *The Body, Dance, and Cultural Theory* and Kasson, *Houdini, Tarzan, and the Perfect Man*. See also Dalcroze, *Rhythm, Music and Education*.
60. This idea is most famously expressed in Wagner's essay, "Jews in Music," 51–59.
61. Ibid.
62. For a detailed treatment of Emile Jacques-Dalcroze, see Michael Golston's *Im Anfang War Der Rhythmu*.

63. See Hall, "The Ideal School as Based on Child Study," and discussion of Hall's work in Baker, *In Perpetual Motion*.
64. See McConathy et al., *The Music Hour*.
65. Anne Dzamba Sessa discusses these aspects of Wagner in American in "British and American Wagnerians."
66. See Mussulman, "Cosmopolitan Nationalism," in *Music in the Cultured Generation*.
67. For a discussion of Dewey's traveling reputation and library of ideas, see Popkewitz, "Preface," in *Inventing the Modern Self and John Dewey*.
68. See Chybowski, "Popularizing Classical Music and Developing American Taste," and Fryberger, *Listening Lessons in Music*.
69. In David Charles Goodman, in his forthcoming book on classical radio broadcasting, *Radio's Public: The Civic Ambitions of 1930s American Radio*, he writes that classical radio's heyday in the early twentieth century conducted a broad campaign to hold out against the tide of mass culture.
70. Chybowski, "Popularizing Classical Music and Developing American Taste."
71. See Sessa, "British and American Wagnerians."
72. See Anon., "*School Lecture Recitals*."
73. See Mussulman, "Wagnerism in America," in *Music in the Cultured Generation*.
74. See Schabas, *Theodore Thomas*.
75. On Dewey and education, see McGerr, *A Fierce Discontent*, 111.
76. See Cundiff and Dykema, *School Music Handbook*.
77. See Vaillant, "Sounds of Whiteness." See also Savage, *Broadcasting Freedom*.
78. See Golston, "Im Anfang War Der Rhythmus."
79. See Victor Talking Machine Company, *Music Appreciation with the Victrola for Children*, 139.
80. The emphasis on recall of musical themes corresponded to an interest in Wagner's work as a signature style of contesting motifs. Following this trend, a small industry of both serious and satiric commentary flourished in the United States. This created a sort of popular familiarity with Wagner's style among music listeners. As one journal put it, "'We have a great deal to learn, and the successful explication or elucidation of Wagnerian theories and practices was a challenge that cultured critics met with eagerness and energy . . . The principle of *Leitmotiv*, for example could be grasped by a sufficient number of cultured readers to make intricate satires on it" (editor of *The Critic*, quoted in Mussulman, 148).
81. This is a gradual development of intellectual thought with respect to music, ranging from Wagner's theories of drama and opera, to Nietzsche's essay on Wagner, to Theodore Adorno. The common theme is how material decadence and hero worship are mirrored in music that fails to integrate the needs of the psyche and society but capitalizes on appeal to bourgeois tastes and simplistic politics.
82. Franklin Dunham mentions Stravinsky as a possibility for instruction in schools in "Can Music Appreciation Be Taught?"
83. See experiments in Schoen, ed., *The Effects of Music*.
84. Wagner, "Jews in Music."
85. Clark, "Festival of the Nations," 14.
86. See Vaillant, "Peddling Noise," "Sounds of Whiteness."
87. See Goodman, "Distracted Listening."

88. Clark, "Music Appreciation of the Future."
89. The Romantic outlook that permeated the nature idylls in nineteenth-century songbooks also made itself felt as "the trope of authenticity." See Radano, "Denoting Difference," 511.
90. See Hooker, "The Invention of American Musical Culture."
91. See Dunham, "Can Music Appreciation Be Taught?" 83.
92. Damrosch, "A Lesson in Music Appreciation," 88.
93. See David C. Goodman's discussion of fragmented attention in his paper, "Distracted Listening," 28–33.
94. The analysis of a fragmented modernism in my argument is borrowed from Crary, *Suspensions of Perception*, chap. 1.
95. See pictures of German socialites performing the "cakewalk" in Blesh and Janis, *They All Played Ragtime*, 82–83.
96. See R, Choate, "Introduction."
97. See, for example, recent curriculum guides that align themselves with national music standards in DPI, "Wisconsin."

Chapter 7

1. See Stallybrass and White, "Introduction," in *The Politics and Poetics of Transgression*.
2. Seashore, "Talent in the Public Schools."
3. See Giddings et al., *Music Appreciation in the Schoolroom*.
4. See Cundiff and Dykema, *School Music Handbook*.
5. Fryberger, *Listening Lessons in Music*.
6. See, for example, Bobbitt, "A City School as a Community Art and Musical Center." More recently, Soderman and Folkestad, in "How Hip Hop Musicians Learn: Strategies in Informal Creative Music Making," *Music Education Research* 6, no. 3 (2004): 314–26, describe the elaborate musical rituals of so-called marginal youth that justify considering popular forms of music as worthy practices.
7. See the biography of conductor Theodore Thomas, *The American Orchestra and Theodore Thomas* by J. Russell (Garden City, NY: Doubleday,1927). See also Vaillant, *Sounds of Reform* for a discussion of Thomas's role in the Chicago Exposition of 1893.
8. Shiraishi, "Calvin Brainerd Cady."
9. See Tröhler, "The 'Kingdom of God on Earth.'" See also Popkewitz, *Cosmopolitanism and the Age of School Reform*.
10. See Baker, "Hear Ye! Hear Ye!"
11. See Fliegelman, *Declaring Independence*, 67, 69, 70.
12. Jefferson, *Notes on the State of Virginia*.
13. See Connor, "What I Say Goes," in *Dumbstruck*.
14. Schmidt's *Hearing Things*, gives a detailed account of this process as does, in another vein, Steven Connor's *Dumbstruck*.
15. See Vaillant, "Sounds of Whiteness."
16. These sounds include peddlers' cries and radio broadcasts that posited competition in airwave space. See Derek Vaillant's description of these phenomena in "Peddling Noise."
17. See, for example, the influential studies of Petzold, *Development of Auditory Perception of Musical Sounds*.

18. Barthes, "The Grain of the Voice."
19. Italian opera had a more ambiguous status in American concert halls.
20. See Pemberton, *Lowell Mason.*
21. See Radano, "Denoting Difference."
22. See Stark, *Bel Canto: A History of Vocal Pedagogy.*
23. See ibid., chap. 7.
24. Exceptions to this rule occur when choral directors call for more feeling in delivery of vocal expression. Julia Eklund Koza, personal communication, 2008.
25. See, for example, Regelski, *Teaching General Music.*
26. Barthes, "The Grain of the Voice."
27. See Stark, *Bel Canto: A History of Vocal Pedagogy,* 86.
28. In personal communications, some vocalists in training and teachers in college and university courses in music have indicated that the discouragement of belting, common in popular and gospel genres, is widespread and based, in their opinion, on anatomical risk.
29. See Bermingham, "The Effects of Performers."
30. Moses, *Developing and Administering a Comprehensive High School Music Program.*
31. Regelski. *Teaching General Music,* 63.
32. For a full treatment of Herbert Spencer's evolutionary typologies covering, but not limited to, music, see his two-volume work, *Principles of Psychology*; also see Spencer, *The Organization and Function of Music.*
33. R. Dunham, "Music Appreciation in the Public Schools."
34. Rhetts, *Outline of a Brief Study of Music Appreciation,* 6. In many Victor Talking Machine publications, "Asiatic" and "oriental" encompassed so-called African or "Negro" music.
35. See Marrocco and Gleason, eds., *Music in America,* 213.
36. See *The Music Hour* series and, specifically, McConathy et al., *The Music Hour.*
37. See Floyd, "The Implications of John Dewey's Theory."
38. See, for example, Mohler, "The Project Method in Teaching Music Appreciation."
39. The chapter titled "Stewart House" in Ruth Crocker's *Social Work and Social Order* provides a detailed account of social institutions and change that were to "rehabilitate" the Southern negro and elevate their condition.
40. Quote is from Tröhler, "The 'Kingdom of God on Earth.'"
41. Lasch-Quinn, *Black Neighbors.*
42. See, for example, Clark, "Festival of the Nations."
43. For a discussion of the relation between progress and degeneration, see "Introduction," in Spadafora, *The Idea of Progress in Eighteenth Century Britain,* and see Pick, *Faces of Degeneration.*
44. Birchard, "Music for Individual and Social Life."
45. Michael McGerr, *A Fierce Discontent.*
46. Expressed in a paper presented at the Music Teachers National Association, "The Mission of Music in Colleges," *Educational Review* 38 (1909): 132–35.
47. See, for example, the series of textbooks for teachers called *The Music Hour,* published by Silver Burdett.
48. Popkewitz and Friedrich, "Professional Development Schools."

49. For a summary of research in rhythm in this period, see Golston, "Im Anfang War Der Rhythmus." A well-known music and dance teacher, Jacques-Dalcroze, led a movement in Europe and the United States to teach rhythm through body movement. See Jacques-Dalcroze, *Rhythm, Music, and Education*.
50. Tröhler, "The 'Kingdom of God on Earth,'" 12.
51. See Sloboda, *The Musical Mind*, for an overview of Carl Seashore's tests for musical ability.
52. See, for example, Petzold, *Development of Auditory Perception of Musical Sounds by Children*.
53. The study Seashore reviewed was by Milton Metfessel (see following note).
54. This study, "Phonophotography in Folk Music," was by Milton Metfessel. See Seashore, *Psychology of Music*, 348–59.
55. Seashore's commentary on Negro singers was to defend primitive art from charges of inaccuracy due to lack of ability. With regard to Metfessel's studies especially, Seashore describes the primitive as a different aesthetic. See Seashore, *Psychology of Music*, 357.
56. For a discussion of coproduction of scientific data, see DeNora's comments on Bruno Latour's work (*Pandora's Hope*) in *After Adorno: Rethinking Music Sociology*, 38.
57. Ronald Radano documents the complex reception of Negro song throughout the nineteenth century, which was uneven and, in some cases, made moves to place the music in the art canon. See Radano, "Denoting Difference," 506–44.
58. See Seashore, *Psychology of Music*, 368.
59. Julia Koza, personal communication, 2008.
60. See Tröhler, "The 'Kingdom of God on Earth.'"
61. Steven Connor, in *Dumbstruck*, and Leigh Schmidt, in *Hearing Things*, offer a wealth of archival work on the suspicions of vocal production and audial technology that permeated the modern era. See Charles Hirschkind, *The Ethical Soundscape*, 106, for an exploration of counterpublic as opposition to the normative notion of public.
62. See Ruth Crocker's *Social Work and Social Order, 1889–1930* for an account of the migration from South to North and several of Indiana's civic notaries' concerns.
63. Clark, "Festival of the Nations," and Vaillant, "Sounds of Whiteness."
64. See Ronald Radano's discussion of African retentions and racial precepts in "Hot Fantasies: American Modernism and the Idea of Black Rhythm." For the historical record, African drumming and dance were allowed to continue in the Caribbean setting after slaves were brought from Africa but this was not the case in the continental United States. See Emery, *Black Dance*. As Emery notes, one reason for this was that the white population looked with suspicion on African religions and voodoo practices; the other was the condemnation of dance by the Protestant churches. What is now termed "black rhythm" is a historical invention that arose from the suppression of particular musical practices, slavery, colonialism, and musical interchange.
65. See Gilliland and Moore, "The Immediate and Long-Time Effects of Classical and Popular Phonograph Selections."
66. The Victor publications of records and curriculum guides specified dancing of the Scandinavian and British traditions, leaving out traditions such as polkas that used paired dance. They also omitted popular dance forms. This follows religious principles from the Calvinist and other Protestant churches. See also

Blesh and Janis, *They All Played Ragtime*. For mention of a physician concerned with the mental health of students at the Philadelphia School for Girls, see Leonard, *Jazz and the White Americans*.
67. Periodically, a type of scientific study appears that correlates musical activity with growth in cognitive ability. Sloboda's Chapter 6 in *The Musical Mind* provides an overview of research in this area. For a recent example of the debate on music and cognitive development, see Rauscher, *Discussion of Research*.
68. See Max Schoen's edited volume of psychoacoustic studies, *The Effects of Music*. For a more recent large-scale study on musical perception among public school students, see Petzold's study on distinctions of auditory perception in children in grades one through six, *Development of Auditory Perception of Musical Sounds by Children in the First Six Grades*.
69. For a glimpse of the use of racial essence in the world of music journalism that corresponded to the blossoming of experimental science on the voice, see the review of Paul Robeson singing "Ol' Man River" in Dorinson's "Paul Robeson and Jackie Robinson: Athletes and Activists at Armageddon," 18.
70. See B. Chinn, "Vocal Self-Identification, Singing Style, and Singing Range," and S. Morrison, "A Comparison of Preferences and Responses of White and African American Students."
71. In *Dumbstruck: A Cultural History of Ventriloquism*, Connor describes the oracle as sitting over a cleft in the earth that doubled as the cleft in the lower body like a mouth (stoma) from which the goddess and gods would speak. The circuit of voice and earth also inscribed a division in space and meaning between the female body and the earth's opening, over which the goddess stood to receive divine communications. In different ages and in different guises, she would "speak" prophecies in various ways, one being the method of dislocating the voice to the "stomach" and having it emanate from the lower body as if from the earth.
72. See Lott, *Love and Theft*.
73. Ronald Radano discusses some of the feminization that went into the construction of "black" music in "Denoting Difference: The Writing of the Slave Spirituals."
74. Nietzsche discusses this historical development in his famous essay, *Genealogy of Morals*. See also Joseph Fontenrose, *The Delphic Oracle*.
75. Ronald Radano cites the German scholar, Forkel, and Jean Jacques Rousseau in relation to making this point. See Rodano, *Lying Up a Nation*, 91.
76. Davis, *I Got the Word in Me and I Can Sing*; Philips, "'Stand by Me': Sacred Quartet Music and Emotionology," 245, 248.
77. See Radano, *Lying Up a Nation*, 247–50.
78. "Hesitating" might refer to syncopated rhythms that compel movements to pause in mid-beat by the standard musical lexicon or it might refer to socializing.
79. Thomas Popkewitz, in "Hopes of Progress and Fears of the Dangerous."
80. Radano, "Denoting Difference." See Judith Butler's description of performance of various states of being and gender in *Bodies That Matter: On the Discursive Limits of "Sex."*
81. One of her biographers, Allan Keiler, makes it clear that there was much physical danger in her appearance and that her fear and self-doubt about the concert revolved around being made use of as a symbol of white liberalism as well

as a valuable commercial commodity to her agent and recording contractors. See Keiler, *Marian Anderson: A Singer's Journey*.
82. For this analysis of Marian Anderson, I have borrowed several ideas from Radano, *Lying Up a Nation*, 138–84.

Chapter 8

1. A radical change in vocal technique, for example, in a summer study with an operatic vocal specialist, Anderson said later, made her lose confidence in her own abilities. See Keiler, *Marian Anderson: A Singer's Journey*, 42.
2. Informal conversations with music educators and African American music students offer some evidence to suggest the phenomenon is widespread and is a key factor in the lack of African American music teachers in public schools.
3. John A. Sloboda's *Exploring the Musical Mind: Cognition, Emotion, Ability, Function* surveys most of the areas of music psychology research. Pedagogical approaches that assert claims of universal talent are ubiquitous and have widespread impact in recent times on new teaching methods, such as the Suzuki approach.
4. See R. James, "A Survey of Teacher-Training Programs in Music."
5. See, for example, Bruce Benward's explanation of an extensive and nationally recognized system for teaching music theory online. http://www.macromusic.org/journal/volume3/06_Musical_Insights.pdf.
6. On musical families, see Francis Galton, "The Comparative Worth of Different Races," in *Hereditary Genius: An Inquiry Into Its Laws and Consequences*; see also pages 291–303.
7. Giddings et al., *Music Appreciation in the Schoolroom*, 29.
8. Derek Vaillant's research on radio stations in Chicago in the early twentieth century notes that there were restrictions. However, according to the teacher conferences and journals, these did not match the agenda of creating a larger national audience for the cultured tradition. See Vaillant, "Sounds of Whiteness." See also David Goodman, "Distracted Listening."
9. Cundiff and Dykema, *School Music Handbook*, 172.
10. See M. Golston, "Im Anfang War Der Rhythmus."
11. As I discuss in a later chapter, activist scholarship on African culture has also had its tendencies to reify identity and rhythm in order to express its opposition to "white" forms of appropriation and European or North American musical superiority. See Agawu, "The Invention of 'African Rhythm.'"
12. See comments on Thomas Bolton in Golston, "Im Anfang War Der Rhythmus."
13. Vaillant, "Peddling Noise."
14. C. M. Tremaine, director of the National Bureau for the Advancement of Music in the early twentieth century, presented the view that music education efforts toward developing the love of music were most profitably done among the young, when tastes are being formed and minds are in an impressionable state. See Tremaine, "The Music Memory Contest," 101.
15. See discussion of aspects of interior or exterior dichotomy in the child in Baker, *In Perpetual Motion*, 441–65.
16. See, for example, Barnes, "The Relation of Rhythmic Exercises to Music," and Gilliland and Moore, "The Immediate and Long-Time Effects of Classical and Popular Phonograph Selections."

17. "Wants Legislature to Stop Jazz as an Intoxicant," *New York Times*, February 12, 1922, 12, cited in Neil Leonard, *Jazz: Myth and Religion*.
18. For similar developments in ascertaining and directing attention, see Sobe, "Challenging the Gaze," and Sobe and Carrie Rackers, "Fashioning Writing Machines."
19. School songbook prefaces commonly discussed this "problem." Judging poor singers could, theoretically, have been strictly on the basis of pitch accuracy, but, as discussed earlier, pronunciation, especially dialect, physiognomy, and bearing, among other things, were part of the overall estimation of the school singer. One author writes, poor singers "should be gotten rid of." See Leslie, *The Cyclone of Song*, ii.
20. See Ruth Crocker's account of settlement house discrimination in *Social Work and Social Order, 1889–1930*. See also Lasch-Quinn, *Black Neighbors*.
21. See Dwight, *Dwight's Journal of Music*.
22. See Fliegelman, *Declaring Independence*.
23. Similar effects were discussed in conjunction with degeneracy, musical taste, and city life (see Chapters 4 through 6). For twentieth-century mapping of the internal and external effects of music, see the collection of studies in Schoen, ed., *The Effects of Music*.
24. Crary, *Suspensions of Perception*.
25. Schoen, *The Effects of Music*, 29.
26. Seashore, *Why We Love Music*, 17. David C. Goodman has dealt extensively with the subject of control over radio listening in his "Distracted Listening: On Not Making Sound Choices in the 1930s."
27. See numerous studies in Schoen, *The Effects of Music*.
28. See Margaret Flo Washburn and George I. Dickinson, "The Source and Nature of the Affective Reaction to Instrumental Music."
29. Victor Talking Machine Company, *Music Appreciation with the Victrola for Children*, 29–41.
30. See, for example, Welch, *The Appreciation of Music*, 18–19.
31. Giddings et al., *Music Appreciation in the Schoolroom*, 31.
32. Clark, "Festival of the Nations," 14.
33. My analysis here is very close to Bourdieu's theory of cultural capital in *Distinctions: A Social Critique of the Judgment of Taste*.
34. See Lempa, "German Body Culture."
35. One dimension of such statuary is often its nudity. François Jullien's *The Impossible Nude: Chinese Art and Western Aesthetics* (Chicago: University of Chicago Press, 2007) explains Western nudity as an expression of the subject's being defined and bounded by the body as opposed to the clothed subject in Eastern art who represents various stages of action and feeling but not a complete persona.
36. See Ruyter, *The Cultivation of Body and Mind*.
37. See Kasson, *Houdini, Tarzan, and the Perfect Man*.
38. See Lempa, "German Body Culture," and McMillan, "Germany Incarnate."
39. See Kasson, *Houdini, Tarzan, and the Perfect Man*, 30–68.
40. Social ambition for the common man was also the prevailing mood in the decades after the Revolution, according to Gordon S. Wood in *The Radicalism of the American Revolution*.
41. See Lott, *Love and Theft*, 115.
42. Stebbins, *Delsarte System of Expression*, 223.

43. Ruyter, *The Cultivation of Body and Mind*, illus. 9 and 10.
44. Mohler, "The Project Method in Teaching Music Appreciation."
45. This incident is noted in Kasson, *Rudeness and Civility*, 283n26.
46. See Popkewitz and Friedrich on Jacques Ranciere's *Hatred of Democracy* in their unpublished paper, "Professional Development Schools: Narratives of Democracy, Theses of Redemption, and Negation of Politics."

Chapter 9

1. See Schneider, *Collecting Lincoln*. See also Comini, *The Changing Image of Beethoven*.
2. Surette, "Musical Appreciation for the General Public."
3. For example, in Derek Vaillant's "Sounds of Whiteness: Local Radio, Racial Formation, and Public Culture in Chicago, 1921–1935," he describes the formation of a nonessentialized whiteness that pervades the cultural tenor and discriminatory practices of broadcasting in Chicago.
4. See David Nye's description of the reception of the telegraph in *The American Technological Sublime*.
5. See Baker, "Hear Ye! Hear Ye!"
6. See Ross, "Listen to This," 128.
7. Reinhardt Kosseleck, "On the Anthropological and Semantic Structure of Bildung."
8. Birge, "The Language Method in Teaching Appreciation," 162.
9. See Comini, *The Changing Image of Beethoven*, 354.
10. See ibid., 56. Ironically, Beethoven's deafness also played with the idea that "tone" deafness refers to having "no ear" for music. See also Baker's account in "Hear Ye, Hear Ye" of the recuperation of deaf individuals from a savage state.
11. See DeNora, *Beethoven and the Construction of Genius*.
12. William Newman, "The Beethoven Mystique in Romantic Art, Literature," 381.
13. Birchard, "Music for Individual and Social Life," 72.
14. For this running commentary on genius, see DeNora, *Beethoven and the Construction of Genius*.
15. This theme is very widespread in the literature related to the music appreciation curriculum. Authors whose work significantly contributed to the dissemination of the use of Beethoven as a national icon in schools are, for example, C. C. Birchard, "Music for Individual and Social Life;" Mabel Glenn and Edith Rhetts, *Reading Lessons in Music Appreciation*; and Frances Elliott Clark, "Music Appreciation of the Future."
16. Nan McMurry's work on whiteness and the consumptive describes part of the discursive atmosphere in which Beethoven and Lincoln's suffering exemplify the superior sensitivity of Caucasian bodies. See her "'And I? I Am in a Consumption.'"
17. In the atmosphere of beliefs current at that time, Beethoven's deafness would have consigned him to a category of "deficient" human types who lacked a capacity for language. See Baynton, "Disability and Justification of Inequality in American History."
18. See also Schneider, *Collecting Lincoln*, movie lobby poster and postcard figs.
19. See Fish, *Lincoln Collections and Lincoln Bibliography*.

20. See Kemp, "The 'Super Artist' as Genius," 49; Kemp quotes Theophile Gautier, who wrote of Dürer's genius, "Your genius . . . taking pity on you, has personified you in your creation. I do not know what could be more admirable in this world, more full of dreaming and deep anguish than this."
21. Kemp, "The 'Super Artist' as Genius," 32.
22. Comini, *The Changing Image of Beethoven*, 27, discusses Schindler's view of Beethoven, 27.
23. See DeNora, *Beethoven and the Construction of Genius*, 147.
24. Starr, *A Bibliography of Lincolniana*.
25. Pre–Nazi era Aryanism is discussed in connection with body culture movements and militias in Germany by Daniel McMillan in "Germany Incarnate: Politics, Gender and Sociability in the Gymnastics Movement, 1811–1871."
26. The portrait of Beethoven most often appearing in schools is one by Josef Karl Stieler in which Beethoven is gazing outward while holding a score. The expression is of great seriousness, but not of suffering—an intense inwardness that characterized his soul in exclusion of considerations of civility. See Comini, *The Changing Image of Beethoven*.
27. See, for example, Thaddeus Giddings in Chapter 8 of this volume, in the section titled "Appreciation, Structural Analysis, and Rhythm."
28. See Grivel, "The Phonograph's Horned Mouth."
29. See, for example, *Journal of the Music Supervisors National Conference* (March 1912), 21.
30. This idea is also expressed in Lott, *Love and Theft*.
31. *Anton Reiser* has sometimes been assessed as an anti-*Bildungsroman* because of the protagonist's failure to reach self-enlightenment. See discussion in La Vopa, *Grace, Talent and Merit*, 100.
32. For an explication of the autobiographical and nonfictional significance of these novels, see Bruford, *The German Tradition of Self-Cultivation: Bildung from Humboldt to Thomas Mann*.
33. Bruford, *The German Tradition of Self-Cultivation*.
34. Here I am not solely referencing National Socialism in Germany, but the early paramilitary organizations of the nineteenth century, such as Jahn's body culture movement (see McMillan, "Germany Incarnate"). For the conjunction of race and nationalism in the United States, among the very large literature on whiteness, see Gail Bederman, *Manliness and Civilization: A Cultural History of Gender and Race, 1880–1917*.
35. Charity board was a regular feature of indentured study. This practice is comparable, in some ways, to public school meal programs for disadvantaged students that have been plagued by participatory problems. See Carol Ann Marples and Diana-Marie Spillman, "Factors Affecting Students' Participation in the Cincinnati Public Schools." In my experiences in the public schools of Madison, Wisconsin, I noted tense interactions between cafeteria staff and students, many of whom were African American, receiving federally funded meals. The students regularly left whole plates of food unfinished. Staff perceived students as "ungrateful" for the free meals complaining of a "waste of taxpayer's money." Students complained that the food was substandard and unappealing. Documentation for minority achievement assistance was provided through the federal meal assistance programs in Madison, Wisconsin, in the 1990s, as a way to identify "students at risk." The effects of this classification were the subjects of reports to the Virginia Henderson, Equity Officer for

the Madison Public Schools (Ruth Gustafson, unpublished manuscript held by Madison Metropolitan School District, 1991–92).
36. See Emery, *Black Dance*.
37. This is a paraphrase from Toni Morrison's critique of the nineteenth-century American novel. See T. Morrison, *Playing in the Dark: Whiteness and the Literary Imagination*, 6.
38. See Bruford, *The German Tradition of Self-Cultivation*.
39. Explicit borrowings are reviewed in general histories of music education and documents such as Michael Mark's *Source Readings in Music Education History* and his *A History of American Music Education*. Other sources include Wilfried Gruhn's "European Methods for American Nineteenth-Century Singing Instruction" and Richard Lee Dunham's "Music Appreciation in the Public Schools, 1887–1930."
40. What it meant to be womanly or manly as future white citizens was often constructed in terms of difference from the black population. See Bederman, *Manliness and Civilization*. Anton Reiser's treatment is reminiscent of the formulation of class stereotyping in Rist's study of the abject condition of poor schoolchildren in the United States. See Ray Rist, "Student Social Class and Teacher Expectations." It is also consistent with Toni Morrison's formulation of disadvantage in *Playing in the Dark* and other writing on racial epistemologies referred to in previous chapters.
41. See Kasson, *Houdini, Tarzan, and the Perfect Man*. See also Kasson, *Rudeness and Civility*, as well as Elias, *The History of Manner*.
42. Moritz, *Anton Reiser*, 50; emphasis added.
43. Marples and Spillman, "Factors Affecting Students' Participation."
44. Quotation from La Vopa, *Grace, Talent, and Merit*, 46. Also see his chapter titled "The Natural Self and the Ethic of Reason," 165–96, for a discussion of the relation between education and the nature and future vocation of the child.
45. For a map of the terrain of racial projections onto slave song, see Radano, "Denoting Difference."
46. See Harrington and Harris, "Letter to the Boston School Committee."
47. See Giddings et al., *Music Appreciation in the Schoolroom*, 39.
48. See Fliegelman, *Declaring Independence*, 68.
49. See Blesh and Janis, *They All Played Ragtime*.
50. See Radano, "Denoting Difference."
51. Rickford and Rickford, *Spoken Soul: The Story of Black English*.
52. Hunter, *Culture and Government: The Emergence of Literary Education*, provides a brilliant analysis of the school subject of English in its historical context in the British Empire.
53. See Rickford, *African American Vernacular English*, chap. 15.
54. University of Wisconsin–Madison, "Spoken Word and Hip Hop in the Classroom."
55. See Marcyliena Morgan, "The African American Speech Community: Reality and Sociolinguistics," in *Language and the Social Construction of Identity in Creole Situations*, ed. Marcyliena Morgan (Los Angeles: Center for Afro-American Studies, UCLA, 1994).
56. Moritz, *Anton Reiser*, 138.
57. Some of the passages in *Anton Reiser* bring out issues previously discussed with regard to Thomas Hastings's *Dissertation on Musical Taste* and Lowell

Mason's comments on the German soprano in his letters. See Pemberton, *Lowell Mason, His Life and Work*. In Mason and Hastings's texts, the European vocal tradition purveys standards of performance that double for the diction and articulation of a particular human type.
58. See Rickford and Rickford, *Spoken Soul*, 143–44.
59. See Fanon, *Black Skin, White Masks*, 140–49.
60. Goethe, *Wilhelm Meisters Wanderjahre*, 200.
61. The instability of inner and outer in the fabrication of the child was an important theme in Goethe's writing. See discussion of pedagogy and *Bildung* in Baker, *In Perpetual Motion*, 372.
62. The traditional low status of musicians, whether as public school teachers, private teachers, or as providers of music for church service or entertainment, is treated in Thomas Tapper's *The Education of the Music Teacher*. His chapters "Music Teaching as Service" and "Music Teaching as a Profession" describe the stigmatizing of the music teacher through a combination of occupational and bodily factors. The worthy musician was one who bore hardship with grace for the sake of his art and provided service for low pay. As a model of professional altruism, Tapper mentions the life of Johann Sebastian Bach and his singing in the streets of Eisenach for alms. The failure to act as if one was engaged in "incessant industry," a lack of "refinement in speech and manner," or a retreat from altruism were characteristics of the music teacher who deserved low status (22–29).
63. The scenes from the novels and autobiographies stage a single drama over and over. As Foucault writes, "[The reiteration] is a play of dominations . . . it is fixed through history, in rituals, in meticulous procedures that impose rights and obligations." See Foucault, "Nietzsche, Genealogy, History," 77.
64. Keiler, *Marian Anderson: A Singer's Journey*.
65. See discussion of relation of racial attitudes to educational strategies in Lisa Delpit, *Other People's Children*, 28–40. Delpit offers a very nuanced argument for specific rule instruction that would be culturally inclusive without denigrating African American culture and family life. The point of agreement here with Delpit is that I found that low expectations of African Americans leads to preemptive discouragement for those seeking to enter advanced level classes and musical ensembles.
66. Quote from Wilson, *Our Nig*, 454. This attitude has survived, historically, in the Jim Crow segregated and de facto segregated educational institutions; it also rides as premise in the procedures of selective college admissions as well as in the idea of ability testing, for example, in Richard Hernnstein and Charles Murray's *The Bell Curve: Intelligence and Class Structure in American Life* (New York: Simon and Schuster, 1996).
67. DeNora, *Beethoven and the Construction of Genius*.
68. Fernold, "The Talking Machine in the Small School," 14.
69. Connerton, *How Societies Remember*.

Chapter 10

1. Music Educators National Conference, *Documentary Report on the Tanglewood Symposium*.
2. See Mark, *Contemporary Music Education*, 1986.
3. Boardman and Landis, *Exploring Music*.

4. See Michael Mark on psychological theory and cognitive growth with respect to the Manhattanville Program in *Contemporary Music Education*.
5. See Bourdieu, *State of Nobility*, 40–41, for discussion of symbolic violence.
6. United States Department of Education, "Arts Education Partnership."
7. Department of Public Instruction, "Winns."
8. J. Anderson, "Still Segregated, Still Unequal."
9. Barry Franklin, *From "Backwardness" to "at Risk"*; Popkewitz, *Struggling for the Soul*.
10. See, for example, Eunice Boardman, "The Relationship of Musical Thinking Learning to Classroom Instruction."
11. See Max Schoen's collection of psychological studies, *The Effects of Music*.
12. Lynn Molenda, personal communication regarding "Arts Portfolio," December 30, 2005, and Julie K. Brown, "Wisconsin Arts Propel Initiative."
13. My informal interviews led me to consider that notions of who is at risk increases in proportion to the list of categories for comparing children, especially as those categories assume a universal cultural context for "all" students.
14. Strandberg, "Listen to My Song, Please!" See also L. Green, *How Popular Musicians Learn*.
15. My informal observations of jazz ensembles, for example, show that students read the chord progressions from printed music; improvisation is worked in as alternating "riffs."
16. See discussion of representation of race in popular genres in Julia Koza's "Rap Music."
17. Bethany Bryson, "'Anything But Heavy Metal': Symbolic Exclusion and Musical Dislikes," *American Sociological Review* 16, no. 5 (1996): 884–99.
18. See Carol Lee, "The Centrality of Culture to the Scientific Study of Learning and Development."
19. Popkewitz, "Hopes of Progress and Fears of the Dangerous."
20. Thomas Popkewitz, in an e-mail, September 27, 2008, remarked that the educational slogan of doing "what works" is an occasion to ask, "What is work?"
21. Kowalczyk and Popkewitz, "Multiculturalism, Recognition and Abjection."
22. See Ginwright, *Black in School*; Koza, "Rap Music"; Soderman and Folkestad, "How Hip Hop Musicians Learn."
23. This debate over hip-hop culture parallels some of the points in the Ebonics debates as well. The issue is the recognition of culture and difference that will give African American Vernacular English a place in the curriculum. See Ginwright, *Black in School*, 114.
24. See one account of a generation gap between Civil Rights adults and youth today, discussed in Shawn Ginwright, *Black in School*.
25. University of Wisconsin–Madison, "Syllabus: Spoken Word and Hip Hop in the Classroom."
26. J. Soderman and G. Folkestad, "How Hip Hop Musicians Learn."
27. University of Wisconsin–Madison, "Syllabus: Spoken Word and Hip Hop in the Classroom."
28. Newman, "The Beethoven Mystique in Romantic Art, Literature."
29. Here, I borrow concepts from Pierre Bourdieu, *Distinctions* and *The State Nobility* as well as *The Field of Cultural Production: Essays on Art and Literature*.
30. See last chapter in Bourdieu, *The Field of Cultural Production*.
31. See Bourdieu, Part 3, "The Pure Gaze" in *The Field of Cultural Production*.

32. Clay, "Keepin' It Real."
33. See Reimer, *A Philosophy of Music Education*.
34. See Bourdieu, *Outline of a Theory of Practice*.
35. See Bourdieu, *The Field of Cultural Production*, 227.
36. Ibid., 219.
37. See Choate, "Music in American Society."
38. Conference participants at Spoken Word spent time elaborating on the cognitive benefits of rap in particular. See University of Wisconsin–Madison, "Syllabus: Spoken Word and Hip Hop in the Classroom."
39. See Shevy, "Music Genre as Cognitive Schema: Extramusical Associations with Country and Hip-Hop Music."
40. Bourdieu, *The State Nobility*, 21. Bourdieu's views on body hexis incorporate music listening as well as speech or dialect. With respect to school music, in general, singers are instructed in "proper" diction. The "correct" accents of phrasing and sound align the singer with Northern Standard American English rather than to Southern Black dialect or African American Vernacular. The debates over Ebonics challenge the general perception that the African American Vernacular is not English and therefore does not garner a position of merit in schooling. In its newer guise, "readiness" stands in for the diction and dispositions of children who are in need of remedial instruction.
41. This is the thesis of Sylvia Wynter in "Towards the Sociogenic Principle" and Franz Fanon's arguments in *Black Skin, White Masks*. For many, double identity is a killing machine of the spirit.
42. Ian Hunter, in *Culture and Government: The Emergence of Literary Education*, discusses the hierarchical principles involved in language teaching in Britain in terms that are especially relevant to music. Also see Bourdieu's findings of academic values as the measure of merit in education in *Distinctions: A Social Critique of the Judgment of Taste*.
43. See discussion of Ranciere's ideas on democracy in Popkewitz and Friedrich, "Professional Development Schools."
44. Carol Lee has presented a similar argument for research on minority cultural epistemologies and their implication for education. See her "The Centrality of Culture to the Scientific Study of Learning and Development."
45. I have gleaned this from observation of racially divided music classrooms. See R. Gustafson, "Theorizing Attrition in School Music Programs." See also Koza, "Multicultural Approaches to Music Education."
46. Another example of similar teaching occurred in Bonnie Green's strings program in Milwaukee public schools. She now directs a strings program in Madison, Wisconsin, outside the public schools.
47. Teachers in training are exposed to a combination of older philosophies of music education and more contemporary views. Texts commonly used to train the professors who teach music education majors are by Suzanne Langer, Bennett Reimer, and Leonard Meyer, among others.
48. See, for example, Linda Jothen, "Music Tells a Story."
49. R. Gustafson, "Stories of Failure or Delight?" 24.
50. Ginwright, *Black in School*.
51. T. Morrison, "Clinton as the First Black President."
52. Thomas Popkewitz provides a deft summary of life constructed around the drama of constant perfection and the problems of research this poses for historians. Popkewitz, *Cosmopolitanism and the Age of School Reform*, 186–87.

53. Koza, "Listening for Whiteness."
54. See King, "Dysconsious Racism."
55. Wynter, "Towards the Sociogenic Principle."
56. See Koza, "Listening for Whiteness."
57. For a discussion of embodied and disembodied aesthetic judgments, see Regelski, "Social Theory and Music Education as Praxis."
58. May Day Group, "Colloquium Schedule."
59. Apple, *Teachers and Texts*.
60. Koza, *Stepping Across*.
61. Ladson-Billings, *Crossing over to Canaan*.
62. See Popkewitz, "Conclusion," in *Cosmopolitanism and the Age of School Reform*.
63. Popkewitz and Friedrich, "Professional Development Schools." See also "Salvation, the Soul and Abjection are Still Projects of Schooling," manuscript submitted to *Harvard Educational Review*, August 2008.
64. For this understanding of intervention's limits, I would like to thank Jinting Wu, my friend and mentor who is searching for a way to represent possibilities. Also see Claudia Ruitenberg, "Teaching So That Democracy May Enter: Jacques Ranciere and the Logic of What If?" *American Educational Research Association* (Paper presented at American Education Research Association, New York, March 26, 2008).
65. Wynter, "Towards the Sociogenic Principle."

Bibliography

Adams, Crosby. "The Meaning of Appreciation." *Journal of the Music Supervisors' National Conference* (March 1929): 85–90.
Adams, John. "A Dissertation on the Canon and Feudal Law." In *The Works of John Adams*, edited by Charles Francis Adams. Boston: Little Brown, 1865.
Adams, John Quincy. *An Oration Delivered at Plymouth*. Boston, 1802. Microfiche.
Agawu, Kofi. "The Invention of 'African Rhythm.'" *Journal of the American Musicological Society* 48, no. 3 (1995): 380–95.
Aiken, Walter. *Aiken's Music Course*. New York: American Book Company, 1908.
Alchin, C. A. *Ear Training and Teacher and Pupil*. Philadelphia: Oliver Ditson, 1904.
Alvarez, Barbara, and Margaret Berg. "Musical Learning and Teaching and the Young Child." In *Dimensions of Musical Learning and Teaching: A Different Kind of Classroom*, edited by Eunice Boardman, 121–39. Reston, VA: Music Educators National Conference, 2002.
Anderson, Benedict. *Imagined Communities: Reflections on the Origin and Spread of Nationalism*. London: Verso, 1991.
Anderson, James D. "Still Segregated, Still Unequal." *Educational Researcher* 35, no. 1 (2006): 30–33.
Anderson, Margo. *The American Census: A Social History*. New Haven: Yale University Press, 1988.
Anderson, William M., and Joy E. Lawrence. *Integrating Music in the Elementary Classroom*. Instructor's 5th ed. Belmont, CA: Wadsworh Thomson Learning, 2001.
Anon. "Albany State Register." *Dwight's Journal of Music* (April 16, 1853): 124.
———. "Conrad O. Johnson." http://www.tmea.org/061_PBM/HOF/html/141_Johnson.htm.
———. "Notes from a Professor's Lecture." *Etude* 8 (January 1895): 19.
———. "Musical Impurity." *Etude* 17, no. 16 (1900).
———. "School Lecture Recitals: Thomas Whitney Surette to Continue Series on Wagner." *New York Times*, November 30, 1912. http://query.nytimes.com/gst/abstract.html?res=9C0DE5DE103CE633A25753C3A9679D946396D6CF.
———. *Silent*. Plano, TX: Music in Motion Catalog, 2004.
Antoine, Sister M. Salome. "The Rhetoric of Jeremy Taylor's Prose: Ornament of the Sunday Sermon." PhD diss., Catholic University of America, 1946.
Apple, M. W. *Teachers and Texts: A Political Economy of Class and Gender Relations in Education*. New York: Routledge and Kegan Paul, 1986.
———. "Foreword." In *White Reign: Deploying Whiteness in America*, edited by Joe Kincheloe. New York: St. Martin's Press, 1998.
Asad, Talal. *Formations of the Secular: Christianity, Islam and Modernity*. Palo Alto, CA: Stanford University Press, 2003.
Bain, Alexander. *The Senses and the Intellect*. New York: Appleton, 1874.
Baker, B. F., and L. H. Southard. *The School Chimes*. Boston: Wilkins, Rice and Kendal, 1852.

Baker, Bernadette M. "'Childhood' in the Emergence and Spread of U.S. Public Schools." In *Foucault's Challenge: Discourse, Knowledge, and Power in Education*, edited by Thomas S. Popkewitz and Marie Brennan, 117–43. New York: Teachers College Press, 1998.

———. *In Perpetual Motion: Theories of Power, Educational History and the Child*. New York: Peter Lang, 2001.

———. "Hear Ye! Hear Ye! Language, Deaf Education, and the Governance of the Child in Historical Perspective." In *Governing Children, Families, and Education: Restructuring the Welfare State*, edited by Marianne Bloch, K. Holmlund, I. Moqvist, and Thomas Popkewitz, 287–312. New York: Palgrave Macmillan, 2003.

Baker, B. F., and L. H. Southard. *The School Chimes*. Boston: Wilkins, Rice and Kendall, 1852.

Barnes, Earl. "The Relation of Rhythmic Exercises to Music in the Education of the Future." Paper presented at the Journal of the Music Supervisors' National Conference, Pittsburgh, March 22–26, 1915.

Barthes, Roland. "The Grain of the Voice." In *Image, Music, Text*, 179–89. New York: Hill and Wang, 1977.

Bascom, E. H. *The School Harp*. Boston: Morris Cotton, 1855.

Baynton, Douglas C. "Disability and the Justification of Inequality in American History." In *Disability History: From the Margins to the Mainstream*, edited by Paul Longmore and Lauri Umansky. New York: New York University Press, 2001.

Bederman, Gail. *Manliness and Civilization: A Cultural History of Gender and Race, 1880–1917*. Chicago: University of Chicago Press, 1995.

Beecher, Lyman. *A Plea for the West*. 1835. Cincinnati: Truman and Smith, 1835/1977.

Benjamin, Walter. *Illuminations*. Translated by Harry Zohn. New York: Schocken Books, 1969.

Bercovitch, Sacvan. *The American Jeremiad*. Madison: University of Wisconsin Press, 1978.

———. *The Rites of Assent: Transformations in the Symbolic Construction of America*. New York: Routledge, 1993.

Bermingham, Gudrun. "The Effects of Performers' External Characteristics on Performance Evaluations." *Update-Applications of Research in Music Education* 18, no. 2 (2000): 3–7.

Bertholot, J. M. "Sociological Discourse and the Body." In *The Body: Social Process and Cultural Theory*, edited by Mike Hepworth, Mike Featherstone, and Bryan S. Turner. London: Sage, 1991.

Bhabha, Homi. *The Location of Culture*. London: Routledge, 1994.

Birchard, C. C. "Music for Individual and Social Life." *Journal of the Music Supervisors National Conference* (April 1923): 68–77.

Birge, Edward Bailey. *History of Public School Music in the United States*. Boston: Oliver Diston, 1928.

———. "The Language Method in Teaching Appreciation." *Journal of the Music Supervisor's National Conference* (1913): 161–68.

Blesh, Rudi, and Harriet Janis. *They All Played Ragtime*. New York: Oak, 1966.

Bloch, Marc. *The Historian's Craft*. 1941. New York: Vintage Books, 1953.

Boardman, Eunice. "The Relationship of Musical Thinking Learning to Classroom Instruction." In *Dimensions of Musical Learning and Teaching: A Different Kind*

of Classroom, edited by Eunice Boardman, 1–20. Reston, VA: Music Educators National Conference, 2002.
Boardman, Eunice, and Beth Landis. *Exploring Music*. New York: Holt, Rinehart, and Winston, 1975.
Bobbitt, John Franklin. "A City School as a Community Art and Musical Center." *School Music Monthly* (January–February 1912): 27–32.
Bohlman, Philip. "The Remembrance of Things Past: Music Race and the End of History in Modern Europe." In *Music and the Racial Imagination*, edited by Ronald Radano and Philip Bohlman, 626–44. Chicago: University of Chicago Press, 2000.
Bolton, Thaddeus L. "Rhythm." *American Journal of Psychology* 6, no. 2 (1894): 145–82.
Bosco, Ronald A., ed. *The Puritan Sermon in America 1630–1750*. Delmar, NY: Scholars Facsimiles and Reprints, 1987.
Boston School Committee. "Report." In *Source Readings in Music Education*, edited by Michael Mark, 134–43. 1837. New York: Schirmer Books, 1982.
Botstein, Leon. "Listening through Reading: Musical Literacy and the Concert Audience." *Nineteenth-Century Music* 16, no. 2 (Fall 1992): 129–45.
Bourdieu, Pierre. *Distinctions: A Social Critique of the Judgment of Taste*. Cambridge: Harvard University Press, 1984.
———. *The Field of Cultural Production: Essays on Art and Literature*. Cambridge, England: Polity Press, 1993.
———. *Outline of a Theory of Practice*. Cambridge: Cambridge University Press, 1977.
———*Pascalian Meditations*. Stanford, CA: Stanford University Press, 2000.
———. *The State Nobility: Elite Schools and the Field of Power*. Palo Alto, CA: Stanford University Press, 1989.
Bowen, George Oscar. "Music Education in Secondary Schools." *Journal of Music Teachers National Association* (1908): 177–79.
Bradbury, William. *Flora's Festival: A Musical Recreation for Schools, Juvenile Singing Classes Etc*. New York: Mark H. Newman, 1847.
Bradbury, William B., and Charles W. Sanders. *The Young Choir*. New York: Ivison and Finney, 1831.
Briggs, Thomas. "Music Memory Contests." *School Music Monthly* (September–October 1925): 5–7.
Broman, Timothy. "The Transformation of Academic Medicine in Germany, 1780–1820." PhD diss., Princeton University, 1987.
Brooks, Van Wyck. *The Flowering of New England*. New York: E. P. Dutton, 1952.
Broudy, Harry S. "The Case for Aesthetic Education." In *Documentary Report on the Tanglewood Symposium. Tanglewood Symposium, 1967*, edited by Robert A. Choate. Washington, DC: Music Educators National Conference, 1968.
Brown, Julie K. "Wisconsin Arts Propel Initiative." http://www.aasd.k12.wi.us/staff/brownjulie/propel/index.html.
Broyles, Michael. *"Music of the Highest Class": Elitism and Populism in Antebellum Boston*. New Haven, CT: Yale University Press, 1992.
Bruford, W. H. *The German Tradition of Self-Cultivation: Bildung from Humboldt to Thomas Mann*. New York: Cambridge University Press, 1975.
Bryant, Willian Cullen. "The Prairies." In *The American Tradition in Literature*, edited by Bradley Sculley, 260–61. 1832. New York: W. W. Norton, 1962.

Burke, Edmund. *Enquiry into the Sublime and Beautiful.* 1757. London: Routledge, Kegan and Paul, 1958.
Burney, Charles. *A General History of Music from the Earliest Ages to the Present Period.* 1776. New York: Dover Press, 1957.
Butler, Charles. *The Silver Bell.* New York: S. T. Gordon, 1869.
Butler, Judith. *Bodies That Matter: On the Discursive Limits of "Sex."* New York: Routledge, 1993.
———. *Theories of Subjection: The Psychic Life of Power.* Palo Alto, CA: Stanford University Press, 1997.
Cady, Calvin. "Exigencies and Possibilities of Secondary Music Education." *Journal of the Music Teachers' National Association* (1908): 148–59.
———. "Music Appreciation and the Correlation of Studies." *Journal of the Music Teachers National Association* (1910): 49–57.
———. *Music Education, an Outline.* Chicago: Clayton F. Summy, 1902.
———. "Preface." In *Music Appreciation with the Victrola for Children*, edited by Victor Talking Machine Company. Camden, NJ: Victor Talking Machine Company, 1923.
California Teachers Association. *Complimentary Souvenir Handbook, Fifty-Third Annual Convention, National Education and International Congress of Education Meeting. August 16–28, 1915.* San Francisco: Arthur Henry Chamberlain, 1915.
Canetti, Elias. *Crowds and Power.* New York: Viking, 1963.
Carlyle, Thomas. "The Nigger Question." In *The Nigger Question/Thomas Carlyle. The Negro Question/John Stuart Mill*, edited by Eugene R. 1849. August. New York: Appleton-Century-Crofts, 1971.
Chamberlin, J. E., and S. L. Gilman, eds. *Degeneration: The Dark Side of Progress.* New York: Columbia University Press, 1988.
Chernoff, John Miller. *African Rhythm and African Sensibility: Aesthetics and Social Action in African Musical Idioms.* Chicago: University of Chicago Press, 1979.
Chinn, B. "Vocal Self-Identification, Singing Style, and Singing Range in Relationship to a Measure of Cultural Mistrust in African American Students Females." *Journal of Research in Music Education* 45, no. 4 (1997): 637–48.
Choate, Robert. "Introduction." *Documentary Report on the Tanglewood Symposium.* Washington, DC: Music Educators National Conference, 1968.
Chybowski, Julia. "Popularizing Classical Music and Developing American Taste: The Role of 1920s Music Appreciation Texts." Master's thesis, University of Wisconsin–Madison, 2004.
Clark, Frances Elliott. "Festival of the Nations." *Journal of the Music Supervisors Conference* (May 1913).
———. "Foreword." In *What We Hear in Music*, edited by Radio Corporation of America (RCA), 9. Camden, NJ: RCA, 1943.
———. "The Interrelation and Interdependence of Records and Radio." *Journal of the Music Supervisors' Conference* (March 1926): 215–18.
———. "Music Appreciation of the Future." *Journal of the Music Supervisors National Conference* (April 1924): 271–78.
———. "Outline of Music History." *The School Music Monthly* (January 1901).
Claxton, P. P. "The Place of Music in National Education." *Journal of the Music Supervisor's Conference* (March 1915): 48–51.
Clay, Andreana. "Keepin' It Real: Black Youth, Hip-Hop Culture, and Black Identity." *American Behavioral Scientist* 46, no. 10 (2003): 1346–58.

Coffin, Leonore. "Report of the Sub-Committee on Music Appreciation in the First Six Grades." *Journal of the Music Supervisors National Conference* (1930): 226–31.
Coleridge, Samuel Taylor. "The Eolian Harp." In *The Norton Anthology of English Literature*, edited by M. H. Abrams. 1795. New York: W. W. Norton, 1968.
———. "Kubla Khan, a Vision in a Dream: A Fragment." In *The Norton Anthology of English Literature*, edited by M. H. Abrams, 1400–1401. 1797. New York: W. W. Norton, 1968.
Combe, George. *The Constitution of Man*. 1828. Boston: Marsh, Capen, Lyon and Webb, 1841.
Comini, Alessandra. *The Changing Image of Beethoven: A Study in Mythmaking*. New York: Rizzoli, 1987.
Connerton, Paul. *How Societies Remember*. Cambridge: Cambridge University Press, 1989.
Connor, Beth, A. Ferri, and J. David. "Tools of Exclusion: Race, Disability, and Re(Segregated) Education." http://www.digitaldivide.net/comm/docs/view.php?DocID=312.
Connor, Steven. *Dumbstruck: A Cultural History of Ventriloquism*. Oxford: Oxford University Press, 2000.
Cooper, Anthony Ashley, Earl of Shaftesbury. *Characteristics of Men, Manners, Opinions, Times*. Edited by Karl Ameriks and Desmond Clarke. Cambridge Texts in the History of Philosophy. 1723. Cambridge: Cambridge University Press, 1999.
Cooper, Grovenor, and Leonard Meyer. *The Rhythmic Structure of Music*. Chicago: University of Chicago Press, 1960.
Crary, Jonathan. *Suspensions of Perception: Attention, Spectacle, and Modern Culture*. Cambridge: MIT Press, 1999.
Crawford, Richard. *The American Musical Landscape*. Berkeley: University of California Press, 1993.
———. "Musical Learning in Nineteenth-Century America." *American Music*, no. 1 (Spring 1983): 144–56.
Crocker, Ruth. *Social Work and Social Order, 1889–1930*. Urbana: University of Illinois Press, 1992.
Cundiff, Hannah M., and Peter W. Dykema. *School Music Handbook: A Guide for Teaching School Music*. Boston: C. C. Birchard, 1927.
Damrosch, Walter. "A Lesson in Appreciation." *Journal of the Music Supervisor's National Conference* (April 1923): 53–59.
———. "Music and the Radio." *Journal of the Music Supervisors National Conference* (April 1928): 55–60.
Davis, Gerald L. *I Got the Word in Me and I Can Sing, You Know: A Study of the Performed African American Sermon*. Philadelphia: University of Pennsylvania Press, 1985.
Delpit, Lisa. *Other People's Children: Cultural Conflict in the Classroom*. New York: New Press, 1995.
DeNora, Tia. *After Adorno: Rethinking Music Sociology*. Cambridge: Cambridge University Press, 2003.
———. *Beethoven and the Construction of Genius: Musical Politics in Vienna, 1792–1803*. Berkeley: University of California Press, 1995.
———. *Music in Everyday Life*. Cambrdige: Cambridge University Press, 2000.

Department of Public Instruction (DPI). "Grading, Instruction, and Assessment in Music." http://dpi.wi.gov/cal/mugrdinstess.html. 2008.
———. "Winns." http://data.dpi.state.wi.us/data/selschool.asp (2002–4).
———. "Wisconsin's Model Academic Standards for Music." Edited by Susan Grady, Thomas Stefonek, and Pauli Nikoly. Madison, WI: Department of Public Instruction, 1997.
DeRogatis, Amy. *Moral Geography: Maps, Missionaries and the New American Frontier.* New York: Columbia University Press, 2002.
Desrosières, Alain. *The Politics of Large Numbers: A History of Statistical Reasoning.* Cambridge, MA: Harvard University Press, 1998.
Dewey, John. *Art as Experience.* New York: Minton, Balch and Co., 1934.
———. *How We Think.* Boston: D. C. Heath, 1933.
Dickinson, Edward. *Music and the Higher Education.* New York: Charles Scribner's Sons, 1915.
Dorinson, Joseph. "Paul Robeson and Jackie Robinson: Athletes and Activists at Armageddon." In *Paul Robeson: Essays on His Life and Legacy,* edited by Joseph Dorinson and William Pencak. Jefferson, NC: McFarland, 2002.
Douglass, Frederick. *Narrative of the Life of Frederick Douglass, An American Slave.* 1845. New Haven: Yale University Press, 2001.
Du Bois, W. E. B. *The Souls of Black Folk.* 1903. New York: Bantam Classics, 1989.
Dumont, Frank. "The Chinese Laundryman." In *Flashes of Merriment: A Century of Humorous Songs in America, 1805–1905,* edited by Lester Levy. Philadelphia: Charles Escher, 1971.
Dumont, Louis. *German Ideology: From France to Germany and Back.* Chicago: University of Chicago Press, 1994.
Dunham, Franklin. "Can Music Appreciation Be Taught?" *Journal of the Music Supervisors National Conference* (March 1929): 83–85.
Dunham, Richard. "Music Appreciation in the Public Schools, 1887–1930." PhD diss., University of Michigan, 1961.
Dwight, John Sullivan. "Introductory." *Dwight's Journal of Music* (April 10, 1852): 2.
Elias, Norbert. "The Civilizing Process." In *The Norbert Elias Reader: A Biographical Selection,* edited by Johann Goudsblom and Steven Mennell. Malden, MA: Blackwell, 1998.
———. *The History of Manners: The Civilizing Process.* 1968. New York: Pantheon, 1982.
Eliot, Samuel. "Third Annual Report for the Boston Academy of Music." *North American Review* 43 (April 1836): 53–85.
Emch-Deriaz, Antoinette, "The Non-Medicals Made Easy." In *The Popularization of Medicine 1650–1850,* edited by Roy Porter. London: Routledge, 1992.
Emerson, L. O. *The Golden Wreath, A Choice Collection of Favorite Melodies, Schools, Seminaries, Select Classes and Etc.* Boston: Oliver Ditson, 1857.
Emery, Lynne Fauley. *Black Dance, 1619 to the Present.* Princeton: Princeton University Press, 1988.
Erskine, John. "Adult Education in Music." *School and Society (Educational Review)* 32, no. 829 (1930): 647–51.
Esterhammer, Angela. *The Romantic Performative: Language and Action in British and German Romanticism.* Palo Alto, CA: Stanford University Press, 2000.
Fanon, Frantz. *Black Skin, White Masks: The Experiences of Black Man in a White World.* New York: Grove Press, 1965.

Fendler, Lynn. "The Educated Subject: Discursive Constructions of Reason and Knowledge in History." PhD diss., University of Wisconsin–Madison, 1999.
Ferguson, Robert. *The American Enlightenment, 1750–1820*. Cambridge: Harvard University Press, 1994.
Fernold, Louise. "The Talking Machine in the Small School—a Few Practical Suggestions," *School Music Monthly* (January–February 1912): 15–17.
Fichte, Johann Gottlieb. *The Vocation of the Scholar*. Translated by William Smith. London: Chapman, 1847.
Fillmore, John. "Traveling Concert Troupes as Educators." *Dwight's Journal of Music* (May 12, 1877): 10.
Fish, Daniel. *Lincoln Collections and Lincoln Bibliography*. New York: Bibliographical Society of America, 1909.
Fitz, Asa. *The American School Songbook*. Boston: Fowle and Capen, 1846.
———. *A Child's Songbook*. Concord, NH: B. Merrill, 1819.
Fitzgerald, Maureen. "Irish Catholic Nuns and the Development of New York City's Welfare System 1840–1900." PhD diss., University of Wisconsin–Madison, 1992.
Fleming, Ada. "Music in the High School—Needs of the Hour—Plans." *School Music Monthly* (January 1908): 24–29.
Fliegelman, Jay. *Declaring Independence: Jefferson, Natural Language, and the Culture of Performance*. Stanford, CA: Stanford University Press, 1993.
Floyd, Samuel A., Jr. "The Implications of John Dewey's Theory of Appreciation for the Teaching of Music Appreciation." PhD diss., Southern Illinois University, 1969.
Fontenrose, Joseph. *The Delphic Oracle, Its Responses and Operations, with a Catalogue of Responses*. Berkeley: University of California Press, 1978.
Foucault, Michel. *Archaeology of Knowledge and the Discourse on Language*. Translated by M. Sheridan Smith. London: Tavistock, 1972.
———. "Different Spaces." In *Aesthetics, Method and Epistemology*, edited by James D. Faubion, 175–85. New York: The New Press, 1998.
———. "Docile Bodies." In *The Foucault Reader*, edited by Paul Rabinow, 179–88. New York: Pantheon Books, 1984.
———. "Nietzsche, Genealogy, History." In *Aesthetics, Method and Epistemology*, edited by James D. Faubion, 369–91. New York: The New Press, 1998.
———. *The Order of Things: An Archaeology of the Human Sciences*. Translator unacknowledged. 1966. New York: Vintage, 1994.
———. "The Subject and Power." In *Michel Foucault: Beyond Structuralism and Hermeneutics*, edited by Hubert L. Dreyfus and Paul Rabinow, 221–26. Chicago: University of Chicago Press, 1982.
———. *Technologies of the Self*. Amherst, MA: Amherst College Press, 1988.
Franklin, Barry. *From "Backwardness" To "At Risk": Childhood Learning Difficulties and the Contradictions of School Reform*. Albany: State University of New York, 1994.
FreeChild Project. "Youth Led Hip Hop Activism." http://www.freechild.org/hiphop.htm.
Froebel, Friedrich. "The Education of Man: Chief Groups of Subjects of Instruction." In *Source Readings in Music Education*, edited by Michael Mark, 95–97. 1850. New York: Schirmer Books, 1982.
———. *Source Readings in Music Education*. New York: Schirmer Books, 1982.

Fryberger, Agnes Moore. *Listening Lessons in Music Graded for Schools*. New York: Silver Burdett, 1925.
Galton, Francis. *Hereditary Genius: An Inquiry into Its Laws and Consequences*. 1869. Cleveland: Meridian Books, 1962.
Gehrkens, Karl. "Rhythm Training and Dalcroze Eurythmics." *Music Supervisors National Conference Yearbook* (1932): 305–10.
Gibling, Sophia. "Types of Musical Listening." *The Musical Quarterly* 3, no. 2 (1917): 385–89.
Giddings, T., W. Earhart, R. Baldwin, and E. Newton. *Manual for Teachers Four Book Course*. Boston: Ginn, 1924.
———. *Music Appreciation in the Schoolroom*. Boston: Ginn, 1926.
Giddings, T. P. *School Music Teaching*. Chicago: C. H. Congdon, 1910.
Giffe, W. T. *The New Favorite*. Indianapolis: H. I. Benham, 1875.
Gilliland, A. R., and H. T. Moore. "The Immediate and Long-Time Effects of Classical and Popular Phonograph Selections." In *The Effects of Music: A Series of Essays*, edited by Max Schoen, 215–21. 1927. London: Routledge, 1999.
Ginwright, Shawn. *Black in School: Afrocentric Reform, Urban Youth, and the Promise of Hip-Hop Culture*. New York: Teachers College Press, 2004.
Ginzburg, Carlos. *The Cheese and the Worms: The Cosmos of a Sixteenth Century Miller*. Baltimore: Johns Hopkins University Press, 1992.
Glenn, Mabel, and Edith Rhetts. *Reading Lessons in Music Appreciation*. Boston: C. C. Birchard, 1923.
Goepp, Philip. "Musical Appreciation in America as a National Asset." *Journal of the Music Teachers' National Association* (December 1910): 27–35.
Goethe, Johann Wolfgang. *Wilhelm Meister's Apprenticeship*. In *Goethe: the Collected Works*, edited and translated by Eric Blackall, vol. 9. 1749. Princeton: Princeton University Press, 1832.
———. *Wilhelm Meister's Journeyman Years*. Translated by Jan Van Huerck, in cooperation with Jane K. Brown. 1825. Princeton, NJ: Princeton University Press, 1989.
Golston, Michael. "Im Anfang War Der Rhythmus: Rhythmic Incubations in Discourses of Mind, Body, and Race from 1850–1944." http://www.stanford.edu/group/SHR/5-supp/text/golston.html.
Goodman, David. "Distracted Listening: On Not Making Sound Choices in the 1930s." In *Sound in the Era of Mechanical Reproduction*, edited by David Suisman and Susan Strasser. Philadelphia: University of Pennsylvania Press, 2008.
———. "Radio's Public: The Civic Ambitions of 1930s American Radio." Unpublished manuscript.
Gould, Stephen Jay. *The Mismeasure of Man*. New York: W. W. Norton, 1996.
Gousouasis, P., M. Guhn, and N. Kishor. "The Predictive Relationship between Achievement and Participation in Music and Achievement in Core Grade 12 Academic Subjects." *Music Education Research* 9, no. 1 (2007): 81–92.
Gove, Philip, ed. *Webster's Seventh New Collegiate Dictionary Based on Webster's Third New International Dictionary of the English Language, Unabridged*. Springfield, MA: Merriam, 1966.
Green, J. W. *School Melodies*. Boston: Morris Cotton, 1852.
Green, Lucy. *How Popular Musicians Learn: A Way Ahead for Music Education*. Burlington, VT: Ashgate, 2001.

Green, Shannon. "'Art for Life's Sake': Music Schools and Activities in U.S. Social Settlement Houses, 1892–1942." PhD diss., University of Wisconsin–Madison, 1998.
Grivel, Charles. "The Phonograph's Horned Mouth." In *Wireless Imagination*, edited by Douglas Kahn and Gregory Whitenead, 31–61. Cambridge: MIT Press, 1992.
Gruhn, Wilfried. "European Methods for American Nineteenth-Century Singing Instruction: A Cross Cultural Perspective in Historical Research." *Journal of Historical Research in Music Education* 23, no. 1 (2001): 3–19.
Gumperz, John J. *Language and Social Identity*. Cambridge: Cambridge University Press, 1982.
Gustafson, James P. *The Great Instrument of Orientation*. Madison, WI: Orion, 2008.
Gustafson, Ruth. "Merry Throngs and Street Gangs: The Fabrication of Whiteness and the Worthy Citizen in Early Vocal Instruction and Music Appreciation, 1830–1930." PhD diss., University of Wisconsin, 2005.
———. "Practicum Notes." In *Practicum C and I, 337 University of Wisconsin–Madison*, Unpublished manuscript. Madison, WI, 2003.
———. "Report of Minority Student Achievement in Madison Metropolitan School District." Unpublished manuscript. Madison, WI: 1991–97.
———. "Stories of Failure or Delight? Reflections on a New Curriculum for Elementary General Music." Master's thesis, University of Wisconsin–Madison, 1991.
———. "Theorizing Attrition in School Music Programs: Bildung's Reverent Body and Good Ears." In *Monografier: Journal of Research in Teacher Education*, edited by Per Olof Erixson, 41–62. Umeå, Sweden: Umeå University, 2004.
Habermas, Jurgen. *The Social Transformation of the Public Sphere*. Translated by Strukturwandel der Offentlichkeit. Cambridge: MIT Press, 1989.
Hall, G. Stanley. "The Ideal School as Based on Child Study." *The Forum* 32, no. 1 (1901): 24–39.
Hammer, Dean. "Puritanism in the Making of a Nation." PhD diss., University of California, Berkeley, 1889.
Hanslick, Edward. *The Beautiful in Music*. 1854. New York: The Liberal Arts Press, 1957.
Harrington, Joseph, and John Harris. "Letter to the Boston School Committee." *Boston Musical Gazette* (May 25, 1838): 7.
Hastings, Thomas. *Dissertation on Musical Taste*. 1822. New York: Da Capo, 1974.
Heath, Shirley. *Ways with Words: Language, Life, and Work in Communities and Classrooms*. New York: Cambridge University Press, 1983.
Henderson, W. J. *What Is Good Music?* New York: Charles Scribner's Sons, 1929.
Hirschkind, Charles. *The Ethical Soundscape: Cassette Sermons and Islamic Counterpublics*. New York: Columbia University Press, 2006.
Hitchcock, H. Wiley. *Music in the United States: A Historical Introduction*. Englewood Cliffs, NJ: Prentice Hall, 1969.
Hooker, Richard. "The Invention of American Musical Culture: Criticism, Musical Acculturation in Antebellum America." In *Keeping Score: Music, Disciplinarity, and Culture*, edited by Anahid Kassabian, David Schwarz, and Lawrence Siegel, 107–28. Charlottesville: University of North Carolina Press, 1997.
Hughes, Ted. *Winter Pollen: Occasional Prose*. New York: Picador, 1995.

Hultqvist, Kenneth. "The Future Is Already Here as It Has Always Been: The New Teacher Subject, the Child and the Technologies of the Self." Paper presented at the Wednesday Group Guest Speaker, University of Wisconsin–Madison, June 25, 2003.

Hunt, Ernest. *The Living Touch in Music and Education: A Manual for Musicians and Others*. London: Kegan Paul, Trench, Trubner and Co., 1924.

Hunter, Ian. *Culture and Government: The Emergence of Literary Education*. Basingstoke, England: Macmillan, 1988.

Hurley, Kelly. *The Gothic Body: Sexuality, Materialism, and Degeneration at the Fin De Siècle*. New York: Cambridge University Press, 1996.

Hutton, Laurence. *Curiosities of the American Stage*. 1891. New York: Random House, 1968.

J. L. Kincheloe, S. R. Steinberg, N. M. Rodriguez, and R. Chennault, ed. *Deploying Whiteness in America*. New York: St. Martin's Griffin, 1998.

Jackson, John, Jr. *Racial Paranoia: The Unintended Consequences of Political Correctness*. New York: Basic Civitas, 2008.

Jacques-Dalcroze, Emile. *Rhythm, Music and Education*. New York: G. P. Putnam and Sons, 1921.

James, Richard. "A Survey of Teacher-Training Programs in Music from the Early Musical Convention to the Introduction of Four-Year Degree Curricula." PhD diss., University of Maryland, 1968.

James, William. *The Principles of Psychology*. Vol. 1. New York: Henry Holt, 1890.

Jefferson, Thomas. *Notes on the State of Virginia*. 1785. New York: Harper Torchbooks, 1964.

Jones, Walter. "An Analysis of Public School Textbooks before 1900." PhD diss., University of Pittsburgh, 1954.

Jothen, Linda. "Music Tells a Story." Madison, WI: Wisconsin Music Educators Association, 2004.

Jsmooth995. "What Is Hip Hop Activism?" http://www.hiphopmusic.com/archives/000147.html.

Jukebox. "Is Hip Hop Bad for Black Culture?" http://www.jukeb0x.com/is-hip-hop-bad-for-black-culture.html.

Jung, Carl. *Civilization in Transition*. Translated by R. F. Hull. London: Routledge and Kean Paul, 1964.

Kant, Immanuel. *Critique of Pure Judgement*. Translated by J. H. Bernard. 1790. London: Macmillan, 1914.

Kasson, John. *Houdini, Tarzan, and the Perfect Man: The White Male Body and the Challenge of Modernity in America*. New York: Hill and Wang, 2001.

———. *Rudeness and Civility*. Toronto: Harper and Collins, 1990.

Keene, James. *A History of Music Education in the United States*. Hanover, NH: University Press of New England, 1982.

Keiler, Allan. *Marian Anderson: A Singer's Journey*. New York: Scribner, 2000.

Keith, Alice. "Music Appreciation Materials." *Music Supervisor's National Conference* (March 1929): 141–45.

Kemp, Martin. "The 'Super Artist' as Genius." In *Genius the History of an Idea*, edited by Penelope Murray, 32–53. Oxford: Basil Blackwell, 1989.

King, Joyce E. "Culture-Centered Knowledge: Black Studies, Curriculum Transformation and Social Action." In *Handbook on Research in Multicultural Education*, edited by James Banks and Cherry A. McGee Banks, 265–90. San Francisco: Jossey-Bass, 1995.

———. "Dysconsious Racism: Ideology, Identity, and the Miseducation of Teachers." *Journal of Negro Education* 60, no. 2 (1991): 133–46.

Kingman, Charles. "The Place and Importance of Music in the High School." *School Music Monthly* (May 1912): 26–30.

Knowlton, Fanny Snow. *Nature Songs for Children*. Springfield, MA: Milton Bradley, 1912.

Koch, Robert. "The History and Promotional Activities of the National Bureau for the Advancement of Music." PhD diss., University of Michigan, 1973.

Koselleck, Rheinhardt. "On the Anthropological and Semantic Structure of Bildung." In *The Practice of Conceptual History: Timing History, Spacing Comments*, edited by Reinhardt Koselleck, 170–207. Palo Alto, CA: Stanford University Press, 2002.

Kowalczyk, Jamie, and Thomas Popkewitz. "Multiculturalism, Recognition and Abjection: (Re)-Mapping Italian Identity." *Policy Futures in Education* 3, no. 4 (2005): 423–35.

Koza, Julia. "Females in 1988 Middle School Music Textbooks: An Analysis of Illustrations. *Journal of Research in Music Education* 42, no. 2 (1988): 145–71.

———. "Listening for Whiteness: Hearing Racial Politics in Undergraduate School Music." *Philosophy of Music Education Review* 16, no. 2 (Fall 2008): 145–55.

———. "Multicultural Approaches to Music Education." In *Making Schooling Multicultural: Campus and Classroom*, edited by Carl Grant and Mary Louis Gomez, 265–87. Englewood Cliffs, NJ: Prentice Hall, 1996.

———. "Rap Music: The Cultural Politics of Official Representation." *Education/Pedagogy/Cultural Studies* 16, no. 2 (1994): 171–96.

———. *Stepping Across: Four Interdisciplinary Studies of Education and Cultural Politics*. New York: Peter Lang, 2003.

Krashen, Stephen. *Second Language Acquisition and Second Language Learning*. Los Angeles: University of Southern California, 1981.

Kraut, Alan. *Silent Travelers: Germs, Genes and the Immigrant Menace*. Baltimore: Johns Hopkins University Press, 1994.

La Vopa, Anthony. *Grace, Talent and Merit: Poor Students, Clerical Careers, and Professional Ideology in Eighteenth-Century Germany*. New York: Cambridge University Press, 1988.

———. *Prussian School Teachers: Profession and Office, 1763–1848*. Chapel Hill: University of North Carolina Press, 1980.

Ladson-Billings, Gloria. *Crossing Over to Canaan: The Journey of New Teachers in Diverse Classrooms*. San Francisco: Jossey-Bass, 2001.

———. "Racialized Discourses and Ethnic Epistemologies." In *Handbook of Qualitative Research*, edited by Norman Denzin and Yvonna S. Lincoln, 257–77. Thousand Oaks, CA: Sage, 2000.

Lasch-Quinn, Elizabeth. *Black Neighbors: Race and the Limits of Reform in the American Settlement House Movement*. Chapel Hill: University of North Carolina Press, 1993.

Lee, Carol D. "The Centrality of Culture to the Scientific Study of Learning and Development: How an Ecological Framework in Education Research Facilitates Civic Responsibility." *Educational Researcher* 37, no. 5 (2008): 267–79.

Lempa, Heikki. "German Body Culture: The Ideology of Moderation and the Educated Middle Class 1790–1850." PhD diss., University of Chicago, 1999.

Lensmire, Timothy. "How I Became White While Punching de Tar Baby." *Curriculum Inquiry*, 38, no. 3 (2008): 299–322.

Leonard, Neil. *Jazz and the White Americans*. Chicago: University of Chicago Press, 1962.
Leppert, Richard, and Susan McClary, eds. *Music and Society: The Politics of Composition, Performance and Reception*. New York: Cambridge University Press, 1987.
Leslie, C. E. *The Cyclone of Song*. Chicago: Chicago Music Company, 1888.
Levine, Lawrence. *Black Culture and Black Consciousness: Afro-American Folk Thought from Slavery to Freedom*. Oxford: Oxford University Press, 1978.
———. *Highbrow and Lowbrow: The Emergence of Cultural Hierarchy in America*. Cambridge: Harvard University Press, 1986.
Levy, Lester. *Flashes of Merriment: A Century of Humorous Songs in America, 1805–1905*. Norman, OK: University of Oklahoma Press, 1971.
Lincoln, Abraham. "The Perpetuation of Our Political Institutions." In *The Political Thought of Abraham Lincoln*, edited by Richard Current, 11–20. 1838. Indianapolis, IN: Bobbs Merrill, 1967.
Lippman, Edward A., ed. *Musical Aesthetics: A Historical Reader*. Vol. 2. New York: Pendragon, 1985.
Litwack, Leon. *North of Slavery: The Negro in the Free States, 1790–1860*. Chicago: University of Chicago Press, 1961.
Locke, John. "Some Thoughts Concerning Education." In *Source Readings in Music Education History*, edited by Michael Mark, 88–90. 1693. New York: Schirmer Books, 1982.
Lott, Eric. *Love and Theft: Blackface Minstrelsy and the American Working Class*. New York: Oxford University Press, 1993.
Macpherson, Stewart. *The Musical Education of the Child*. Boston: Boston Music, 1915.
Mann, Horace. *Sixth Annual Report to the Boston School Committee*. Edited by Horace Mann League and the National Educational Association. 1843. Boston: Dutton and Wentworth, 1950.
———. "Report for 1844: Vocal Music in the Schools." In *Source Readings in Music Education*, edited by Michael Mark, 144–54. 1844. New York: Schirmer Books, 1982.
Mantle, Burns. "Mantle Hopes Othello Does Not Come to the US." *Chicago Tribune*, August 3, 930, E4.
Mark, Michael. *Contemporary Music Education*. New York: Schirmer Books, 1986.
———. *A History of American Music Education*. New York: Schirmer Books, 1999.
———. *Source Readings in Music Education History*. New York: Schirmer Books, 1982.
Marrocco, Thomas, and Harold Gleason, eds. *Music in America: An Anthology from the Landing of the Pilgrims to the Close of the Civil War*. New York: W. W. Norton, 1964.
Marples, Carol Ann, and Diana-Marie Spillman. "Factors Affecting Students' Participation in the Cincinnati Public Schools." *Adolescence* 30 (Fall 1995): 745–54.
Mason, Lowell. *The Elements of Vocal Music on the System of Pestalozzi Published for the Boston Academy of Music*. Boston: Wilkins and Carter, 1834.
———. "Manual of the Boston Academy of Music." In *Source Readings in Music Education*, edited by Michael Mark, 127–33. 1834. New York: Schirmer Books, 1982.
———. *Musical Letters from Abroad*. 1854. New York: Da Capo Press, 1967.
———. *The Normal Singer*. New York: Mason Brothers, 1856.

———. *A Songbook of the Schoolroom*. Boston: Wilkins and Carter, 1847.
Mason, Lowell, and Thomas Hastings. *Spiritual Songs for Social Worship*. Utica, NY: G. Tracy, Robinson, and Pratt, 1831.
Mason, Lowell, and Elam Ives. *The Juvenile Lyre or Hymns and Songs, Religious, Cheerful Set to Appropriate Music for Primary and Common School*. Boston: Richardson, Lord and Holbrook, 1831.
Mathews, Charles. *The London Mathews Containing an Account of This Celebrated Comedian's Trip to America, Being an Annual Lecture on Peculiarities, Characters, and Manners, Founded on His Own Observations and Adventures*. Philadelphia: Morgan and Yeager, 1824.
Mathews, W. S. B. *How to Understand Music: A Concise Course of Musical Culture by Object Lessons and Essays*. Philadelphia: Theodore Presser, 1888/96.
May Day Group. "Colloquium Schedule." http://www.maydaygroup.org/.
McClary, Susan. *Conventional Wisdom: The Content of Musical Form*. Berkeley: University of California Press, 2000.
———. *Feminine Endings: Music, Gender and Sexuality*. Minneapolis: University of Minnesota Press, 1991.
McConathy, Osbourne. "High School Music." *Music Supervisors National Conference* (1910): 70–77.
———. "Introduction." In *Listening Lessons in Music*, edited by Anne Moore Fryberger. New York: Silver, Burdett, 1925.
McConathy, Osbourne, Otto Meissner, Edward Bailey Birge, and Mabel Bray. *The Music Hour: Elementary Teacher's Book*. New York: Silver Burdett, 1929a.
———. *The Music Hour: Intermediate Teacher's Book*. New York: Silver Burdett, 1930.
———. *The Music Hour: Kindergarten and First Grade*. New York: Silver Burdett, 1929b.
McGerr, Michael. *A Fierce Discontent: The Rise and Fall of the Progressive Movement in America, 1870–1920*. New York: Free Press, 2003.
McMillan, Daniel Alexander. "Germany Incarnate: Politics, Gender and Sociability in the Gymnastics Movement, 1811–1871." PhD diss., Columbia University, 1997.
McMurry, Nan Marie. "'And I? I Am in a Consumption': The Tuberculosis Patient, 1780–1930." PhD diss., Duke University, 1985.
McWhorter, John. "How Hip-Hop Hold Blacks Back." http://www.city-journal.org/html/13_3_how_hip_hop.html.
Ment, David Martin. "Racial Segregation in the Public Schools of New England and New York, 1840–1940." PhD diss., Columbia University, 1975.
Miller, Kiri. "Americanism Musically: Nation, Evolution, and Public Education at the Columbian Exposition, 1893." *Nineteenth-Century Music* 27, no. 2 (2003): 137–55.
Miller, Perry. *The Errand into the Wilderness*. Cambridge: Harvard University Press, 1961.
———, ed. *The Transcendentalists: An Anthology*. Cambridge: Harvard University Press, 1950.
Mohler, Louis. "The Project Method in Teaching Music Appreciation." *Journal of the Music Supervisors National Conference* (April 1924): 261–64.
Moritz, Karl. *Anton Reiser*. 1785. London: Penguin Books, 1997.
Morris, R. L. *Cholera, 1832: The Social Response to an Epidemic*. Chicago: University of Chicago Press, 1976.

Morrison, S. J. "A Comparison of Preferences and Responses of White and African American Students to Musical Vs. Musical/Visual Stimuli." *Journal of Research in Music Education* 46, no. 2 (1998): 208–22.

Morrison, Toni. "Clinton as the First Black President." New Yorker, October 5, 1998. http://ontology.buffalo.edu/smith/clinton/morrison.html.

———. *Playing in the Dark: Whiteness and the Literary Imagination.* New York: Vintage Books, 1992.

Moses, Harry. *Developing and Administering a Comprehensive High School Music Program.* West Nyack, NY: Parker, 1970.

Murray, James E. *Dainty Songs for Little Lads and Lasses.* Cincinnati: S. Brainerd, 1887.

Music Educators National Conference. *Documentary Report on the Tanglewood Symposium.* Edited by Robert Choate. Reston, VA: Music Educator's National Conference, 1967.

Mussulman, Joseph. *Music in the Cultured Generation: A Social History of Music in America, 1870–1900.* Evanston: Northwestern University, 1972.

Nason, Elias. *Our National Song.* Albany, NY: Joel Munsell, 1869.

Nelson, Jennifer, and Bradley Behrens. "Spoken Daggers, Deaf Ears, and Silent Mouths." In *The Disabilities Studies Reader*, edited by Lenard J. Davis, 29–74. New York: Routledge, 1997.

Nichols, Roger and Richard Langham Smith. Debussy: Pelléas and Mélisande. Cambridge: Cambridge University Press, 1994.

Nietzsche, Friedrich. *The Birth of Tragedy and the Genealogy of Morals.* New York: Anchor Books, 1990.

Newman, William. "The Beethoven Mystique in Romantic Art, Literature." *Musical Quarterly* 69 (Summer 1983): 354–87.

Nye, David. *The American Technological Sublime.* Cambridge, MA: MIT Press, 1999.

Oberndorfer, Anne Faulkner. *What We Hear in Music.* Edited by Victor Talking Machine Company. Camden: RCA Victor, 1913, 1921, 1928, 1939, 1943.

Orsini, G. N. G. *Coleridge and German Idealism.* Carbondale: University of Southern Illinois Press, 1969.

Paine, Thomas. "The Rights of Man." In *Complete Writings of Thomas Paine*, edited by Philip S. Foner, 243–458. 1792. New York: Citadel, 1945.

Painter, Nell. *Standing at Armageddon: The United States 1877–1919.* New York: W. W. Norton, 1987.

Parry, Sir Hubert. *The Evolution of the Art of Music.* 1884. London: Kegan Paul, Trench and Trübner, 1901.

Peabody, Augustus. *The Child's Songbook for the Use of Schools and Families.* Boston: Richardson, Lord and Holbrook, 1830.

Pemberton, Carol. *Lowell Mason: His Life and Work.* Ann Arbor: University of Michigan Research Press, 1992.

Pencak, William. "Paul Robeson and Classical Music." In *Paul Robeson: Essays on His Life and Legacy*, edited by Joseph Dorinson and William Pencak, 152–79. Jefferson, NC: McFarland, 2002.

Perkins, Charles, and John S. Dwight. *History of the Handel and Haydn Society.* Boston: Mudge and Son, 1883.

Perkins, W. O., and H. S. Perkins. *The Nightingale: Songs, Chants and Hymns Designed for the Use of Juvenile Classes, Public Schools and Seminaries.* Boston: Oliver Ditson, 1866.

Petzold, Robert. *Development of Auditory Perception of Musical Sounds by Children in the First Six Grades. Cooperative Research Project No. 766 (Sae 8411)*. Madison, WI: University of Wisconsin, 1960.
Philips, Kimberley E. "'Stand by Me': Sacred Quartet Music and Emotionology." In *An Emotional History of the United States*, edited by Peter Stearns and Jan Lewis, 241–59. New York: New York University Press, 1998.
Pick, Daniel. *Faces of Degeneration: A European Disorder c. 1848–1918*. New York: Cambridge University Press, 1989.
Pitts, Lilla Belle, Mabel Glenn, and Lorrain Waters. *The Kindergarten Book*. Boston: Ginn, 1949.
Pontious, Melvin. "A Guide to Curriculum Planning." Madison, WI: Department of Public Instruction Wisconsin, 1997.
Popkewitz, Thomas. "The Alchemy of the Mathematics Curriculum: Inscriptions and the Fabrication of the Child." *American Educational Research Journal* 41, no. 1 (Spring 2004): 3–34.
———. *A Political Sociology of Educational Reform: Power/Knowledge in Teaching, Teacher Education, and Research*. New York: Teachers College Press, Columbia University, 1991.
———. *Struggling for the Soul: The Politics of Schooling and the Construction of the Teacher*. New York: Columbia University Teachers College Press, 1998.
———, ed. *Cosmopolitanism and the Age of School Reform: Science, Education, and Making Society by Making the Child*. New York: Routledge, 2008.
———. "Education Sciences, Schooling, and Abjection: Recognizing Difference in the Making of Inequality." (forthcoming).
———. "Hopes of Progress and Fears of the Dangerous: Research, Cultural Theses, and Planning Different Human Kinds." In *Education Research in the Public Interest: The Place for Advocacy in the Academy*, edited by Gloria Ladson-Billings and William Tate 119–40. New York: Teachers College Press, 2006.
———. *Inventing the Modern Self and John Dewey: Modernities and the Traveling of Pragmatism*. New York: Palgrave MacMillan, 2005.
Popkewitz, Thomas, and Daniel S. Friedrich. "Professional Development Schools: Narratives of Democracy, Theses of Redemption and the Negation of Politics." Unpublished manuscript. 2008.
Popkewitz, Thomas, and Ruth Gustafson. "Standards of Music Education and the Easily Administered Child/Citizen: The Alchemy of Pedagogy and Social Inclusion/Exclusion." *Philosophy of Music Education Review* 10, no. 2 (2002): 80–91.
Radano, Ronald. "Denoting Difference: The Writing of the Slave Spirituals." *Critical Inquiry* 22, no. 3 (1996) 506–44.
———. "Hot Fantasies: American Modernism and the Idea of Black Rhythm." In *Music and the Racial Imagination*, edited by Ronald Radano and Philip Bohlman, 459–82. Chicago: University of Chicago Press, 2000.
———. *Lying Up a Nation: Race and Black Music*. Chicago: University of Chicago Press, 2003.
Radano, Ronald, and Philip Bohlman. "Introduction." In *Music and the Racial Imagination*, edited by Ronald Radano and Philip Bohlman, 1–53. Chicago: University of Chicago Press, 2000.
Rancière, Jacques. *Hatred of Democracy*. Translated by Steve Corcoran. London: Verso, 2007.
———. *The Ignorant Schoolmaster: Five Lessons in Intellectual Emancipation*. Translated by Kristin Ross. Stanford, CA: Stanford University Press, 1991.

Rauscher, Frances. "Discussion of Research." Educational Cyber Playground. http://www.edu-cyberpg.com/Music/nprmozart.html.
Reagan, Ronald. "A Nation at Risk." Speech delivered in London at Westminster, June 8, 1982.
Redfield, Marc. *The Politics of Aesthetics: Nationalism, Gender, and Romanticism.* Stanford, CA: Stanford University Press, 2003.
Reese, William. *The Origins of the American High School.* New Haven: Yale University Press, 1995.
Regelski, Thomas. "Social Theory and Music Education as Praxis." ACT: Mayday Group, December, 2004. http://www.nyu.edu/education/music/mayday.
———. *Teaching General Music: Action Learning for Middle and Secondary Schools,* 178–256. New York: Schirmer Books, 1981.
———. *Teaching General Music in Grades 4–8: A Musician's Approach.* New York: Oxford University Press, 2004.
Rehding, Alexander. "Nature and Nationhood in Hugo Rieman's Dualistic Theory of Harmony." PhD diss., University of Cambridge, 1998.
Reimer, Bennett. *A Philosophy of Music Education.* Englewood Cliffs, NJ: Prentice Hall, 1989.
Reuben, Julie. *The Making of the Modern University: Intellectual Transformation and the Marginalization of Morality.* Chicago: University of Chicago Press, 1996.
Rhetts, Edith. *Outlines of a Brief Study of Music Appreciation for High Schools.* Camden, NJ: Victor Talking Machine Company, 1923.
Rickford, John. *African American Vernacular English: Features, Evolution and Educational Implications.* Malden, MA: Blackwell, 1999.
Rickford, John, and Russell Rickford. *Spoken Soul: The Story of Black English.* New York: John Wiley, 2000.
Rist, Ray. "Student Social Class and Teacher Expectations: The Self-Fulfilling Prophecy in Ghetto Education." *Harvard Educational Review* 40, no. 3 (1970): 411–51.
Ritter, Frederic Louis. *Musical Dictation.* London: Novello, Ewer, 1887.
Rosenberg, Charles. *The Cholera Years: 1832, 1849, 1866.* Chicago: University of Chicago Press, 1962.
Ruskin, John. "On the Relation of National Ethics to National Arts." In *Source Readings in Music Education,* edited by Michael Mark, 98–103. 1867. New York: Schirmer Books, 1982.
Ruyter, Nancy Lee Chalfa. *The Cultivation of Body and Mind in Nineteenth-Century American Delsartism.* Westport, CT: Greenwood, 1999.
Rydell, Robert W. *All the World's a Fair: Visions of Empire at American Expositional Exhibitions, 1876–1916.* Chicago: University of Chicago Press, 1984.
Said, Edward. *Culture and Imperialism.* New York: Vintage Books, 1994.
Savage, Barbara Dianne. *Broadcasting Freedom.* Chapel Hill: University of North Carolina Press, 1999.
Schabas, Ezra. *Theodore Thomas.* Urbana: University of Illinois Press, 1989.
Schama, Simon. *Landscape and Memory.* New York: A. A. Knopf, 1995.
Schelling, Friedrich. "The Special Part of the Philosophy of Art." In *Musical Aesthetics: A Historical Reader,* vol. 2. Edited by Edward Lippman, 67–84. 1800. Stuyvesant, NY: Pendragon Press, 1985.
Schiller, Friedrich. *On the Aesthetic Education of Mann.* 1795. New Haven: Yale University Press, 1954.

Schmidt, Leigh. *Hearing Things: Religion, Illusion and the American Enlightenment.* Cambridge: Harvard University Press, 2000.
Schneider, Stuart. *Collecting Lincoln.* Atglen, PA: Schiffer, 1997.
Schoen, Max, ed. *The Effects of Music.* 1927. London: Routledge, 1999.
———. "Psychological Problems in Musical Art." *Journal of Research in Music Education* 3, no. 1 (1955): 27–39.
Scholes, Percy. *The Puritans and Music in England and New England: A Contribution to the Cultural History of the Two Nations.* London: Oxford University Press, 1934.
Schultz, Stanley K. *The Culture Factory, Boston Public Schools, 1789–1860.* New York: University of Oxford Press, 1973.
Schulze, Hagen. *The Course of German Nationalism from Frederick the Great to Bismarck, 1763–1867.* Translated by Sarah Hanbury-Tenison. Cambridge: Cambridge University Press, 1991.
Seashore, Carl. "The Measurement of Musical Talent." *Journal of the Music Teachers National Association* (1913): 210–12.
———. *Psychology of Music.* New York: McGraw Hill, 1938.
———. "Talent in the Public Schools." *Journal of the Music Supervisor's National Conference* (January 1916): 10–11.
———. *Why We Love Music.* Philadelphia: Oliver Ditson, 1940.
Serres, Michel, and Latour, Bruno. *Conversations on Science, Culture and Time.* Translated by Roxanne Laspidus. Ann Arbor: University of Michigan Press, 1995.
Sessa, Anne Dzamba. "British and American Wagnerians." In *Wagnerism in European Culture and Politics,* edited by William Large and David S. Weber, 246–77. Ithaca, NY: Cornell University Press, 1984.
Shelemay, Kay Kaufman. *Let Jasmine Rain Down: Song and Remembrance among Syrian Jews.* Chicago: University of Chicago Press, 1998.
Shevy, Mark. "Music Genre as Cognitive Schema: Extramusical Associations with Country and Hip-Hop Music." *P9sychology of Music* 36, no. 4 (October 2008): 477–98.
Shiraishi, Fumiko. "Calvin Brainerd Cady: Thought and Feeling in the Study of Music." *Journal of Research in Music Education* 47, no. 2 (1999): 150–62.
Sloboda, John A. *Exploring the Musical Mind: Cognition, Emotion, Ability, Function.* New York: Oxford University Press, 2005.
———. *The Musical Mind: The Cognitive Psychology of Music.* Oxford: Clarendon, 1985.
Sobe, Noah. "Challenging the Gaze: The Subject of Attention and a 1915 Montessori Demonstration Classroom." *Educational Theory* 54, no. 3 (2004): 281–97.
Sobe, Noah, and Carrie Rackers. "Fashioning Writing Machines: Typewriting and Handwriting Exhibits at U.S. World's Fairs, 1893–1915." In *International Expositions and the Materiality of Education: Modeling the Future,* edited by Martin Lawn and Ian Grosvenor. London: Symposium Books, forthcoming.
Soderman, Johan, and Goran Folkestad. "How Hip Hop Musicians Learn: Strategies in Informal Creative Music Making." *Music Education Research* 6, no. 3 (2004): 314–26.
Southey, Robert. *The Doctor and Etc.* New York: Harper Brothers, 1836.
Sovetov, Vladimir. "Aunt Jemima." http://www.arf.ru/Notes/Uncle/eaj.html.
Spadafora, David. *The Idea of Progress in Eighteenth-Century Britain.* New Haven: Yale University Press, 1990.

Spalding, Walter. *Music: An Art and a Language*. Boston: Arthur Schmidt, 1920.
Spear, Allan H. *Black Chicago: The Making of a Negro Ghetto*. Chicago: University of Chicago Press, 1967.
Spencer, Herbert. "The Origin and Function of Music." In *Herbert Spencer on Education*, edited by Andreas Kazamias, 210–17. 1857. New York: Teachers College Press, Columbia University, 1966.
———. *Principles of Psychology*. Vols. 1, 2. 1855. New York: Appleton, 1899.
Stallybrass, Peter, and Allon White. *The Politics and Poetics of Transgression*. New York: Cornell University Press, 1986.
Stark, James. *Bel Canto: A History of Vocal Pedagogy*. Toronto: University of Toronto Press, 1999.
Starr, John W. *A Bibliography of Lincolniana*. Madison WI: Madison Historical Society, 1926.
Stebbins, Genevieve. *Delsarte System of Expression*. New York: Edgar S. Werner, 1902.
Steedman, Carolyn. *Strange Dislocations: Childhood and the Idea of Human Interiority, 1780–1930*. Cambridge: Harvard University Press, 1995.
Stellings, Alan. "Music Cognition Theory: The Legacy of the Formalist Aesthetic." Paper presented at the Second International Symposium on the Philosophy of Music Education, Toronto, 1994.
Stewart, N. Coe. *Merry Voices*. Cleveland: S. Brainerd, 1873.
Strandberg, Tommy. "'Listen to My Song, Please!' Understanding of Composing in the Classroom." In *Monografier: Journal of Research in Teacher Education*, edited by Per Olof Erixson, 163–76. Umeå, Sweden: Umeå University, 2004.
Subotnik, Rose. *Deconstructive Variations: Music and Reason in Western Society*. Minneapolis: University of Minnesota Press, 1996.
Surette, Thomas. "Musical Appreciation for the General Public." *Journal of the Music Teachers' National Association* (1906): 109–14.
Surette, Thomas, and Daniel Gregory Mason. *The Appreciation of Music*. Vol. 1. New York: H. W. Gray, 1907.
Swales, Martin. *The German Bildungsroman from Wieland to Hesse*. Princeton, NJ: Princeton University Press, 1978.
Tapper, Thomas. *The Education of the Music Teacher*. Philadelphia: Theodore Presser, 1914.
Tharp, Louise. *Until Victory: Horace Mann and Mary Peabody*. Boston: Little Brown, 1953.
Thomas, Helen. *The Body, Dance, and Cultural Theory*. New York: Palgrave Macmillan, 2003.
Thomas, Rose Fay. *Memoirs of Theodore Thomas*. New York: Moffat, Yard, 1911.
Thompson, Randall. *College Music: An Investigation for the Association of American Colleges*. New York: Macmillan, 1935.
Tremaine, C. M. "The Music Memory Contest, Etc." *Journal of the Music Supervisors' National Conference* (April 1918): 99–107.
Tröhler, Daniel. "Geschichte und Sprache der Pädagogik." *Zeitschrift für Pädagogik* 51 (2005): 218–35.
———. "Philosophical Argument, Historical Contexts and Theory of Education." *Educational Philosophy and Theory* 39, no. 1 (2007): 11–19.
———. "The 'Kingdom of God on Earth' and Early Chicago Pragmatism." *Educational Theory* 6, no. 1 (2006): 89–106.
Tuveson, Ernest. *Redeemer Nation*. Chicago: University of Chicago Press, 1968.

United States Department of Education. "The Place of Music in the Scheme of Modern Democratic. "Arts Education Partnership." 2007–8. http://www.aep-arts.org/.
University of Wisconsin–Madison. "Syllabus: Spoken Word and Hip Hop in the Classroom." Vilas Hall Madison, Wisconsin 2006.
Vaillant, Derek. "Peddling Noise: Contesting the Civic Soundscape of Chicago, 1890–1913." *Journal of the Illinois State Historical Society* 96, no. 3 (2003): 257–87.
———. *Sounds of Reform: Progressivism and Music in Chicago, 1873–1935*. Chapel Hill: University of North Carolina Press, 2003.
———. "Sounds of Whiteness: Local Radio, Racial Formation, and Public Culture in Chicago, 1921–1935." *American Quarterly* 54, no. 1 (2002): 25–64.
Victor Talking Machine Company. *Music Appreciation with the Victrola for Children: Designed to Meet the Needs of the Child Mind During the Period of Development, from First to Sixth Grade, Inclusive*. Camden, NJ: Educational Department, Victor Talking Machine Company, 1923.
———. "My Master's Voice." Keokuk, IA: School Music Monthly, 1912.
Von Humboldt, Wilhelm. "A General Introduction to Language." In *An Anthology of the Writings of Wilhelm Von Humboldt. Humanist without Portfolio*, 252–98. 1799. Detroit: Wayne State University Press, 1963.
Wagner, Richard. "Artwork of the Future." In *Wagner on Music and Drama: A Compendium of Richard Wagner's Prose Works*, edited by Albert Goldman and Evert Sprinchorn, 179–235. 1849. New York: Da Capo, 1964.
———. "Jews in Music." In *Wagner on Music and Drama: A Compendium of Richard Wagner's Prose Work*, edited by Albert Goldman and Evert Sprinchorn, 179–235. 1850. New York: Da Capo Press, 1964.
Warren, J. T. "Whiteness and Cultural Theory: Perspectives on Research and Education." *Urban Review* 31, no. 2 (1999): 185–203.
Washburn, Margaret Flo, and George I. Dickinson. "The Source and Nature of the Affective Reaction to Instrumental Music." In *The Effects of Music: A Series of Essays*, edited by Max Schoen, 121–51. 1927. London: Routledge, 1999.
Weber, William. "Wagner, Wagnerism, and Musical Idealism." In *Wagnerism in European Culture and Politics*, edited by Large and William Weber David S, 28–71. Ithaca, NY: Cornell University Press, 1984.
Weekley, Ernest. *An Etymological Dictionary of Modern English*. Edited by John Simpson and Edmund Weiner. Vol. I, Oxford English Dictionary. New York: Dover, 1993.
Welch, Roy Dickinson. *The Appreciation of Music*. New York: Harper and Brothers, 1927.
Welsbacher, Betty, and Elaine Bernstorf. "Musical Thinking Amongst Diverse Students." In *Dimensions of Musical Learning and Teaching*, edited by Eunice Boardman, 155–68. Reston, VA: Music Educators National Conference, 2002.
White, Charles. *An Account of the Regular Gradation in Man*. London: C, Dilly, 1799.
Whittier, John G. "The Farewell of a Virginia Slave Mother to Her Daughter Sold into Southern Bondage." 2008. http://www.readbookonline.net/readOnLin.
Williams, Sudie L. "The Music Memory Contest." *Journal of the Music Supervisors National Conference* (April 1921): 147–54.
Wilson, Harriet E. "Our Nig; or Sketches from the Life of a Free Black in a Two-Story White House, North." In *The Norton Anthology of African American Literature*,

edited by Henry Louis Gates and Nellie Y. McKay, 439–58. 1859. New York: W. W. Norton, 1997.

Wingren, Gustaf. *The Christian's Calling: Luther's Vocation*. Edinburgh: Oliver and Boyd, 1958.

Withers, John. "The Place of Music in the Scheme of Modern Democratic Education." *Journal of the Music Supervisor's National Conference* (March 1916): 25–29.

Wood, Gordon S. *The Radicalism of the American Revolution*. New York: A. A. Knopf, 1992.

Wynter, Sylvia. "Towards the Sociogenic Principle: Fanon, the Puzzle of Conscious Experience, of Identity and What It Is Like to Be Black." In *National Identity and Sociopolitical Change: Latin America between Marginalization and Integration*, edited by Mercedes Duran-Cogan and Antonio Gomez-Moriana. New York: Garland Press, 1999.

York, Francis. "Report of Appreciation Conference." *Journal of the Music Teachers' National Association* (1915): 68–74.

Index

Adams, Crosby, 73, 149, 175
Adams, Henry, 215
Adams, John, 8, 9, 207
Adams, John Quincy, 9, 207
Agawu, Kofi, 230
Aiken, Walter, 218
Alchin, C. A., 223
Anderson, Margo, 24
Anderson, Marian, 141–45, 180, 193, 195, 203, 223, 224, 230, 235
Anton Reiser, 172–82, 233, 234
Apple, Michael, 238

Baker, B. F., 29, 33, 40, 50–56
Baker, Bernadette M., 4, 156, 168, 208, 212, 213, 219, 220–22, 225, 226, 230, 232, 235, 239
ballad, 1, 3, 7–10, 24, 29, 35, 36, 55, 59, 61, 64, 68, 70, 74, 75, 77, 104, 146, 207, 215, 218
Barnes, Earl, 100, 105, 120, 131–33, 140, 153,154, 157, 222, 230
Barthes, Roland, 123, 227
Bascom, E. H., 26, 64, 214
Baynton, Douglas, 232
Beacon Hill, 14
Bederman, Gail, 15, 67, 212, 215, 233, 234
Beecher, Lyman, 11, 43
Beethoven, 54, 55, 73, 95, 96, 103, 108, 115–21, 151, 155, 163–70, 178, 191–94, 198, 217, 220, 221, 232, 233, 235, 236
bel canto, 124–29, 140, 146, 201, 227
Bell Curve, The, 235
Bercovitch, Sacvan, 11, 206, 211
Bermingham, Gudrun, 227
Berthelot, J. M., xii

Bildung, 89, 90, 161, 206, 221, 232, 233, 235
Billings, William, 65–67, 79, 80, 106, 123, 169, 215, 216
Birchard, C. C., 118, 131, 168, 175
Birge, Edward Bailey, 168, 178, 215, 232
blackface, xi, 7, 10, 11, 28, 29, 37, 42–44, 46, 63, 74, 75, 78, 84, 85, 92, 110, 18, 127, 139, 206, 211
Blasius, Leslie, 213
Boardman, Eunice, 235, 236
Bobbitt, Franklin, 97, 226
Bohlman, Philip, 206, 213
Boston, 3–12, 14, 17–25, 35, 39–50, 55, 65, 66, 71, 79, 97, 125, 152, 205, 210, 212, 213, 215, 217, 234, 239
Boston Musical Gazette, 21
Boston School Committee, 4, 5, 12, 17, 19, 35, 40–45
Bourdieu, Pierre, 1, 161, 185, 192–94, 205, 211, 236, 237
Bowen, George Oscar, 112, 129
Bradbury, William, 14, 24, 28, 29, 41, 43, 50, 62, 64, 68, 212, 213
Briggs, Thomas, 73, 91, 107, 149, 162, 223
Broyles, Michael, 43, 66, 206, 297, 209, 212, 214, 216, 217
Burney, Charles, 86, 220
Butler, Charles, 31, 40, 208, 213
Butler, Judith, 53, 206, 207, 214, 229

Cady, Calvin, 95–98, 103, 126, 148, 159, 178, 226
Calvinist, 9, 90, 105, 228
Carlyle, Thomas, 49, 50, 213
Cartwright, S. A., 26
Celtic, xv, 43

census, 23, 24, 28, 209, 210, 217
Chamberlin, J. E., 36, 7
Chicago, 81, 85, 86, 93, 96, 97, 108, 118, 125, 126, 132, 146, 161, 220–24, 230–32
Chicago World's Columbian Exposition, 85, 86, 108, 109, 226
childhood, 14, 25, 35, 37, 40,50, 55, 56, 59, 62, 64, 77, 115, 116, 215, 219
cholera, 17, 22–31, 208–10
Chybowski, Julia, 224
Civil War, 2–4, 15, 16, 30, 39, 43, 44, 75, 160, 165, 206, 211
Clark, Frances E., 70, 91–98, 112–16, 119–21, 131, 134, 148, 150, 152, 157, 159, 168, 177–79, 218, 224, 225, 228, 231, 232, 226–28, 231, 232
classroom, xi–xvii, 1–6, 14, 15, 57, 63, 70, 76, 83, 97, 109–12, 121–25, 134, 146, 147, 150, 154, 162, 163, 171, 185, 189, 195, 200, 234, 236, 237
Claxton, P. P., 70, 105, 109, 112, 125, 126
Coleridge, Samuel T., 54–58, 211, 214, 215
Combe, George, 208–10
comportment, xvi, 1, 3, 4, 8, 15, 25, 30, 32, 33, 37, 44, 46–48, 54, 65, 66, 71, 79, 83, 110, 115, 128, 131, 140, 153, 160, 171, 172, 200, 211
Connor, Steven, 139, 226, 228, 229
consumption, 18, 19, 25–31, 168, 208, 210, 211, 232
contagion, 16, 27, 28, 33, 34
Cooper, Grovenor, 213
Crary, Jonathan, 154, 222, 226, 231
Crawford, Richard, 207, 216, 223
Crocker, Ruth, 118, 160, 221, 227, 228, 231
Cundiff, Hannah, 84, 101, 150, 225, 230
curriculum, xi–xvii, 2, 4–17, 35, 37, 38, 43, 53, 62, 67, 71, 76, 81, 82, 84, 86, 87, 90, 95, 98, 101, 103, 104, 107–16, 121, 123, 124, 128, 132, 133, 134, 138, 140, 141, 146, 147, 151, 156, 158, 162, 163, 165, 168, 172, 176, 177, 181, 182, 183, 186–94, 198–202, 216, 220, 221, 222, 226, 228, 232, 236

Dalcroze, 91, 115, 119, 150, 221, 224, 225, 228
Damrosch, Walter, 121, 165, 166, 168, 222, 224
Declaration of Independence, 29, 57, 106, 117, 153, 214
degeneracy, 7, 16, 21, 34–38, 42, 43, 52, 61, 63, 66–68, 70, 71, 75, 77, 80, 104, 123, 189, 211, 212, 215, 217
Delpit, Lisa, 177, 195, 196, 235
DeNora, Tia, 206, 207, 217, 220, 222, 232, 233, 235
Dewey, John, 82, 83, 98, 116–18, 125, 126, 130, 132, 137, 148, 153, 184, 193, 225, 227
Dickinson, Edward, 93, 105
Douglass, Frederick, 3, 32, 41, 48, 75, 172, 174, 181, 206, 212, 214, 244
Dubois, W. E. B., 140
Dunham, Richard Lee, 86, 120, 162, 221, 223–27, 234
Dwight, John Sullivan, 39–43, 49–57, 60, 66, 68–71, 153, 207, 215–17, 224, 231
Dykema, Peter, 84, 101, 150, 225, 230

ear training, 146, 147, 223
elementary school, xi, 101, 149, 154, 196, 199
Elias, Norbert, 211, 214, 234
Eliot, Samuel, 66, 69, 215
Ellison, Ralph, 193
Emerson, L. O., 52, 58, 61, 207, 213, 216
Emerson, Ralph Waldo, 115
Emery, Lynne Fauley, 222, 228, 234

entrainment, xi–xvi, 12, 37, 82, 95, 122, 162, 200
ethnology, 15, 43, 85, 223
Eve, 42, 43

Fanon, Frantz, 205, 214, 235, 237
Fendler, Lynn, 218
Fichte, Johann, 45, 46, 90, 213
Fitz, Asa, 2, 29, 41, 205
Fitzgerald, Maureen, 207, 217, 219
Fliegelman, Jay, 106, 214, 220, 222, 226, 231, 234
Foucault, Michel, xvii, 17, 33, 205, 235
Froebel, Friedrich, 212
Fryberger, Agnes Moore, 90, 91, 95, 134, 157, 176, 177, 217, 241, 243, 244

Galton, Francis, 220, 230
Gear, Joseph, 10
Giffe, W. T., 23, 24, 27
Gilman, Sandor, 36, 37
Glenn, Mabel, 117, 167, 168, 177, 192, 194, 224, 232
Goethe, Johann, 10, 44
Golston, Michael, 150, 220, 222, 225, 228, 230
good ear, xiii, xv, 6, 68, 72, 82, 116, 145, 147, 153, 166, 170, 187, 189, 199
Goodman, David, 104, 105, 108, 222, 225, 226, 230, 231
Gould, Stephen Jay, 28, 47, 207, 209, 210, 212, 213
Green, Lucy, 236, 237
Green, Shannon, 222
Grivel, Charles, 233
Gruhn, Wilfried, 234
Gustafson, James Paul, v, 203
Gustafson, Ruth, xii, xiii, 123, 145–47, 165, 183, 197, 207, 223, 234, 237

Hall, G. Stanley, 113, 125, 150, 152, 156, 184, 224
Handel and Haydn Society, 8

Hansel and Gretel, 102
Hanslick, Edward, 73, 149, 161, 218, 220, 222
Hastings, Thomas, 8, 67, 68, 73, 128, 206, 215, 234, 235
Heath, Shirley, xii, 146, 147
Henderson, W. J., 74, 218, 220, 233
high school, 73, 84, 86, 87, 96, 100, 109, 110, 114, 134, 138, 147, 151, 157, 189, 197, 199, 207, 217, 222, 227
Hitchcock, H. Wiley, 11
Hoffman, Duwayne, 165, 198
Hultqvist, Kenneth, 208
Hunt, Ernest, 158, 159, 162, 234
Hutton, C., 36, 206

Irish Catholic, 2, 7, 8, 15, 16, 24–25
Ives, Elam, 4, 14

Jackson, John L., Jr., 82
James, Richard Lee, 110, 216, 220, 223, 230
James, William, 155
Jefferson, Thomas, 2, 48, 106, 117, 126, 142, 153, 211–13, 222, 226
Jung, Carl, 137, 220

Kant, Immanuel, 14, 73, 214, 217, 220
Kasson, John, 108, 223, 224, 231, 232, 234
Keene, James, 155, 184, 207, 220, 223, 224
Keiler, Allan, 142–45, 180, 223, 224, 229, 230, 235
kindergarten, 86, 149, 177, 178, 183, 184, 199, 212
King, Joyce E., xi, xvi,
Kingman, Charles, 221
Knowlton, Fanny Snow, 218
Koselleck, Rheinhardt, 90, 221
Koza, Julia Eklund, xvii, 120, 187, 202, 219, 223, 227, 228, 236, 237
Kraut, Alan, 210

Ladson-Billings, Gloria, 173, 197, 238
Lasch-Quinn, Elizabeth, 223, 227, 231
La Vopa, Anthony, 90, 174, 178, 206, 221, 233, 234
Lee, Carol D., 205, 236, 237
Lempa, Heikki, 32, 208–11, 214, 231
Lensmire, Timothy, 193
Leppert, Richard, 218
Leslie, C. E., 27, 28
lesson, 3, 24, 42, 84, 99, 102, 116–19, 135, 147–50, 156, 162, 165, 184, 226, 243
Levine, Lawrence, 216, 217, 220
Levy, Lester, 10, 28, 74
Lincoln, Abraham, 11, 12, 141–43, 163, 165, 169, 203, 217, 231, 233
Lippman, Edward, 40, 45, 221
listening ears, xiii, xv, 6
Litwack, Leon, 213, 217, 219
Locke, John, 9
Lott, Eric, xi, xvii, 8, 28, 37, 49, 63, 65, 79, 107, 116, 131, 201, 206, 207, 210–13, 215, 218, 220, 223, 229, 231, 233
Lucy, 1

Maeterlinck, Maurice, 119
Mann, Horace, 3–21, 25–29, 48, 49, 58, 61, 62, 66, 70, 110, 205, 208, 210, 212, 214, 215
Mann, Thomas, 233
Mark, Michael, 38–40, 42, 117, 118, 199, 207, 210, 212, 215, 234–36
Mason, Daniel Gregory, 71, 111, 112, 120, 130, 135, 222, 223
Mason, Lowell, 3, 4, 5, 7–9, 12, 14, 18–26, 28, 31, 35, 41, 42, 47, 52, 65–70, 122, 123, 128, 205, 206, 208, 213, 215, 218, 227, 235, 238
Mathews, Charles, 36
Mathews, W. S. B., 85–87, 90, 104, 109, 146, 217, 220
May Day Group, 238
McClary, Susan, 206, 218

McConathy, Osbourne, 97, 105, 148–50, 167, 225, 227
McMillan, Daniel Alexander, 209, 213, 231, 233
McMurry, Nan, 31, 208, 210, 211, 232
Metfessel, Milton, 127, 135, 136, 228
Meyer, Leonard, 213, 237
Miller, Kiri, 97, 103, 220, 223
Miller, Perry, 214, 215
minstrelsy, xi, 7–9, 36, 37, 41–43, 49, 50, 61–68, 74–78, 84–88, 104, 109, 118, 124, 127, 139, 151, 160, 161, 211, 218
Mohler, Louis, 94–98, 102, 132, 152, 222, 227, 232
Moritz, Karl, 172, 174, 178, 180, 234
Morrison, Toni, 28
mortality rates, 25
Murray, James, 51
Murray, Penelope, 248
music appreciation, 67, 71–77, 79–121, 124, 124, 125, 130–36, 140–43, 146–49, 151–57, 159, 161, 162, 165–71, 176–82, 192–94, 199, 218, 220–28, 230–34, 239, 240, 242–48, 251, 254, 256–58

Nason, Elias, 74, 75
New England, 9
New York, 8, 17, 36, 42, 63, 65, 70, 74, 106, 115, 162, 165, 193, 207, 209, 210, 219, 224
Newman, William, 170, 191, 232, 236
Nietzsche, Friedrich, 205, 225, 229, 235
Nutcracker, The, xvi
Nye, David, 211, 214, 217, 223, 232

Oberndorfer, Anne Faulkner, 134, 135, 141
opera, 31, 39, 46, 47, 63, 71, 89, 96, 100, 101, 102, 108, 115–17, 125, 128, 135, 136, 143, 145–49, 162, 163, 175, 180, 225, 227, 230

orchestra, 46, 89, 96, 107, 109–11, 115, 161, 165, 186, 197, 200

Paine, Thomas, 9
Panama Pacific International Exposition, 125
Parry, Sir Hubert, 104, 111, 208, 218, 224
Peanuts, 1
Pemberton, Carol, 213, 227, 233
Pestalozzi, 45, 46, 240
Philadelphia, 17, 42, 64, 145, 151, 229
phrenology, 15, 19, 20, 26, 209, 210, 212
physiology, 20, 28, 44, 155, 208, 213
polygenist, 26
Popkewitz, Thomas, xvii, 102, 124, 187, 201, 202, 203, 205, 206, 210, 214, 222, 226, 227, 229, 232, 236, 237, 238
Prussia, 3, 4, 172, 206, 209, 221

Quincy, Josiah, 23

race theory, xvi
racial essence, xiv
Rackers, Carrie, 231
Radano, Ronald, xiv, 119, 138–40, 187, 205, 206, 207, 212, 215, 216, 217, 218, 219, 220, 222, 224, 226, 227, 228, 229, 230, 234
Rancière, Jacques, 232, 237, 238
Redfield, Marc, 40
Reese, William, 207
Regal, Mary, 87, 210, 217, 220
Regelski, Thomas, 130, 147, 199, 200, 202
Rhetts, Edith, 76, 116, 131, 134, 167, 168, 192, 224, 227, 232
rhythm, xiii–xvi, 9, 35, 39–42, 50–59, 68, 83, 84, 91–101, 106, 113–16, 119, 120, 128–40, 147–57, 161, 162, 184, 186, 187, 191, 199, 200, 203, 215, 220–22, 224, 225, 228–30, 233

Rice, Thomas D., 37, 75, 218
Rickford, John, 221, 234, 235
Rickford, Russell, 234, 235
Rist, Ray, 234
Ritter, Frederick, 113, 146
Robeson, Paul, 84, 128, 141, 224, 229
Rodin, Auguste, 1
Rosenberg, Charles, 209, 210
Rush, Benjamin, 18
Ruskin, John, 38–40, 50, 212
Ruyter, Nancy, 231, 232
Rydell, Robert W., 220, 222, 223

Sanders, Charles W., 24, 28, 29
Schelling, Friedrich, 39, 40, 214
Schmidt, Leigh, 127, 137, 226, 228
Schoen, Max, 121, 122, 225, 229, 231, 236
Scholes, Percy, 213, 219
Schroeder, 1
Schultz, Stanley K., 4, 22, 26, 206, 207, 209
Schulze, Hagen, 213
Seashore, Carl, 120, 125, 127, 134–37, 154, 176, 206, 220, 226, 228, 231
segregation, xiii, xvi, 15, 43, 48, 71, 79, 92–95, 110, 132, 140–42, 146, 206, 208, 212, 219, 224, 235, 236
self-cultivation, 16, 34, 105–10, 116, 174, 184, 189–90, 209, 229, 251, 252
Sessa, Anne Dzamba, 162, 224
Shaftesbury, Earl of, 9
Shelemay, Kay Kaufman, 207, 213
Sloboda, John, 228–30
Sobe, Noah, 231
songbook, 2–4, 6–10, 12–16, 22–37, 40–46, 50–53, 58, 63–69, 72, 77, 123, 128, 174, 203, 211, 212, 218, 227, 231
Spear, Allan H., 221–23
Spencer, Herbert, 111, 133, 140, 224, 227
Stallybrass, Peter, 79, 123, 169, 217, 220, 226

Stark, James, 227
Stebbins, Genevieve, 159, 182, 231
Steedman, Carolyn, 213
Stellings, Alan, 205
Strandberg, Tommy, 236
Subotnik, Rose, 217
Surette, Thomas, 71, 111, 112, 130, 135, 220, 222, 224, 232

Tanglewood Symposium, 121, 122, 183–85
Tapper, Thomas, 152, 175, 180, 216, 235
Tchaikovsky, Piotr I., xvi
teacher training, 107, 109–11, 122, 130, 145, 146, 148, 188, 201, 209, 216, 221, 223, 227, 230, 237
temperance, 13, 14, 22–25, 61, 70–73
Thomas, Theodore, 96, 104, 161, 222, 225, 226
Tröhler, Daniel, 221
Twain, Mark, 115, 162, 221

Victor Talking Machine, 76, 79, 92, 94, 96, 104, 112–14, 121, 125, 130, 131, 133, 134, 137, 152–56, 167, 170, 177, 179, 218, 223–26, 231
vocal, 2, 3–8, 16–26, 28, 33, 35, 41–47, 53, 56–58
Von Humboldt, Wilhelm, 90, 214, 221

Wagner, Richard, 94, 102, 115–20, 161, 162, 169, 191, 213, 218–21, 224, 225, 239
Welch, Roy Dickinson, 147, 231
White, Allon, 79, 123, 169, 217, 220, 226
White, Charles, 28
whiteness, xiv, xv, 2, 16, 21, 27, 31–33, 37, 43, 52, 54, 56, 59, 75, 82–84, 90, 101, 104, 110, 114, 118, 122, 124, 130, 140, 154, 160, 166, 171, 182, 188, 191, 193, 196, 201, 210, 220, 221, 223, 224–26, 228, 232–34, 237, 238
Wilhelm Meister, 44, 179, 212, 213, 227, 233, 235
Wilson, Harriet E., 181, 235
Wissenschaft, 19, 20, 208
Wood, Gordon S., 9, 206, 211, 214, 216, 231
Wynter, Sylvia, 214, 237, 238

GPSR Compliance
The European Union's (EU) General Product Safety Regulation (GPSR) is a set of rules that requires consumer products to be safe and our obligations to ensure this.

If you have any concerns about our products, you can contact us on

ProductSafety@springernature.com

In case Publisher is established outside the EU, the EU authorized representative is:

Springer Nature Customer Service Center GmbH
Europaplatz 3
69115 Heidelberg, Germany

www.ingramcontent.com/pod-product-compliance
Lightning Source LLC
LaVergne TN
LVHW011807060526
838200LV00053B/3695